The Physics of Particle Detectors

This text provides a comprehensive introduction to the physical principles and design of particle detectors, covering all major detector types in use today. The book begins with a reprise of the size and energy scales involved in different physical processes. It then considers non-destructive methods, including the photoelectric effect, photomultipliers, scintillators, Cerenkov and transition radiation, scattering and ionization and the use of magnetic fields in drift and wire chambers. A complete chapter is devoted to silicon detectors. In the final part of the book, the author discusses destructive measurement techniques including Thomson and Compton scattering, Bremsstrahlung and calorimetry. Throughout the book, emphasis is placed on explaining the physical principles on which detection is based, and showing, by considering appropriate examples, how those principles are best utilized in real detectors. This approach also reveals the limitations that are intrinsic to different devices.

DAN GREEN received his Ph.D. from the University of Rochester in 1969. He was a post-doc at Stony Brook from 1969 to 1972 and worked at the Intersecting Storage Rings (ISR) at CERN. His next appointment was as an Assistant Professor at Carnegie Mellon University from 1972 to 1978 during which time he was also Spokesperson of a BNL Baryonium Experiment. He has been a Staff Scientist at Fermilab from 1979 to the present, and has worked in a wide variety of roles on experiments both at Fermilab and elsewhere. He worked on the D0 Experiment as Muon Group Leader from 1982 to 1990 and as B Physics Group Co-Convener from 1990 to 1994. He led the US Compact Muon Solenoid (CMS) Collaboration as Spokesperson for the US groups working at the Large Hadron Collider (LHC) at CERN. At Fermilab, he was Physics Department Deputy Head from 1984 to 1986 and Head from 1986 to 1990. From 1993 to present he has served as the CMS Department Head in the Particle Physics Division.

CAMBRIDGE MONOGRAPHS ON PARTICLE PHYSICS, NUCLEAR PHYSICS AND COSMOLOGY

12

General Editors: T. Ericson, P. V. Landshoff

THE PHYSICS OF PARTICLE DETECTORS

DAN GREEN

Fermilab

CAMBRIDGE
UNIVERSITY PRESS

CAMBRIDGE UNIVERSITY PRESS
Cambridge, New York, Melbourne, Madrid, Cape Town, Singapore, São Paulo

Cambridge University Press
The Edinburgh Building, Cambridge CB2 2RU, UK

Published in the United States of America by Cambridge University Press, New York

www.cambridge.org
Information on this title: www.cambridge.org/9780521662260

First published 2000
This digitally printed first paperback version 2005

A catalogue record for this publication is available from the British Library

Library of Congress Cataloguing in Publication data

Green, Dan.
The physics of particle detectors / Dan Green.
p.   cm. – (Cambridge monographs on particle physics,
nuclear physics, and cosmology ; 12)
Includes bibliographical references and index.
ISBN 0 521 66226 5 (hardback)
1. Nuclear counters. I. Title. II. Series.
QC787.C6 G67 2000
539.7′7–dc21   99-053251

ISBN-13 978-0-521-66226-0 hardback
ISBN-10 0-521-66226-5 hardback

ISBN-13 978-0-521-67568-0 paperback
ISBN-10 0-521-67568-5 paperback

Contents

Acknowledgments

This book represents a distillation of 30 years of experimental experience. The author cannot possibly individually acknowledge all the colleagues who 'taught him the business' of experimental high energy physics. Suffice it to say that he is indebted to a multitude. The input of the students who were subjected to lectures consisting of parts of this text was often incisive and thought provoking. The enthusiasm and dedication of Ms. Terry Grozis in assembling the final document from inaudible tapes, scraps of paper and marginal digressions were also of inestimable value. Dr. John Womersley and Dr. Adam Para are thanked for a critical reading of the text and for valuable suggestions. Finally, the students subjected to a full course of lectures in the summer of 1997 gave very valuable criticism.

Part I

Introduction

The subject of particle detectors covers those devices by which the existence and attributes of particles in a detecting medium is made manifest to us. The full and complete understanding of these devices requires a good understanding of basic physics. Without that knowledge we are simply 'mechanics' without the capacity to advance the state of the art of detector technology. On the other hand, a rigorous understanding from first principles is also not an optimal approach. The 'useful' understanding of a given device proceeds from an understanding of what approximations to full rigor are possible. That understanding can only come from experience and it is the purpose of this volume to communicate that experience. The aim of this text is to steer a perilous course between the purely descriptive and the purely theoretical.

The role of detectors can be visualized by assuming that an interesting interaction occurs at a point in space and time. From that point several secondary particles of different masses are emitted with various angles and momenta as shown in Fig. I.1. It is the job of the detector designer to measure the time of interaction, t, and the vector momenta, $\mathbf{p}_i$, and masses, M_i, of those emitted particles. The text is organized so as to show the ensemble of tools available to the designer. Typically, mathematical detail and topics outside the main scope of the text are relegated to the Appendices.

A list of the topics covered is given in the Table of Contents. The first chapter is an introduction devoted to a numerical description of the appropriate size, energy scales, and cross sections for different processes. The numerical data given in the tables of Chapter 1 will constantly be referred to in later chapters.

There then follow eight chapters on 'non-destructive' measurements, or those which do not appreciably change the measured particle's position or momentum. The first subtopic concerns the measurement of time and velocity. Chapter 2 starts with the basic physics of the photoelectric effect and leads into photomultiplier tubes, scintillator and time of flight measurement.

Introduction

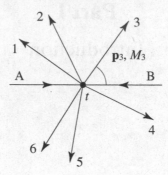

Fig. I.1.   A schematic representation of the reaction $A + B \rightarrow 1 + 2 + 3 + 4 + 5 + 6$. The reaction is specified when the vector momentum and mass of each particle is determined.

Chapter 3 describes velocity measurement by way of the emission of optical photons in the Cerenkov process. Chapter 4 follows with a discussion of the closely related emission of x-ray photons in transition radiation which also determines the velocity of a particle.

The second subtopic within non-destructive readout first lays the physics groundwork of elastic scattering in Chapter 5. Chapter 6 then covers the application of single scattering to scattering off atomic electrons and the resulting energy loss. Detection of the energy lost is the physical basis of many of the techniques used in charged particle detection. The third subtopic then follows up with the non-destructive measurement of the position and momentum of charged particles. Chapter 7 contains a derivation of the particle trajectory in a magnetic field and the consequent measurement of momentum. Those measurements have intrinsic limitations which are explored first in Chapter 8 in studying diffusion in gases and wire chambers. Chapter 9 looks at faster and higher spatial resolution silicon detectors for more accurate position measurements.

The text then switches to destructive measurements, where the particle to be measured loses a significant fraction of its energy or is fully absorbed in the detector. First, in Chapter 10, the physics foundation is laid by exploring radiation and photon scattering. Then these concepts are applied in exploring the topic of destructive energy measurements. Chapter 11 describes measurements of electron and photon energy, while Chapter 12 describes the measurement of the energy of strongly interacting particles.

Finally, a general-purpose high energy physics detector using all the previously described techniques is sketched in Chapter 13. The concept of multiple redundant measurements is introduced and several examples are given.

The full set of material in the text is suitable for a one year course. If a one semester course is desired, the algebraic details in the Appendices can be skipped as well as Chapter 1, the first half of Chapter 2 and Chapters 5, 6, and 10 assuming that no supplementary physics background was required.

Note that the subjects covered in the text are strongly limited to detectors themselves. Exceptions are a brief description of coincidence circuits in Chapter 2 and front end noise processing in Chapter 9. These brief forays were made since these special topics were tightly connected to the detectors themselves. However, there is no other discussion of front end electronics, trigger systems, data acquisition, or computer programming. In addition, the vital area of detector modeling and Monte Carlo techniques is only sketched in Appendix K. Probability theory and statistical analysis appear only briefly in Appendix J. References to these vital areas are given at the end of the text for readers who want to go beyond the scope of this volume.

The aim of this text is to describe the full ensemble of particle detectors from first principles. The goal is to strike a balance between simply presenting the final result and a full and rigorous derivation and thus to extract the relevant physics in a clear fashion. Intuition and order of magnitude numerical estimates are stressed throughout in an attempt to communicate the insights garnered from experience.

> Ah, but a man's reach should exceed his grasp, or what's a heaven for?
>
> *Robert Browning*

> Curiosity is, in great and generous minds, the first passion and the last.
>
> *Dr Samuel Johnson, 1750*

General references – A

Mechanics, electricity and magnetism and quantum mechanics

[1] *The Feynman Lectures in Physics*, R. Feynman, R. Leighton, and M. Sands, Addison-Wesley Publishing Co., Inc. (1963).
[2] *Classical Mechanics*, H. Goldstein, Addison-Wesley Publishing Co., Inc. (1950).
[3] *Classical Electricity and Magnetism*, W.K.H. Panofsky and M. Phillips, Addison-Wesley Publishing Co., Inc. (1962).
[4] *Quantum Mechanics*, E. Merzbacher, John Wiley & Sons, Inc. (1961).

General references – B

Textbooks on particle detectors

[1] *Detectors for Particle Radiation*, K. Kleinknecht, Cambridge University Press (1987).

[2] *Experimental Techniques in High Energy Physics*, T. Ferbel, Addison-Wesley Publishing Co., Inc. (1987).

[3] *Instrumentation in High Energy Physics*, Ed. F. Sauli, World Scientific Publishing Co. (1992).

[4] *Instrumentation in Elementary Particle Physics*, J.C. Anjos, D. Hartill, F. Sauli, and M. Sheaf, Rio de Janeiro, 1990, World Scientific Publishing Co. (1992).

[5] *Instrumentation in Elementary Particle Physics*, C.W. Fabjan and J.E. Pilcher, Trieste 1987, World Scientific Publishing Co. (1988).

[6] C.W. Fabjan and H.F. Fisher, 'Particle detectors', *Rep. Prog. Phys.* **43** 1003 (1980).

1
Size, energy, cross section

Beauty depends on size as well as symmetry.

Aristotle

Energy is eternal delight.

William Blake

Textbooks on detectors often jump directly into a description of the devices themselves. The relevant descriptive formulae are then simply given without derivation and readers are instead referred to the relevant texts. A complementary approach is taken here. A 'derivation' of the relevant physics is always attempted first. Armed with the derivations, the reader is then introduced to the detector where the approximations which are made are explained along with the reasons why they are valid. In order to contain the length of the text, all 'derivations' are heuristic and thus either compressed or left to the Appendices. Numerical examples are given at regular intervals in order that the reader be firmly connected to real devices and have a firm grasp of the appropriate orders of magnitude. We note that 'intuition' is largely the result of experience. The judgement that allows a simplifying approximation to be made usually comes with an appreciation of orders of magnitude of the quantities involved. In this text, that hard won 'intuition' is, it is hoped, passed on to the student.

Detectors function by causing a particle to interact with some detecting medium. For example, a charged particle might ionize a gas in a device and the freed charge might be collected as an electrical signal localized in time, a 'pulse', on a detector electrode. To characterize the detector it is fundamental to understand the probability of interaction of the particle with the device. The aim of Chapter 1 is to provide the basic numerical data needed to later characterize the interaction probability of the different particles which we wish to detect.

1.1 Units

It is traditional in high energy physics to work in dimensionless units where $\hbar$ and c are defined to be equal to 1. In those units momentum (pc), energy (ε), mass (mc^2), inverse time ($\hbar/t$) and inverse length ($\hbar c/x$) all have the same dimensions, which we take to be energy. Units for energy are taken to be the electron volt, where one electron volt, eV, is the energy, ΔU, gained by an electron of charge e, in dropping through a potential difference, ΔV, of 1 volt, $\Delta U = e\Delta V$. A tabulation of many of the physical quantities used throughout the text is given in Table 1.1[1]. In that table, the speed of light, the Planck constant, the electron charge, the masses of elementary particles, the fine-structure constant, the classical electron radius, the Compton wavelength of the electron etc. are gathered together in electron volt and in MKS units. This table and Table 1.2 contain sufficient numerical information for the needs of this text. Kinematics, the constraints imposed by energy–momentum conservation, are worked out in Appendix A.

1.2 Planck constant

In going from the dimensionless calculations of high energy physics to dimensional quantities, it is necessary to know the Planck constant and be able to use it. (See Table 1.1.)

$$\hbar c = 0.2 \text{ GeV fm} = 2000 \text{ eV Å}$$

$$1 \text{ GeV} = 10^9 \text{ eV}$$

$$1 \text{ Å} = 10^{-8} \text{ cm} = 10 \text{ nm} \qquad (1.1)$$

$$1 \text{ fm} = 10^{-13} \text{ cm}$$

The Planck constant is given for two different distance scales, the angstrom and the fermi. Those two distance scales are characteristic of the size of an atom and the size of a nucleus respectively. The conversion of size to energy leads us to the immediate conclusion that nuclear energy scales are GeV whereas atomic energy scales are of order a few electron volts. The ratio of one angstrom to one fermi is 100 000 to 1.

1.3 Electromagnetic units

In the body of the text we will most often use CGS units in symbolic manipulations and we will freely convert back and forth to MKS units. It is a fact of life that practicing physicists must acquire a facility with both systems of units since experimentalists have connections to both theory and engineering. (See the beginning of Chapter 3 for a full explanation.) We will later explicitly give a prescription for converting between MKS and CGS units.

Table 1.1. *Fundamental physical constants*

Quantity	Symbol, equation	Value	Uncert. (ppm)
Speed of light in vacuum	c	$299\,792\,458$ m s⁻¹	exact[a]
Planck constant	h	$6.626\,075\,5(40)\times10^{-34}$ J s	0.60
Planck constant, reduced	$\hbar = h/2\pi$	$1.054\,572\,66(63)\times10^{-34}$ J s	0.60
		$= 6.582\,122\,0(20)\times10^{-22}$ MeV s	0.30
Electron charge magnitude	e	$1.602\,177\,33(49)\times10^{-19}$ C $= 4.803\,206\,8(15)\times10^{-10}$ esu	0.30, 0.30
Conversion constant	$\hbar c$	$197.327\,053(59)$ MeV fm	0.30
Conversion constant	$(\hbar c)^2$	$0.389\,379\,66(23)$ GeV² mbarn	0.59
Electron mass	m_e	$0.510\,999\,06(15)$ MeV/$c^2 = 9.109\,389\,7(54)\times10^{-31}$ kg	0.30, 0.59
Proton mass	m_p	$938.272\,31(28)$ MeV/$c^2 = 1.672\,623\,1(10)\times10^{-27}$ kg	0.30, 0.59
		$= 1.007\,276\,470(12)$ u $= 1836.152\,701(37)\,m_e$	0.012, 0.020
Deuteron mass	m_d	$1875.613\,39(57)$ MeV/c^2	0.30
Unified atomic mass unit (u)	(mass ¹²C atom)/12 $= (1\text{g})/(N_A$ mol)	$931.494\,32(28)$ MeV/$c^2 = 1.660\,540\,2(10)\times10^{-27}$ kg	0.30, 0.59
Permittivity of free space	$\epsilon_0\ \}\ \epsilon_0\mu_0 = 1/c^2$	$8.854\,187\,817\ldots \times10^{-12}$ F m⁻¹	exact
Permeability of free space	$\mu_0\ \}$	$4\pi\times10^{-7}$ N A⁻² $= 12.566\,370\,614\ldots\times10^{-7}$ N A⁻²	exact
Fine-structure constant	$\alpha = e^2/4\pi\epsilon_0\hbar c$	$1/137.035\,989\,5(61)^b$	0.045
Classical electron radius	$r_e = e^2/4\pi\epsilon_0 m_e c^2$	$2.817\,940\,92(38)\times10^{-15}$ m	0.13
Electron Compton wavelength	$\lambdabar_e = \hbar/m_e c = r_e\alpha^{-1}$	$3.861\,593\,23(35)\times10^{-13}$ m	0.089
Bohr radius ($m_{\text{nucleus}}=\infty$)	$a_\infty = 4\pi\epsilon_0\hbar^2/m_e e^2 = r_e\alpha^{-2}$	$0.529\,177\,249(24)\times10^{-10}$ m	0.045
Wavelength of 1 eV/c particle	hc/e	$1.239\,842\,44(37)\times10^{-6}$ m	0.30
Rydberg energy	$hcR_\infty = m_e e^4/2(4\pi\epsilon_0)^2\hbar^2 = m_e c^2\alpha^2/2$	$13.605\,698\,1(40)$ eV	0.30
Thomson cross section	$\sigma_T = 8\pi r_e^2/3$	$0.655\,246\,16(18)$ barn	0.27
Bohr magneton	$\mu_B = e\hbar/2m_e$	$5.788\,382\,63(52)\times10^{-11}$ MeV T⁻¹	0.089
Nuclear magneton	$\mu_N = e\hbar/2m_p$	$3.152\,451\,66(28)\times10^{-14}$ MeV T⁻¹	0.089
Electron cyclotron freq./field	$\omega^e_{\text{cycl}}/B = e/m_e$	$1.758\,819\,62(53)\times10^{11}$ rad s⁻¹ T⁻¹	0.30
Proton cyclotron freq./field	$\omega^p_{\text{cycl}}/B = e/m_p$	$9.578\,830\,9(29)\times10^{7}$ rad s⁻¹ T⁻¹	0.30

Table 1.1. *(cont.)*

Quantity	Symbol, equation	Value	Uncert. (ppm)
Gravitational constant	G_N	$6.672\,59(85)\times10^{-11}\ \mathrm{m^3\,kg^{-1}\,s^{-2}}$	128
		$=6.707\,11(86)\times10^{-39}\ \hbar c\ (\mathrm{GeV}/c^2)^{-2}$	128
Standard grav. accel., sea level	g	$9.806\,65\ \mathrm{m\,s^{-2}}$	exact
Avogadro constant	N_A	$6.022\,136\,7(36)\times10^{23}\ \mathrm{mol^{-1}}$	0.59
Boltzmann constant	k	$1.380\,658(12)\times10^{-23}\ \mathrm{J\,K^{-1}}$	8.5
		$=8.617\,385(73)\times10^{-5}\ \mathrm{eV\,K^{-1}}$	8.4
Molar volume, ideal gas at STP	$N_A k(273.15\ \mathrm{K})/(101\,325\ \mathrm{Pa})$	$22.414\,10(19)\times10^{-3}\ \mathrm{m^3\,mol^{-1}}$	8.4
Wien displacement law constant	$b=\lambda_{\max}T$	$2.897\,756(24)\times10^{-3}\ \mathrm{m\,K}$	8.4
Stefan–Boltzmann constant	$\sigma=\pi^2k^4/60\hbar^3 c^2$	$5.670\,51(19)\times10^{-8}\ \mathrm{W\,m^{-2}\,K^{-4}}$	34
Fermi coupling constant	$G_F/(\hbar c)^3$	$1.166\,39(2)\times10^{-5}\ \mathrm{GeV^{-2}}$	20
Weak mixing angle	$\sin^2\hat\theta(M_Z)\ (\overline{\mathrm{MS}})$	$0.2319(5)$	2200
$W^\pm$ boson mass	m_W	$80.22(26)\ \mathrm{GeV}/c^2$	3200
Z^0 boson mass	m_Z	$91.187(7)\ \mathrm{GeV}/c^2$	77
Strong coupling constant	$\alpha_s(m_Z)$	$0.116(5)$	43 000

$\pi=3.141\,592\,653\,589\,793\,238$ $e=2.718\,281\,828\,459\,045\,235$ $\gamma=0.577\,215\,664\,901\,532\,861$

$1\ \mathrm{in}=0.0254\ \mathrm{m}$	$1\ \mathrm{G}\equiv10^{-4}\ \mathrm{T}$	$1\ \mathrm{eV}=1.602\,177\,33(49)\times10^{-19}\ \mathrm{J}$
$1\ \text{Å}\equiv10\ \mathrm{nm}$	$1\ \mathrm{dyne}\equiv10^{-5}\ \mathrm{N}$	$1\ \mathrm{eV}/c^2=1.782\,662\,70(54)\times10^{-36}\ \mathrm{kg}$
$1\ \mathrm{barn}\equiv10^{-28}\mathrm{m^2}$	$1\ \mathrm{erg}\equiv10^{-7}\ \mathrm{J}$	$2.997\,924\,58\times10^9\ \mathrm{esu}=1\mathrm{C}$

kT at $300\ \mathrm{K}=[38.681\,49(33)]^{-1}\ \mathrm{eV}$

$0°\mathrm{C}\equiv273.15\ \mathrm{K}$

$1\ \mathrm{atmosphere}=760\ \mathrm{torr}\equiv101\,325\ \mathrm{Pa}$

Notes:

[a] The meter is defined to be the length of path traveled by light in vacuum in $1/299\,792\,458$ s. See B. W. Petley, *Nature* **303**, 373 (1983).

[b] At $Q^2=0$. At $Q^2\approx m_w^2$, the value is approximately $1/128$.

Source: From Ref. 1.1.

Table 1.2. *Atomic and nuclear properties of materials*

Material	Z	A	Nuclear total cross section σ_T (barn)	Nuclear inelastic cross section σ_I (barn)	Nuclear collision length λ_T (g/cm²)	Nuclear interaction length λ_I (g/cm²)	$dE/dx\|_{min}$ ($\frac{MeV}{g/cm^2}$) ( ) is for gas	Radiation length X_0 (g/cm²) ( ) is for gas	(cm)	Density (g/cm³) (g/ℓ) ( ) is for gas	Refractive index n ( ) is $(n-1)\times10^6$ for gas
H₂ gas	1	1.01	0.0387	0.033	43.3	50.8	(4.103)	61.28	865	(0.0838)[0.090]	[140]
H₂ (B.C., 26K)	1	1.01	0.0387	0.033	43.3	50.8	4.045	61.28	865	0.0708	1.112
D₂	1	2.01	0.073	0.061	45.7	54.7	(2.052)	122.6	757	0.162[0.177]	1.128
He	2	4.00	0.133	0.102	49.9	65.1	(1.937)	94.32	755	0.125[0.178]	1.024[35]
Li	3	6.94	0.211	0.157	54.6	73.4	1.539	82.76	155	0.534	—
Be	4	9.01	0.268	0.199	55.8	75.2	1.594	65.19	35.3	1.848	—
C	6	12.01	0.331	0.231	60.2	86.3	1.745	42.70	18.8	2.265	
N₂	7	14.01	0.379	0.265	61.4	87.8	(1.825)	37.99	47.0	0.808[1.25]	1.205[300]
O₂	8	16.00	0.420	0.292	63.2	91.0	(1.801)	34.24	30.0	1.14[1.43]	1.22[266]
Ne	10	20.18	0.507	0.347	66.1	96.6	(1.724)	28.94	24.0	1.207[0.900]	1.092[67]
Al	13	26.98	0.634	0.421	70.6	106.4	1.615	24.01	8.9	2.70	—
Si	14	28.09	0.660	0.440	70.6	106.0	1.664	21.82	9.36	2.33	—
Ar	18	39.95	0.868	0.566	76.4	117.2	(1.519)	19.55	14.0	1.40[1.782]	1.233[283]
Ti	22	47.88	0.995	0.637	79.9	124.9	1.476	16.17	3.56	4.54	—
Fe	26	55.85	1.120	0.703	82.8	131.9	1.451	13.84	1.76	7.87	—
Cu	29	63.55	1.232	0.782	85.6	134.9	1.403	12.86	1.43	8.96	—
Ge	32	72.59	1.365	0.858	88.3	140.5	1.371	12.25	2.30	5.323	—
Sn	50	118.69	1.967	1.21	100.2	163	1.264	8.82	1.21	7.31	—
Xe	54	131.29	2.120	1.29	102.8	169	(1.255)	8.48	2.77	3.057[5.858]	[705]
W	74	183.85	2.767	1.65	110.3	185	1.145	6.76	0.35	19.3	—
Pt	78	195.08	2.861	1.708	113.3	189.7	1.129	6.54	0.305	21.45	—
Pb	82	207.19	2.960	1.77	116.2	194	1.123	6.37	0.56	11.35	—
U	92	238.03	3.378	1.98	117.0	199	1.082	6.00	≈0.32	≈18.95	—

Table 1.2. (cont.)

Material	Nuclear collision length λ_T (g/cm²)	Nuclear interaction length λ_I (g/cm²)	$dE/dx\vert_{min}$ ($\frac{MeV}{g/cm^2}$) ( ) is for gas	Radiation length X_0 (g/cm²) ( ) is for gas	(cm) ( ) is for gas	Density (g/cm³) ( ) is for gas (g/ℓ)	Refractive index n ( ) is $(n-1)\times10^6$ for gas (273) [293]
Air, (20 °C, 1 atm), [STP]	62.0	90.0	(1.815)	36.66	[30 420]	(1.205) [1.29]	(273) [293]
H₂O	60.1	84.9	1.991	36.08	36.1	1.00	1.33
CO₂	62.4	90.5	(1.819)	36.2	[18 310]	[1.977]	[410]
Shielding concrete	67.4	99.9	1.711	26.7	10.7	2.5	—
Borosilicate glass (Pyrex)	66.2	97.6	1.695	28.3	12.7	2.23	1.474
SiO₂ (fused quartz)	67.0	99.2	1.697	27.05	11.7	2.32	1.458
Methane (CH₄)	54.7	74.0	(2.417)	46.5	[64 850]	0.423 [0.717]	[444]
Ethane (C₂H₆)	55.73	75.71	(2.304)	45.66	[34 035]	0.509 (1.356)	(1.038)
Propane (C₃H₈)	—	—	(2.262)	—	—	(1.879)	
Isobutane ((CH₃)₂CHCH₃)	56.3	77.4	(2.239)	45.2	[16 930]	[2.67]	[1900]
Octane, liquid (CH₃(CH₂)₆CH₃)	—	—	2.123	—	—	0.703	
Paraffin wax CH₃(CH₂)$_n$CH₃, $\langle n\rangle\approx25$	—	—	2.087	—	—	0.93	
Nylon, type 6	—	—	1.974	—	—	1.14	
Polycarbonate (Lexan)	—	—	1.886	—	—	1.200	
Polyethylene terephthlate (Mylar) (C₅H₄O₂)	60.2	85.7	1.848	39.95	28.7	1.39	—
Polyethylene (monomer CH₂=CH₂)	56.9	78.8	2.076	44.8	≈47.9	0.92–0.95	—
Polyimide film (Kapton)	—	—	1.820	—	—	1.420	
Polymethylmethacrylate (Lucite, Plexiglas) (monomer (CH₂=C(CH₃)CO₂CH₃))	59.2	83.6	1.929	40.55	≈34.4	1.16–1.20	≈1.49
Polystyrene, scintillator (monomer C₆H₅CH=CH₂)	58.4	82.0	1.936	43.8	42.4	1.032	1.581

Polytetrafluoroethylene (Teflon) (monomer $CF_2=CF_2$)	—	—	1.671	—	—	2.20	—
Polyvinyltoluene, scintillator (monomer 2-$CH_3C_6H_4CH=CH_2$)	—	—	1.956	—	—	1.032	—
Barium fluoride (BaF_2)	92.1	146	1.303	2.05	9.91	4.89	1.56
Bismuth germanate (BGO) ($Bi_4Ge_3O_{12}$)	97.4	156	1.251	1.12	7.98	7.1	2.15
Cesium iodide (CsI)	—	—	1.243	—	—	4.51	—
Lithium fluoride (LiF)	62.00	88.24	1.514	14.91	39.25	2.632	1.392
Sodium fluoride (NaF)	66.78	97.57	1.69	11.68	29.87	2.558	1.336
Sodium iodide (NaI)	94.8	152	1.305	2.59	9.49	3.67	1.775
Silica Aerogel	65.5	95.7	1.83	≈150	29.85	0.1–0.3	$1.0+0.25\rho$
NEMA G10 plate	62.5	90.2	1.87	19.4	33.0	1.7	—

Note: Table revised June 1994. Gases are evaluated at 20 °C, 1 atm (in parentheses), or at STP [square brackets].

Source: From Ref. 1.1.

1.4 Coupling constants

What about the characteristic strength of the forces between particles? These are defined by dimensionless coupling constants whose size gives us a measure of the strength of the interaction. For electromagnetism we have the familiar fine-structure constant, α, which has the strength of roughly 1/137, Table 1.1. The strong interactions have a coupling constant α_s which is larger because the forces are stronger. Why is α dimensionless? The dimension of $U = e^2/r$, indicated by brackets, is energy so $[e^2] = [\text{energy} \cdot \text{length}]$. The dimension of $\hbar$ is [energy · time] and of $\hbar c$ is [energy · length]. Thus $\alpha = e^2/\hbar c$ is indeed dimensionless.

$$\alpha = e^2/\hbar c \sim 1/137$$

$$\alpha_s = g_s^2/\hbar c \sim (1/10 - 1) \text{ (Table 1.1)} \tag{1.2}$$

The electromagnetic force is very familiar in everyday life. The existence of a strong force may be inferred by the observation that nuclei are bound systems containing one or more protons. Coulomb repulsion of the protons must then be overcome by another, stronger force. The strong interaction is responsible for binding the hadronic, or strongly interacting, particles like the proton together. Nuclei are then bound together by means of a residual 'strong interaction' in analogy to the van der Waals molecular binding of neutral atoms. Our treatment of the strong interactions will be completely phenomenological and contained only in the last two chapters. We will not discuss the weak interaction which is responsible for radioactive decay nor the gravitational interaction. Detectors using non-destructive methods usually use only the electromagnetic interaction of a particle with the detecting medium.

1.5 Atomic energy scales

Let us now turn towards a description of the energy scales of systems, beginning with the atomic energy scale. The atom is held together by the electromagnetic attraction between the nucleus and the 'orbiting' electrons. Since the proton is so much heavier than the electron, $m_e/m_p \sim 1/2000$ (Table 1.1), it is essentially at rest and there is only one dynamic mass scale in the problem which is the electron mass, as illustrated in Fig. 1.1. The bound state lowest Bohr energy level is E_o.

$$E_o = -mc^2 \, \alpha^2/2 \text{ (Table 1.1, Rydberg)}$$

$$m_e c^2 = 0.51 \text{ MeV}, \quad E_o = -13.6 \text{ eV} \tag{1.3}$$

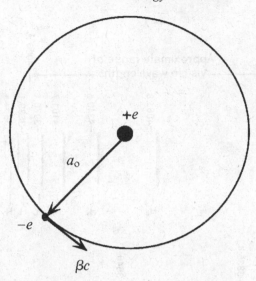

Fig. 1.1. Definition of orbital radius and velocity of the electron.

We know experimentally from spectroscopy that the energy levels of bound systems are quantized. The hydrogen atom is characterized by a quantum number n defining an energy $E_n = E_o/n^2$. An example of data for atomic hydrogen is shown in Fig. 1.2 where the discrete atomic spectrum of the emitted light is obvious and the $1/n^2$ behavior of the lines is evident. Let us compute the H_α wavelength to give us confidence. This line is due to a transition from $n = 3$ to $n = 2$ which causes emission of a photon with energy (Eq. 1.3) of 1.89 eV. Using the Planck constant to convert energy to frequency, $\lambda = 2\pi(\hbar c)/1.89 \ \text{eV} = 6645$ Å as seen in Fig. 1.2. The ionization continuum where the atom is broken apart and the electron is freed is also seen in Fig. 1.2.

It is easy to 'derive' the Bohr energies by appealing to the minimization of the total system energy in the ground, or lowest energy, state. The kinetic, T, plus the potential energy, U, is the total system energy, E, which is a conserved constant. There is a balance between the negative attractive potential, whose magnitude increases as the radius of the system, a, decreases and the positive kinetic energy which also increases as the radius decreases. Since T goes as $1/a^2$, while U goes as $1/a$, there is a stable minimum. The minimum occurs at the first Bohr radius, a_o, which defines the atomic size scale, 0.54 Å, Table 1.1. The Heisenberg uncertainty principle for position and momentum uncertainties $\Delta p, \Delta x, \Delta p \, \Delta x \sim \hbar$ supplies the crucial element of this 'derivation'. Throughout this text p represents particle momentum, and we assume that the uncertainty in p, Δp, is of order p while the uncertainty in position, Δx, is of order a.

$$T + U = E, \ \Delta p \Delta x \sim \hbar \sim pa$$

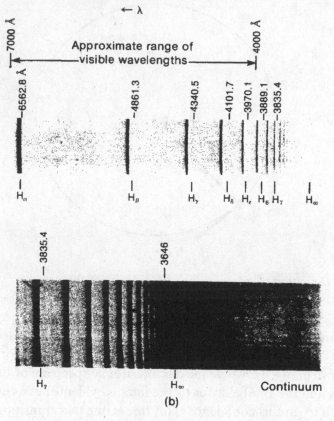

Fig. 1.2. Discrete emission spectrum for atomic hydrogen. (From Ref. 1.7, with permission.)

$$\frac{(\Delta p)^2}{2m} + U = E = \frac{\hbar^2}{2ma^2} - e^2/a$$

$$\partial E/\partial a = 0 \text{ at } a = a_\mathrm{o} \qquad (1.4)$$

The other characteristic length scale is the Compton wavelength of the electron, $\lambdabar = \hbar/mc$, which is much smaller than a_o (see Table 1.1). The characteristic size of the hydrogen atom is the Compton wavelength divided by α.

$$a_\mathrm{o} = \lambdabar/\alpha = 0.54\,\text{Å}, \quad E = E_\mathrm{o} = -e^2/2a_\mathrm{o}$$

$$\lambdabar = \hbar/mc = 0.004\,\text{Å (Table 1.1)} \qquad (1.5)$$

$$\beta = \alpha, \quad \beta \equiv v/c$$

For example, if the coupling, α, went to zero the characteristic size for the system would become infinite, which means that the system is unbound. A larger value of α would mean tighter binding or a smaller bound state radius. Plugging the Bohr radius back into the expression for E we can show that the electron velocity with respect to that of light, β, is in fact equal to the fine-structure constant. Therefore, the atom is fairly loosely bound and its motion is non-relativistic, $\beta \ll 1$.

Perhaps this should not be surprising because we know that, for example, batteries are driven by processes with atomic (or chemical) energy scales. Hence we expect them to produce of the order of 1 volt and they do. As another example, since the atomic energy scales are electron volts, we expect the light which is emitted in making transitions between those energy levels to be a few thousand angstroms which is indeed the wavelength of visible light, Eq. 1.1. Thus, if we knew only that human beings see at the thousand angstrom scale of wavelength, we could infer that they live in a world with atomic energy levels that were of the order of electron volts.

1.6 Atomic size

The structure of multielectron atoms is determined by the Fermi exclusion principle which states that one and only one electron can occupy a quantum state. Recall that the energy levels in the hydrogen atom are $E_n = E_0/n^2$, $n = 1$, $2 \ldots$. Higher level atomic states, with larger n, have more momentum, and are thus less deeply bound. As n increases the electrons are less tightly bound, $a_n = a_0 n^2$ (see Appendix B). Therefore in atoms lower n values are filled by electrons first.

The angular momentum, L, is quantized in units of $\hbar$, $L^2 = \ell(\ell+1)\hbar^2$, $\ell < n$ and the number of projections of angular momentum along an arbitrary axis is $L_z = m\hbar$, $-\ell < m < \ell$ or $(2\ell+1)$ projections. The electron spin, $s_z = s\hbar/2$ can be $s = +1$ or or $s = -1$. Thus there are $2(2\ell+1)$ electron states for a given ℓ. These are called 'shells'. Since the centrifugal force pushes us away from the origin, the effective potential is positive (repulsive) and lower ℓ states lie lower in energy and hence are filled first (see also Appendix B). The fact that lower n fill first and that within a given n lower ℓ fill first is sufficient to understand much of the periodic table of the elements.

The atomic size, a, and ionization potential I (the energy needed to free a single electron from the element) of the first, $Z \leqslant 36$, elements are shown in Fig. 1.3. Note that the atomic 'shell structure' (labeled in the figure) is clearly mirrored in the behavior of I, which rises until the shell is filled at the location of a noble gas. That is $n = 1$, $\ell = 0$, 2 states (1s). For $n = 2$, $\ell = 0$, 2 states (2s) plus $\ell = 1$ or 6 states (2p). For $n = 3$, $\ell = 0$, 2 states (3s), plus $\ell = 1$ or 6 states (3p). At

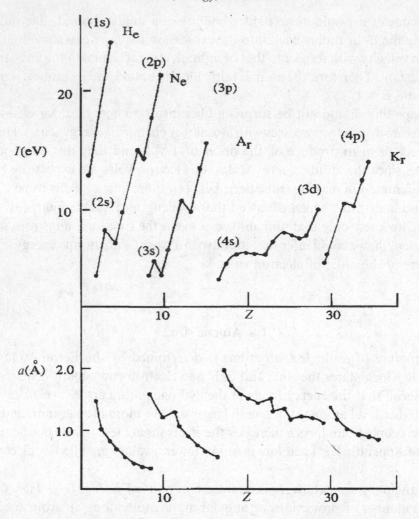

Fig. 1.3.   Ionization potential and atomic size for atoms with $Z \leq 36$. The scales are ~10 eV for I and ~1 Å for a.

that point $n=4$ $\ell=0$, 2 states (4s) intervenes prior to $n=3$, $\ell=2$ or 10 states (3d).

The behavior of the size is to decrease with the filling of the shell. The largest atoms therefore correspond to metals with a single electron loosely bound outside a closed (noble gas) shell. The variation of a and I is less than a factor of about 4 for $Z<36$. Thus, to lowest approximation, all atoms have the same size because as Z increases the force on the electrons becomes stronger but the Pauli exclusion principle forces the occupation of higher quantum states, n, which are more loosely bound. The inner closed shells of electrons are tightly bound and largely chemically inert and only the outer or valence electrons are chemically active.

The small size of atoms means that we perceive macroscopic bodies as continuous. For example, for atoms of 10 Å size, a $1\,m \times 1\,m \times 1\,m$ object has 1 billion $\times$ 1 billion 'pixels'. Compared to a computer screen of high resolution this number is enormous. A dust speck 0.1 mm on a side, just visible to us, contains $100\,000 \times 100\,000$ atomic 'pixels'.

1.7 Atomic spin effects

In what follows we will essentially ignore spin effects. The reason we can do this is that the energy of an electron magnetic moment, $\mu = e\hbar/2m_e$, in a magnetic field, B, is, $E_B \sim \mu B$. Numerically $\mu \sim 6 \times 10^{-6}$ eV/kG, Table 1.1. For $B = 20$ kG, $E_B \sim 1.2 \times 10^{-4}$ eV. Hence, the spin energies are normally tiny with respect to E_o which allows us to ignore them for the purposes of describing detectors.

1.8 Cross section and mean free path

Let us now consider the cross section, σ, which describes the probability of collision. It gets its name from the fact that in some cases it is related to the geometric cross sectional area of the system which is scattered from. The kinematic definitions of circular frequency, ω, wavelength, λ, and wave number, k, for an incident wave packet scattering off an extended object of radius a are shown in Fig. 1.4. The linear frequency is f and $\omega = 2\pi f$. Assuming the velocity of propagation is c, then $f\lambda = c$. The wave number k is related to ω as, $k = \omega/c$, $k = 1/2\pi\lambda$.

The geometric cross section of the scattering centers is classically simply the cross sectional area of a sphere of characteristic size a.

$$\sigma_{atom} \sim \pi a_o^2 \sim 3 \times 10^8 \, b, \quad a_o \sim 1\,\text{Å}$$

$$\sigma_{nuc} \sim \pi a_N^2 \sim 31 \, mb, \quad a_N \sim 1\,\text{fm}$$

$$(\hbar c)^2 = 0.4 \, \text{GeV}^2 \text{mb}$$

$$1\,mb = 10^{-27} \, cm^2, \quad 1\,b = 10^{-24} \, cm^2$$

(1.6)

The geometric cross section for atoms is expected to be hundreds of millions of barns, while the cross section for nucleons is expected to be on the order of tens of millibarns. That ratio of 10^{10} is simply the square of the ratios of the characteristic sizes of the systems. For reference, the dimensionless quantity $(\hbar c)^2$ is 0.4 (GeV)^2 mb which means that a system bound with typical energies

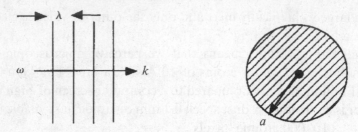

Fig. 1.4.   Kinematic quantities for wave packet scattering off an extended object of size $\sim a$.

of GeV has a geometric cross section of order millibarns. The cross section will be a fundamental concept in what follows because it quantifies the probability of a particle to interact with the detecting medium.

It is useful to define a quantity related to σ which is called the mean free path, $\langle L \rangle$. The quantity $\langle L \rangle$ is defined to be the average distance between scatterings. The scatterings are distributed in distance x of material traversed as $\exp(-N_o \rho x \sigma / A)$, where σ is the cross section to scatter off a nucleus, N_o is Avogadro's number, (see Table 1.1), ρ is the mass density of scattering centers and A is the atomic weight. Thus $N_o \rho / A$ is the number of nuclei per unit volume of the medium having density ρ.

This equation can be easily visualized. Suppose we throw N objects at a slab of material of thickness dx. We cannot aim at the nuclei, so the chance to scatter is probabilistic. The number of nuclei per unit area transverse to the incident x direction is $(N_o \rho dx / A)$. Therefore, if each nucleus has an effective transverse area of σ, the probability to scatter is $dN = -N(N_o \rho \sigma / A)dx$ which has the solution

$$N(x) = N(0)\exp(-N_o \rho x \sigma / A) \qquad (1.7)$$

The mean free path $\langle L \rangle$ is then

$$\langle L \rangle^{-1} = N_o \rho \sigma / A \text{ cm}^{-1}$$
$$\langle L\rho \rangle^{-1} = N_o \sigma / A \text{ (g/cm}^2)^{-1} \qquad (1.8)$$

We define the mean free path in centimeters or in grams per centimeter squared by using the mass density ρ. The latter definition is more compact in that different systems will typically have mean free paths which are fairly constant in g/cm^2 units. See, for example, the column of entries in Table 1.2 for λ_T which varies by only a factor of about 2 over the full periodic table. As defined, $dN/N = -dx/\langle L \rangle$, so that $\langle x \rangle = \langle L \rangle$ where the bracket notation denotes an average over some statistical distribution. For example, taking σ_{atom} from Eq. 1.6, assuming a gas with $\rho = 10^{-3}$ g/cm^3 (Table 1.2) and $A = 10$ we find an atomic mean free

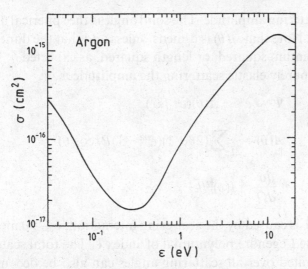

Fig. 1.5.   Atomic cross section off argon as a function of projectile energy. (From Ref. B.1, with permission.)

path of $\langle L \rangle_{\text{atom}} \sim 5 \times 10^{-5}$ cm $= 0.5$ μm. In contrast, using σ_{nuc}, we find $\langle L \rangle_{\text{nuc}} \sim 5.5 \times 10^5$ cm in the gas, or ~ 550 cm in a solid of density equal to 1 g/cm^3. Thus the atomic mean free path in a gas is of order μm, which we refer to in Chapter 8, while the nuclear mean free path in a solid is of order meters. Clearly detectors of atoms can be rather compact.

The atomic cross section for scattering on argon as a function of projectile energy is shown in Fig. 1.5. It should be noted that, although there is an energy dependent structure, the estimate given in Eq. 1.6 is certainly a reasonable representation of the typical size of the atomic cross section. The two decades shown span the region from 10^7 to 10^9 barns.

1.9 Partial waves and differential cross section

In order to motivate the geometric cross section given in Eq. 1.6 let us very briefly review the quantum mechanical theory of scattering [6]. Some additional detail is given in Appendix B.

The wave function of a scattering state, Ψ, consists of an incoming plane wave and an outgoing spherical wave with a scattering amplitude, A, which can be decomposed into individual partial waves since central forces conserve angular momentum. The wave number is k which is related to the momentum p as $\mathbf{p} \equiv \hbar \mathbf{k}$. In the spinless case the conservation of angular momentum means that the quantum number ℓ is conserved. The elastic differential cross section, or the distribution function for the scattering angle, $d\sigma/d\Omega$, is related to the square of the scattering amplitude. The solid angle is the spherical area element $d\Omega = d(\cos \theta)\, d\phi$. Note that $A(\theta)$ is dimensionless and that the dimensions of σ are inverse momentum squared, or length squared, as expected.

In the case of purely elastic scattering the amplitude is

$$\Psi \rightarrow e^{ikz} + A(\theta)(e^{ikr}/kr)$$

$$A(\theta) = \frac{1}{2i}\sum_{\ell}(2\ell+1)(e^{i\delta\ell}-1)P_{\ell}(\cos \theta) \tag{1.9}$$

$$\frac{d\sigma}{d\Omega} = |A(\theta)|^2/k^2$$

The phase shift δ_{ℓ} is caused by the scattering interaction; $\delta_{\ell} \rightarrow 0$ means $A(\theta) \rightarrow 0$. $P_{\ell}(\cos \theta)$ is the Legendre polynomial of index ℓ. The total scattering cross section, σ, integrated over all scattering angles can also be decomposed as a sum over partial waves.

$$\sigma = \frac{4\pi}{k^2}\sum_{\ell}(2\ell+1)\sin^2\delta_{\ell}$$

$$= \frac{4\pi}{k^2}[\mathrm{Im}(A(0))] \tag{1.10}$$

The optical theorem relates the imaginary part of the forward scattering amplitude, $\mathrm{Im}(A(0))$, to the total cross section, σ. We can relate the total cross section to the index of refraction in order to make the connection to optics. As we will see in Chapter 3, a wave propagating in a medium is characterized by an index of refraction n. Since the phase factor of the associated field is $i(\mathbf{k}\cdot\mathbf{x}-\omega t)$ and $\omega/k = c/n$, the attenuation of the wave is $\exp[-(\mathrm{Im}k)x]$. Since the probability goes as the square of the wave function, the mean free path is $\langle L\rangle = 1/2\, \mathrm{Im}k = c/[2\omega\mathrm{Im}(n)]$. (See Chapter 3.)

Unitarity is another incarnation of the conservation of probability. Each partial wave, labeled by ℓ, has a unitary upper limit which, as can be seen from Eq. 1.10, occurs when the sine of the phase shift δ_{ℓ} is a maximum. The center of mass (CM) wave number is k, while the CM energy is $\sqrt{s}$. (See Appendix A.)

$$\sigma_{\ell} < \frac{4\pi}{k^2}(2\ell+1)$$

$$k = p^* \cong \sqrt{s}/2, \quad p = \hbar k, \quad \hbar = 1 \tag{1.11}$$

$$s = (\text{CM energy})^2$$

If there are absorptive processes then the phase shift becomes complex, and we must distinguish between elastic scattering, σ_{EL}, inelastic scattering, σ_{I}, and the

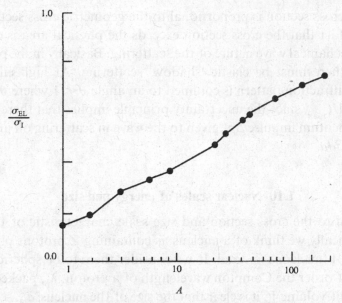

Fig. 1.6. Ratio of elastic to inelastic cross section for proton scattering off nuclei of different atomic weight A. A 'transparent' nucleus has $\sigma_{EL}/\sigma_I \to 0$ while a 'black' nucleus has $\sigma_{EL}/\sigma_I \to 1$.

total cross section, σ_T. In the absence of absorption $\sigma_{EL} = \sigma_T$, $\sigma_I = 0$. A classical example with non-zero inelastic cross section is scattering off a completely absorbing body. In that case $\sigma_{EL} = \sigma_I$, $\sigma_T \equiv \sigma_{EL} + \sigma_I = 2\sigma_{EL}$. The intermediate case of partial absorption is discussed in the references given at the end of this chapter, e.g. Ref. 6, and in Appendix B.

The ratio of σ_{EL}/σ_I is shown in Fig. 1.6 for proton scattering off nuclei (see Table 1.2). As $A \to 0$, $\sigma_{EL}/\sigma_I \to 0$ indicating little absorption, while heavy nuclei look like 'black' fully absorbing objects, since as $A \to \infty$, $\sigma_{EL}/\sigma_I \to 1$. These facts will be applied in our description of calorimetry in Chapter 12.

A geometric interpretation of the cross section is possible if the incoming wave suffers no scattering when it is outside an absorbing object of radius a but is totally absorbed when it is inside radius a. In that case there is an absorption of all angular momenta, $\mathbf{L}$, up to some maximum which is proportional to the radius.

$$\ell_{max} \sim ka(\mathbf{L} \equiv \mathbf{r} \times \mathbf{p}, \ \mathbf{L} \sim \hbar\ell, \ \mathbf{p} = \hbar\mathbf{k})$$

$$\sigma_T \sim \sum_0^{\ell_{max}} \sigma_\ell = \frac{4\pi}{k^2} \sum_0^{(ka)} (2\ell+1) \tag{1.12}$$

$$\sim 4\pi a^2$$

Indeed, the cross section is proportional to the geometric cross section of the object. The fact that the cross section exceeds the physical area is due to the quantum mechanical wave nature of the scattering. Basically in the presence of absorption there must be elastic 'shadow' scattering. At high energies the absorptive diffraction pattern is confined to an angle $\theta \ll 1$ where $\theta \sim \Delta p_T / p \sim \lambda/a \sim \hbar/ap \sim 1/\ell_{max}$, since the uncertainty principle implies that there is an irreducible momentum impulse Δp_T given to the wave in scattering off an object of size a, $\Delta p_T \sim \hbar/a$.

1.10 Nuclear scales of energy and size

We now turn to the cross section and size scale characteristic of the atomic nucleus. Basically we think of a nucleus as containing Z protons plus $(A - Z)$ neutrons or A 'nucleons' in total. If we consider them to be spherical objects with a size of order the Compton wavelength of a proton, λ_p, packed together in a nucleus of volume $\underline{V}$, it is clear that the size of the nucleus, a_N, scales as the atomic weight, A, to the 1/3 power.

$$\underline{V} = \frac{4\pi}{3} a_N^3 = A \frac{4\pi}{3} \lambda_p^3$$

$$a_N \sim \lambda_p A^{1/3}$$

(1.13)

Notice that the Compton wavelength of a proton, λ_p, is about 0.2 fermi. Referring to Eq. 1.4, $a = \lambda/\alpha$, and knowing that the strong coupling constant α_s is large, we expect a proton size of about λ_p. Using the derived geometrical interpretation of the total cross section, we expect scaling of the nuclear cross section, σ_N, as the 2/3 power of the atomic weight, and the scaling of the mean free path, $\langle L \rangle$ as the 1/3 power of the atomic weight. In Table 1.2 $\langle L \rangle$ is given, in g/cm^2 units, as the column labeled λ_I.

$$\sigma_N \sim \pi a_N^2$$

$$\sim A^{2/3}$$

$$\langle L \rangle \sim A^{1/3} \qquad \text{(Eq. 1.8)}$$

(1.14)

$$\lambda_I \sim (35 \text{ g/cm}^2) A^{1/3}$$

For example, a detection device which absorbed all the energy in an incident proton might take roughly $10\lambda_I$ since the secondary debris may itself interact. Thus, a steel device (hadron calorimeter – Chapter 12) could be expected to be roughly 1.6 m 'deep'. This device is less compact than a detector of atoms.

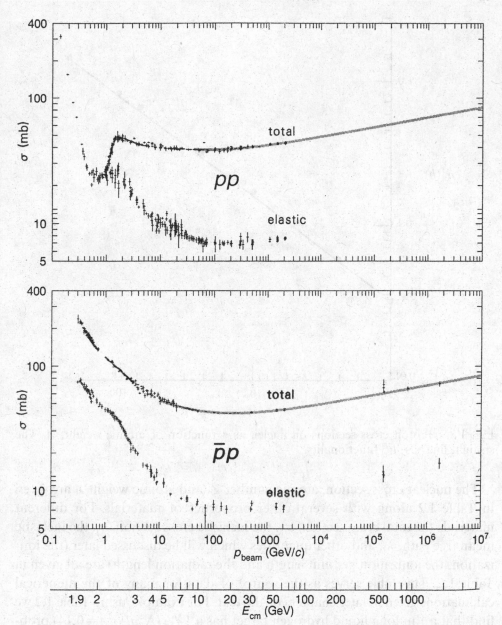

Fig. 1.7. Proton/antiproton cross section on proton as a function of projectile energy. (From Ref. 1.1.)

1.11 Nuclear cross section

In Fig. 1.7 are shown the proton–protron and antiproton–proton scattering cross section as a function of energy. The geometric cross section is about 30 millibarns which is comfortably within an order of magnitude of the observed total cross section over a substantial range of energy.

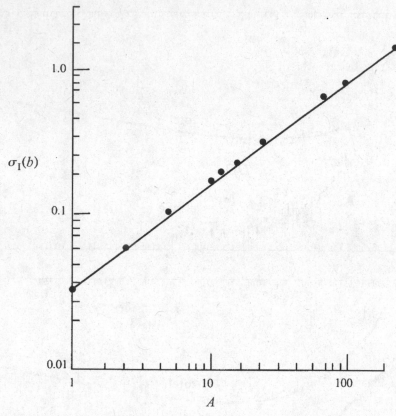

Fig. 1.8.   Proton cross sections on nuclei, as a function of atomic weight, A. The straight line has $A^{2/3}$ functionality.

The nuclear cross section, atomic number Z, and atomic weight A are given in Table 1.2 along with several other properties of materials. For different materials the total cross section, σ_T, the inelastic cross section, σ_I, the inelastic mean free path, λ_I, and other quantities which will be discussed later (the ionization, the ionization per unit length and the radiation length) are all given in Table 1.2. This table serves as the source of data for many of the numerical calculations given in later chapters of the text. For example, using Table 1.2 we find that a 1 m long liquid hydrogen target has a 14% ($N_o \rho L \sigma / A = 0.14$) probability that an incident proton will interact in traversing it.

In order to display the expected functional dependence of the inelastic cross section, that quantity is plotted as a function of the atomic weight in Fig. 1.8. It is clear that the expected power law dependence of the cross section going as $A^{2/3}$ describes the data well.

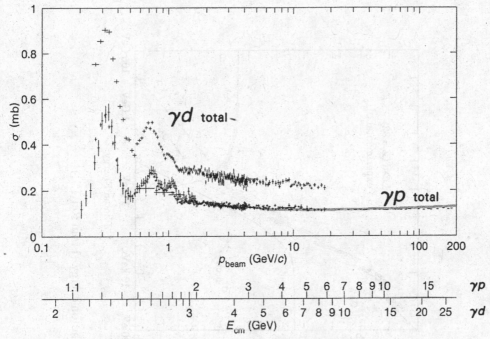

Fig. 1.9. Photon cross section on nucleons as a function of photon energy. (From Ref. 1.1.)

1.12 Photon cross section

In the case of photons scattering off protons, a compilation of the energy dependence of the total cross section is shown in Fig. 1.9. Note that the scale is about 0.2 millibarns which, compared to the nucleon scattering case, is reduced by a factor of about 100. As shown in Eq. 1.2, this is roughly the ratio of the strong to the electromagnetic coupling constant. The factor of α can be thought of as the probability that a photon will virtually disassociate into a strongly interacting particle.

The energy dependence of the cross section after some low energy structure at $\sqrt{s} \sim 2$–3 GeV is reasonably constant over several decades in energy. In analogy to the transition from nucleon–nucleon to nucleon–nucleus cross sections, in Fig. 1.10 we show the cross sections of photons on nuclei.

The high energy behavior is fairly straightforward. The energy dependence above about 1 GeV is very slight. However, the low energy behavior of the photon–nucleus cross section is rather complicated. At energy scales of a few keV the photoelectric effect, which will be discussed in Chapter 2, is the dominant process. In the intermediate range, at energy scales of order 1 MeV, the Compton cross section, discussed in Chapter 10, becomes the dominant

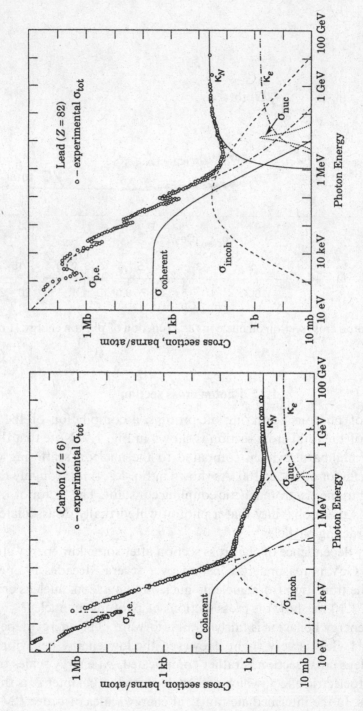

Fig. 1.10. Photon cross sections on carbon and lead as a function of photon energy. (From Ref 1.1.)

process. At high energies radiative processes such as pair production, which are constant in energy, dominate. We will discuss all of these processes in subsequent chapters. The cross section for photons on nuclei above a few MeV is of order 100 b which is intermediate between atomic (Fig. 1.5) and nuclear (Fig. 1.8) cross sections.

Some interaction rates are remarkably small. For example, we are all bathed in a sea of photons of temperature 2.3 K with a blackbody frequency spectrum which are a relic of the 'big bang' creation of the universe. The number density of these photons is about 400 photons/cm^3 which implies a flux of roughly 1.2×10^{13} photons/(cm^2 s). However, since these photons are below the energy threshold to excite any bound atoms (photon energy $\sim kT = 1/40$ eV at $T = 300$ or $kT \sim 1/4000$ eV, $k = $ Boltzmann constant, see Table 1.1), our atoms are transparent to the photons. Therefore, we do not directly experience them; they only appear in sensitive detectors. For example, they can be seen as the 'snow' on our personal photon detectors – our TV sets. If you tune to a blank channel, some of the speckles you see are relic photons from the birth of the universe.

Exercises

1. How far does light travel in a time interval of 1 nanosecond (ns) $= 10^{-9}$ s?
2. How many electrons cross the plane of a wire carrying a current of 1 ampere in 1 s? In 1 ns?
3. What is the size of the dimensionless quantity ka for the Hα line shown in Fig. 1.2? The assumption that the wavelength of the light is much greater than the size of the atom is often made. Is it valid in this case?
4. Fill in the steps to calculate a at the energy minimum for the hydrogen atom. Show that at $a = a_o$, $U = -mc^2\alpha^2/2$ and $T = -U/2$. Thus show that $T = mv^2/2 = p^2/2m = mc^2\beta^2/2$ or $\beta = \alpha$.
5. Compute the number of quantum states for $n = 1$, 2 and 3. Identify where they appear in Fig. 1.3. Identify the 'metal' which has a single loosely bound electron outside a shell at $Z = 1, 3, 11, 19$.
6. What are the radius and binding energy of an innermost electron compared to a_o and E_o for an element with atomic number Z? Ignore the effect of the other electrons. Modify the derivation of Eq. 1.4.
7. Show that $(\hbar c)^2$ is equal to 4×10^{14} (eV^{2}b).
8. Show that for a nuclear cross section of 0.23 b on carbon the mean free path is about 38 cm using the data given in Table 1.2.
9. Consider the collision of an electron with 100 eV energy with an atom of radius 10 angstroms. Show that the maximum partial wave excited has

$l\sim 50$ and that the diffraction pattern is confined to forward angles of about 1 degree.

10. Suppose an electron is bound by 1 eV in an atom of silicon. Consider a cube 1 cm on a side. Estimate how many electrons are thermally ionized in this sample at room temperature using Boltzmann statistics.

References

[1] 'Review of particle properties', *Phys. Rev D. Particles and Fields* **50** (1994).
[2] *Twentieth Century Physics*, J. Norwood, Jr., Prentice-Hall, Inc. (1976).
[3] *Elementary Modern Physics*, A.P. Arya, Addison-Wesley (1974).
[4] *Principles of Modern Physics*, R.B. Leighton, McGraw-Hill Book Company, Inc. (1959).
[5] *Atomic Physics*, M. Born, Hafnor Publishing Company (1935).
[6] *Quantum Mechanics*, L.I. Schiff, McGraw-Hill Book Company (1955).
[7] *An Introduction to Quantum Physics*, A.P. French and Edwin F. Taylor, W.W. Norton & Company, Inc. (1978).

Part II

Non-destructive measurements

The particles emitted in a collision, Fig. I.1, have position, momentum, charge, and other properties (e.g. lifetime) which we wish to measure. If the interaction with a detecting medium transfers but little energy to that medium, the measurement is called non-destructive. Within this category we distinguish between measurements of time and velocity (Chapters 2–4) and measurements using ionization (Chapter 6) but limited by scattering (Chapter 5) of the trajectory of the particles in the detecting medium (Chapters 7–9).

In the first case physical processes which depend on velocity are used, while in the latter the trajectory in electric and magnetic fields is used (Chapters 7–9) to infer the position, momentum, and charge of a particle.

Part IIA

Time and velocity

In Chapter 2, we examine the photoelectric effect whereby light is converted to an electrical signal. Devices using this effect are commonplace; for example photocells which are used to control the doors of elevators. By utilizing photo-multiplier tubes and scintillating materials, precise time measurements can be made. These 'clocks' are then used to measure velocity as the time to go to a fixed distance. In Chapter 3 we turn to the Cerenkov radiation which is emitted at a velocity dependent angle when a charged particle moves with a velocity which exceeds the velocity of light in a medium. The emitted light can be converted to an electrical signal using the techniques of Chapter 2. Chapter 4 concerns x-ray radiation emitted by a charged particle in a medium with index of refraction <1. This 'sub-threshold' Cerenkov photon emission is used to measure particles moving at velocities very near to the speed of light.

2

The photoelectric effect, photomultipliers, scintillators

> There is a crack in everything, that's how the light gets in.
> *Leonard Cohen*
> Time is nature's way of keeping everything from happening at once.
> *Anon*

Of the physical scattering processes introduced in Chapter 1, the photoelectric effect is our starting point. The effect is due to the absorption of a photon by an atom with subsequent electron emission. We first derive the photoelectric cross section and examine the regime (low photon energy) where it dominates. This chapter then takes up our first extended discussion of a detector, the photomultiplier tube. Applications of the device include time of flight measurements with fast scintillators, and coincidence logic. A discussion of light collection in scintillation counters explores both classical 'light pipes' and 'wavelength shifting' techniques. Other devices also require a knowledge of the photoelectric effect. For example see Chapter 4 where the transition radiation detector performance depends critically on photoelectric absorption of the emitted x-rays.

2.1 Interaction Hamiltonian

As we saw from the data shown in Chapter 1, the low energy interaction of a photon is dominated by the photoelectric effect. This effect was first explained quantitatively by Einstein, using the new quantum theory and treating the photon as a particle. A schematic of the process in which there is an incident plane wave with wave number k, wavelength λ, and frequency ω, is shown in Fig. 2.1. Also indicated is the energy level diagram appropriate to the kinematics. The atom is initially in a bound state with negative energy $-\varepsilon$ and is described by a wave function $u_o(r)$. The final state for the electron has momentum p and exists in the ionization continuum as a free particle.

The free particle Hamiltonian H_o, is modified in the presence of an electromagnetic field by an interaction term arising from a replacement of the momentum, $\mathbf{p} \rightarrow \mathbf{p} - ie\mathbf{A}$ where $\mathbf{A}$ is the vector potential of the electromagnetic field. This replacement is valid in both classical mechanics and quantum

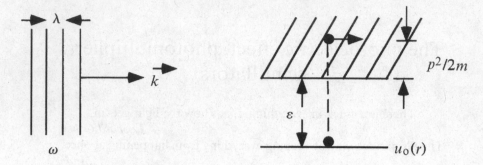

Fig. 2.1.   Energy level diagram for the kinematics of the photoelectric effect.

mechanics. It leads to an interaction term in the Hamiltonian, H_I, which is proportional to the vector potential, $\mathbf{A}$, of the photon and the momentum $\mathbf{p}$ of the electron.

$$H_0 = p^2/2m$$

$$\mathbf{p} \rightarrow \mathbf{p} - ie\mathbf{A} \tag{2.1}$$

$$H_I \sim \left(\frac{ie\mathbf{p}\cdot\mathbf{A}}{m} \right)$$

The kinematics for classical energy and momentum conservation are

$$-\varepsilon + \hbar\omega = p^2/2m$$

$$\underline{\mathbf{p}}_e + \hbar\mathbf{k} = \mathbf{p} \tag{2.2}$$

Ignoring the small energy of the recoil atom, the final state energy is simply the kinetic energy of the ionized electron. The wave vector $\mathbf{k}$ refers to the initial photon state, while $\mathbf{p}$ refers to the final state free electron. The initial electron momentum $\mathbf{p}_e$ is not observed. Therefore this free variable is 'integrated over' in the sense that the initial state wave function, $u_o(r)$, contains the probability to obtain different initial electron momenta.

2.2 Transition amplitude and cross section

In non-relativistic perturbation theory the lowest order transition amplitude is the matrix element, A, of the interaction potential between the unperturbed initial and final states.

$$A = \langle f | H_I | i \rangle$$

$$\sim \frac{e}{m} \int e^{i\mathbf{p}\cdot\mathbf{r}/\hbar} (\mathbf{A}\cdot\mathbf{p}) u_0(r) d\mathbf{r} \tag{2.3}$$

Details of the calculation are given in Appendix C. Here we simply sketch the steps.

We need an expression for the bound state wave function of the inner shell electrons in the lowest angular momentum state. Recall that the square of the wave function is the probability density. Since the characteristic size for the state is the Bohr radius, a, the wave function is contained in a volume $< \pi a^3$ or $|u_0(r)|^2 \sim 1/\pi a^3$. Referring to Appendices B and C, the bound state wave function is:

$$u_0(r) \sim \frac{1}{\sqrt{\pi a^3}} e^{-r/a}$$

$$(2.4)$$

$$a_0 = \lambda/\alpha, \quad a \sim a_0/Z$$

The inner electrons in an atom 'see' the full charge, Z, of the nucleus unscreened by the other electrons and thus are bound tightly. Therefore, their radius is reduced with respect to the hydrogenic Bohr radius by a factor $1/Z$.

The matrix element A is proportional to $(\lambda_{DB}/a)^4$ where λ_{DB} is the outgoing electron de Broglie wavelength $\lambda_{DB} \equiv \hbar/p$. The smallness of the amplitude is due to the poor overlap integral for A since $a \gg \lambda_{DB}$. Basically, it is very unlikely to find a high momentum in the initial state. Thus the photoelectric cross section, σ_{PE}, scales in a complicated way with photon energy. The main physics contribution is the $(\lambda_{DB}/a)^8 = (\hbar/pa)^8$ behavior coming from the square of the matrix element. As before, λ is the Compton wavelength of the electron.

$$\sigma_{PE} \sim \alpha \lambda^2 \left[\frac{mc^2}{\hbar \omega} \right] [\lambda_{DB}/a]^5$$

$$\sim 1/\omega^{7/2} \tag{2.5}$$

$$\hbar \omega \sim p^2/2m \quad \text{(Eq. 2.2)}$$

The cross section falls rapidly with incident photon energy; in this approximation as $1/\omega^{7/2}$. Therefore the photoelectric effect is only important at low photon energies as is, indeed, clear looking back at Chapter 1.

If the photon has an energy which is equal to the energy of one of the inner electrons, an enhanced absorption will occur. The simple approximate generalization of the Bohr result for different principle quantum numbers, n, and for inner electrons bound to atoms with atomic number Z is given below. The effect for E_n is taken care of by the replacement $U = e^2/r \rightarrow U = (Ze)(e)/r$ or $\alpha \rightarrow Z\alpha$.

$$E_n = - \left[\frac{mc^2}{2} (Z\alpha)^2 \right]/n^2$$

$$(2.6)$$

$$= -13.6 eV [Z^2/n^2]$$

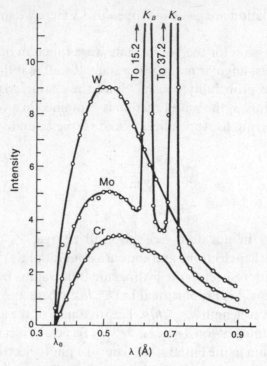

Fig. 2.2. Data on the intensity of emitted x-rays as a function of wavelength in various heavy elements. The 'quantized' lines corresponding to the binding energies of the inner electrons are very evident. (From Ref. 1.7, with permission.)

Some representative data on the intensity of the emitted x-ray for the inverse process in various heavy elements is shown in Fig. 2.2. In this inverse process, $e + A \rightarrow \gamma + A^*$, elements under electron bombardment emit x-ray photons. The photon emission occurs when the atom, A^*, de-excites by an inner electron transition accompanied by the emission of an x-ray with a quantized wave length. The scale for λ is Å, or keV energies, as expected from Eq. 2.6.

　　Data on the relationship between the square root of the emitted frequency and the atomic number is shown in Fig. 2.3. As expected from Eq. 2.6, the bound state energies, and therefore the emitted x-ray frequencies are proportional to Z^2, which is called Moseley's law. The mechanism of x-ray emission by electron bombardment is the basis of medical and dental radiology and is now a commonplace.

　　The inverse mean free path, in $(g/cm^2)^{-1}$, as a function of the photon energy incident on lead is shown in Fig. 2.4. We calculate the inner electron binding energy for lead for the first two principle quantum numbers. The computed energies from Eq. 2.6 are 91 and 23 keV respectively. As seen in Fig. 2.4, there is pronounced structure at an incident photon energy corresponding to these

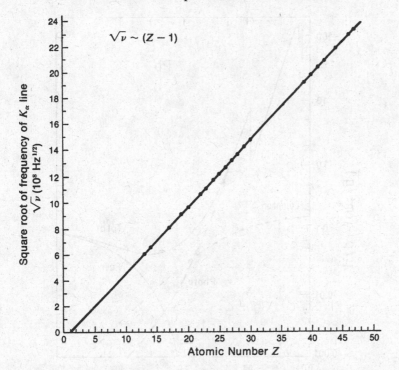

Fig. 2.3.   Data on the relationship between the x-ray wavelength and the atomic number. The $\hbar\omega = E \sim Z^2$ behavior expected from Eq. 2.6 is observed. (From Ref. 1.7, with permission.)

two energies. We compare the photoelectric cross section to the Thomson cross section, σ_T, for photon elastic scattering off free electrons given in Table 1.1 and derived later in Chapter 10. Assuming that Thomson scattering is incoherent, the sum over Z electrons is simply Z times the cross section off a single electron.

$$\sigma_{PE} \sim \frac{32\pi}{3} \sqrt{2}(Z\alpha)^4 Z\left(\frac{m}{\hbar\omega}\right)^{7/2}(\alpha\lambdabar)^2$$

$$\sigma_T \sim \frac{8\pi}{3}Z(\alpha\lambdabar)^2 \qquad (2.7)$$

$$\sigma_{PE}/\sigma_T \sim 4\sqrt{2}(Z\alpha)^4(m_e c^2/\hbar\omega)^{7/2}$$

Therefore, for heavy atoms where $Z\alpha \sim 1$, we expect the photoelectric effect to dominate over Thomson scattering for energies, $\hbar\omega < mc^2 \sim 0.51$ MeV. Indeed, this is the observed behavior and the approximate energy region where the two cross sections are comparable. For example, in lead, a photon with 10 keV

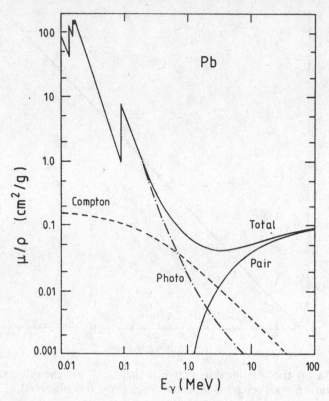

Fig. 2.4.   The inverse mean free path, $1/\langle L \rangle \rho$, for photons incident on lead. The distinct physical mechanisms of photoelectric effect, Compton scattering and pair production are evident, along with the 'resonant' behavior near the bound state energies indicated by Eq. 2.6. (From Ref. B.1, with permission.)

energy has $\sigma_{PE}/\sigma_T \sim 6.8 \times 10^5$ or $\sigma_{PE} \sim 0.46$ Mb. This estimate from Eq. 2.7 is reasonably close to the value which can be read off from Fig. 1.10.

At photon energies below about 1 MeV in lead the photoelectric effect dominates. Above 1 MeV the Compton effect (relativistic elastic photon scattering), see Chapter 10, then dominates briefly before pair production rises rapidly to be the major physical effect at high energy. At low energies a power law behavior is observed, as expected from Eq. 2.5 which gives a straight line on the log–log plot of Fig. 2.5. The mean free path, $\langle L \rangle \rho$, for photon attenuation is shown, as a function of photon energy, in different materials in Fig. 2.5 for energies from 1 keV to 100 GeV. At high energies the constant value of the pair production cross section is evident as is strong dependence on the atomic number, $\langle L \rangle \sim X_o \sim 1/Z$, where X_o is the radiation length, to be defined later.

At intermediate energies the incoherent nature of the elastic Compton scattering implies that $\langle L \rangle$ is independent of Z, as is seen for photons of 1 to 10 MeV energy.

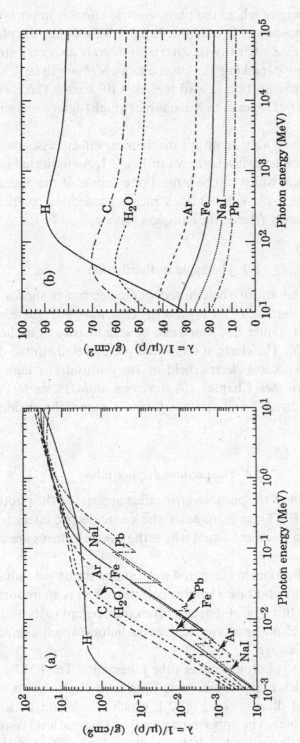

Fig. 2.5. Mean free path, $\langle L \rangle \rho$, as a function of photon energy for different materials. At low energies the photoelectric Z^5 behavior is displayed, while at high energies the energy independent $X_0 \sim 1/Z$ pair production behavior is observed. (From Ref. 1.1.)

At still lower energies where the photoelectric effect is important we see a complicated Z dependence. Looking at Eq. 2.7 we see that the photoelectric cross section goes as Z^5. Clearly the overlap integrals also vary strongly with energy. As an example, picking a photon energy of $\hbar\omega = 10$ keV, we find that the photoelectric cross section in lead is $\sim 2.0 \times 10^5$ barns. The corresponding mean free path is 0.0015 g/cm^2, in reasonable (rough) agreement with the data given in Fig. 2.5.

Note that there are also quantum mechanical effects associated with the emission due to the energy levels shown in Fig. 2.1. As indicated in Appendix B, there is a reflection which can be large. For example, if the potential well of the metal is deep, $U_o = 10$ eV, $E = 0.10$ eV means a reflection coefficient of $R \sim 0.67$ which significantly retards the emission process.

2.3 The angular distribution

The angular distribution of the emitted photoelectron is shown for several photon energies in Fig. 2.6. In the low energy case, $\hbar\omega < mc^2$, there is a classical dipole pattern (see Chapter 10), which can be seen to arise from the **p·A** factor in Eq. 2.3 for $\langle f|A|i \rangle$. The electron is emitted preferentially in the direction of the accelerating transverse electric field of the photon. For high energies, a 'searchlight' pattern (see Chapter 10) develops, $d\sigma/d\Omega < \sin^2\theta/(1 - \beta_e\cos\,\theta)^4$ leading to the electron being thrown forward along the incident photon direction.

2.4 The photomultiplier tube

A major application of the photoelectric effect appears in the photomultiplier tube (PMT). In a PMT use is made of the photoelectric effect to convert a photon signal to an electrical signal where the freed electrons are collected as a current.

As a second application, in Chapter 4 we examine transition radiation detectors which emit x-ray photons. The photoelectric effect is an important source of absorption since this type of detector is not transparent to its own emissions. Thin foils of low Z material are used as radiators which significantly self absorb the emitted x-rays

Let us now turn to photomultiplier tube kinematics. The PMT have special photocathodes which efficiently convert light into electrons. In a solid, the atomic energy level diagram of Fig. 2.1 which is relevant to single atoms becomes more complex. The uppermost populated atomic level is smeared into a continuum (the 'valence band', VB) by the mutual interaction of the electrons

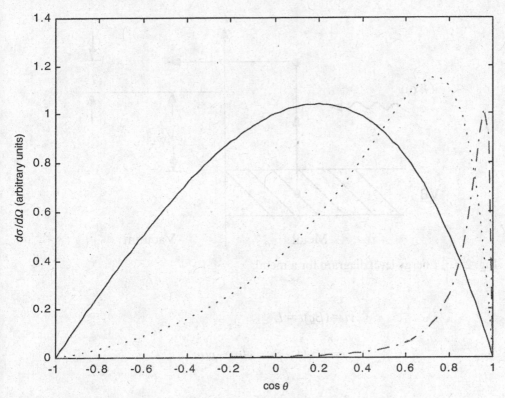

Fig. 2.6. Angular distribution of the photoelectron with respect to the incident photon direction for several recoil electron velocities. —, $\beta=0.1$; $\cdots$, $\beta=0.5$; $-\cdot-$, $\beta=0.9$.

in the solid. The electrons remain bound to the solid. The energy diagram relevant in a metal is shown in Fig. 2.7. Electrons in the metal are bound by energies $\geq eV_{WF}$ (the 'work function'). Absorption of a photon of energy $\hbar\omega$ leads to ionized electrons with kinetic energy T, $T \leq (\hbar\omega - eV_{WF})$. As the reaction time for this process is very short, PMT are very useful devices for the accurate measurement of time.

2.5 Time of flight

A first application of the PMT is the direct measurement of the time of particle passage between two fixed locations. If a system of particles of different masses has been prepared to possess a unique momentum, for example by collimation in a magnetic field (see Chapter 8), then time of flight gives us velocity and hence 'particle identification' or the mass of a given particle. Assume a fixed distance L and a time of passage t. In Eq. 2.8 the relativistic expressions for energy ε and momentum p are used. (See Appendix A.)

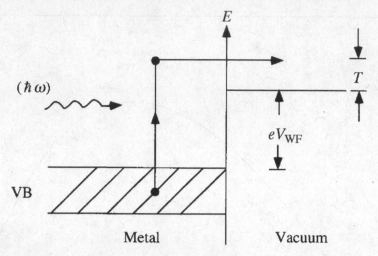

Fig. 2.7. Energy level diagram for a metal.

$$vt = (\beta c)t = L$$

$$\gamma = 1/\sqrt{1-\beta^2} = \varepsilon/m \sim p/m$$

$$\beta = \sqrt{1 - \frac{1}{\gamma^2}} \sim 1 - \frac{1}{2\gamma^2}, \quad \beta \rightarrow 1 \qquad (2.8)$$

$$t \sim \frac{L}{c}\left(1 + \frac{1}{2\gamma^2}\right)$$

$$\Delta t \sim \frac{L}{c}\left(\frac{m_1^2 - m_2^2}{2p^2}\right), \quad \beta \rightarrow 1$$

For a given flight path, L, the difference in time of flight for particles of different mass, Δt, is proportional to the difference of the squares of the masses of particles. The time separation becomes worse as the inverse of the square of the particle momentum. Therefore time of flight is typically a useful technique only for slow, almost non-relativistic, particles.

The time of flight method collapses at high energies because, by virtue of the special theory of relativity, $\beta \rightarrow 1$ independent of the mass. The required flight path at fixed time resolution goes like p^2. For example, to achieve a four standard deviation π/K separation in Δt, $m_\pi \sim 0.14$ GeV, $m_K \sim 0.49$ GeV, with a detector with a resolution of 300 ps, requires a 3 meter flight path, L, at 1 GeV and a 12 meter flight path at 2 GeV. As this example shows, we are normally limited by geometric and fiscal constraints to rather low momenta.

2.6 Scintillators and light collection

So far we have not specified how particle passage is detected and made into a light signal which impinges on the PMT. We now consider the use of plastic 'scintillators' for this purpose. These plastics consist of a 'base' structural plastic which has small additives of 'fluors' dissolved in the base. The real physics of scintillation light production is rather complicated. An incident charged particle first interacts with the atoms of the scintillator exciting a 'primary' fluorescent material which de-excites and emits photons in the ultraviolet. This light is absorbed within a mean free path of ~1 mm by a 'secondary' fluorescent material also dissolved in the 'base' plastic which emits in the visible part of the spectrum where the 'base' plastic is fairly transparent.

For example, the primary fluor might be *p*-terphenyl (PTP) which emits at 4400 Å with a 5 ns decay time, while the secondary fluor might be 'POPOP' which emits at 4200 Å in 1.5 ns. A typical base plastic is polystyrene. Suffice it to say that operationally a plastic scintillator emits blue light with a time constant for defluorescence $\lesssim 8$ ns.

Plastic scintillator is a basic workhorse detector used in many applications requiring charged particle detection. The plastic is easy to machine and can be cut and polished into shapes specific to a given application.

A scintillator plus photomultiplier tube system can be used to measure time of flight since all of the components of this system are very fast. Let us work through a typical application. A typical plastic scintillator such as NE102 has a density of about 1 g/cm³, and an index of refraction n of 1.58. It emits blue light with a characteristic decay time constant of only a few ns.

$$\tau \sim 1.6 \text{ ns}$$
$$\lambda_{max} \sim 4250 \text{ Å} (\sim 3.0 \text{ eV}) \tag{2.9}$$

As a 'rule of thumb' we get roughly one scintillation photon for every 3000 eV of deposited ionization (see Chapter 6) energy (0.1% conversion efficiency). For the case of plastic scintillator, looking in Table 1.2 (see also Chapter 6), this means a yield of about 500 photons per centimeter.

$$1\gamma/3000 \text{ eV}$$
$$500\gamma/\text{cm} \tag{2.10}$$

The emitted light needs to be captured and collected. As it is emitted isotropically inside the plastic, there is a critical angle for total internal reflection, θ_c. In plastic the critical angle is about 39° so that of order $1/3$ of the emitted photons are captured within the plastic and move towards one end. A schematic representation of the capture of the light is shown in Fig. 2.8.

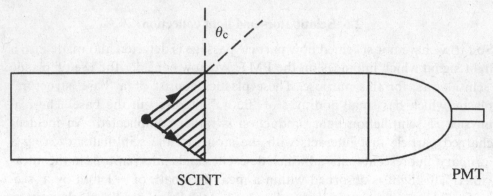

Fig. 2.8. Schematic of a scintillator with phototube readout showing the critical angle for light collection.

$$\sin \theta_c = 1/n$$

$$\theta_c \sim 39° \qquad (2.11)$$

$$\frac{\Delta\Omega}{4\pi} = \left(\frac{1 - \sin\theta_c}{2}\right) \sim 0.18$$

As a concrete example, a 0.5 cm thick scintillation counter would capture 45 photons within the critical angle. If the optical collection efficiency is only $1/2$ due to poor surface quality of the plastic, we still have 22 photons arriving at the phototube and impinging on the photocathode. The photocathode has some probability to convert the photon into a photoelectron. That probability, the 'quantum efficiency', at the wavelengths emitted by the scintillator may be of order 25%, see Fig. 2.9. We now have 6 photoelectrons coming off the photocathode. Since this is a stochastic or random process, the number of independently emitted photoelectrons is Poisson distributed and the inefficiency, or probability to observe no photoelectrons, at a mean, $\langle N \rangle$, of 6 is 0.2%, i.e. $e^{-6} \approx 0.002$. (See Appendix J.)

$$1 - \varepsilon = e^{-\langle N \rangle}, \quad \varepsilon = \text{efficiency} \qquad (2.12)$$

A plot of the window transmission coefficient for different PMT as a function of λ is shown in Fig. 2.10. Clearly we can extend the range of photon energies passed by choosing a window made of MgF_2. This ability to pass UV light will be mentioned again in our discussion of Cerenkov counters in Chapter 3. Fiscal constraints may limit us to the use of glass windows which only pass $\lambda > 3000$ Å photons. Since MgF_2 windows are expensive, one other option is to put a layer of 'wavelength shifter' on a cheap glass window. This material

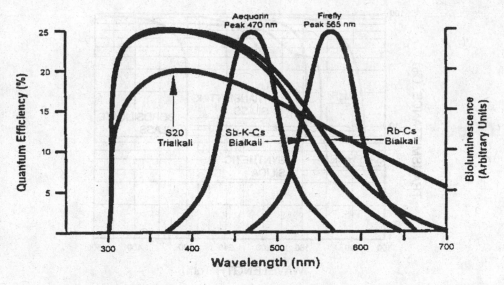

Fig. 2.9.   Photocathode 'quantum efficiency' as a function of λ of the light impinging on a PMT. (From Ref. 2.10.)

absorbs high energy photons and re-emits them isotropically at a λ matched to the peak photocathode efficiency. We will see examples of the wavelength shifting (WLS) of light later. Clearly better transmission means more photoelectrons and thus higher detection efficiency.

The photocathode layer is thin because the photoelectric cross section is very large at low photon energies. For typical PMT cathode materials the mean free path for a 1 to 3 eV photon is 10^{-5} cm or 1000 Å. Thus the cathode material may only be about 100 atoms thick and can be vacuum deposited on the glass window. (See Fig. 2.5 and extrapolate in energy as $1/\omega^{7/2}$.)

2.7 Gain and time structure

The rest of the photomultiplier consists of electrodes called 'dynodes'. At each dynode the electrons from the previous stage strike the material after falling through an accelerating voltage. A schematic of a typical PMT electrode structure is shown in Fig. 2.11. Due to secondary electron emission there is electron multiplication by a factor δ, typically of order 3, at each dynode. Therefore, the gain, G, for a tube with 14 dynode 'stages', N_d, is of order 5×10^6. This being a geometric multiplication, the total gain is given by the gain per stage, δ, raised to the power, N_d.

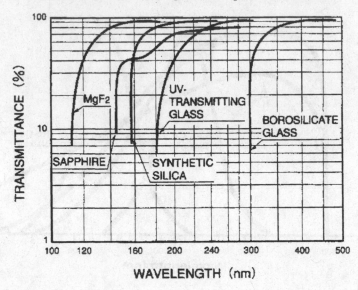

Fig. 2.10.   Transmission of different PMT window materials as a function of λ. (From Ref. 2.10.)

$$\delta = \text{gain/dynode} \sim 3$$
$$G = (\delta)^{N_d} = \text{PMT gain} \qquad (2.13)$$

Note that the variation in gain with applied voltage, which sets δ, is rapid, which means we need to be careful to regulate the voltage well. Note also that small manufacturing variations in the gain per stage lead to large changes in the overall gain. Thus we expect PMT to have variable gains from tube to tube, which we can equalize by supplying a slightly different voltage to each tube.

For example, the approximate cross section for electron multiplication is given in Eq. 2.14 where $I_o =$ ionization energy and $\varepsilon =$ incident electron energy. For $I_o = 10$ eV and $\varepsilon = 100$ eV, $\sigma \sim 6.7 \times 10^{-17}$ cm^2 or $\langle L \rangle = 1.7 \times 10^{-7}$ cm in a metal such as copper. Thus the dynodes can be very thin.

$$\sigma(e^- \rightarrow e^- + e^-) < \pi \alpha^2 / \varepsilon I_o \qquad (2.14)$$

A contribution to the rise time of the current pulse comes from the spread in kinetic energy of the ejected photoelectrons. The spread is partially caused by the spread of energies in the valence band, as is obvious from looking at Fig. 2.7. Hence there is a spread in transit time to the first dynode, dt. For an initial kinetic energy spread from zero to T, the time spread, dt, for non-relativistic motion in an electric field E is, $dt = \sqrt{2mT}/(eE)$, which for $T = 1.2$ eV in a field of $E = 150$ V/cm is $dt \sim 0.2$ ns.

Yet another contribution to the rise time is the path length difference from

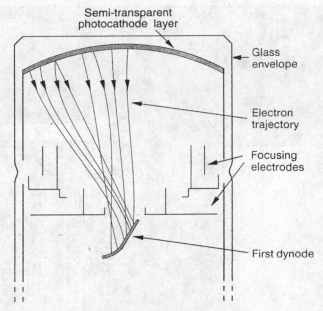

Fig. 2.11. Schematic of the electrode structure of a typical photomultiplier. (From Ref. 2.10.)

the first dynode to different locations on the cathode, as is visible in Fig. 2.11. The resultant spread in arrival times scales with the size of the cathode. Typical values of intrinsic phototube rise time (for 5 cm diameter tubes) are ~ 2 ns. These times are well matched to the scintillator decay lifetimes. (Eq. 2.9.)

Thus our six photoelectrons, having dropped through a 14 stage tube, become 4.6×10^{-12} C = 4.6 pC of anode signal charge. They are delivered in a rather short time because the photocathode response is effectively instantaneous and the PMT rise time and the fluorescence time of the scintillator are very rapid, $\tau < 3$ ns. If we assume that the total signal charge is delivered in 10 ns, it corresponds to a current of about 0.46 mA. The photomultiplier output is effectively a current source. The PMT current delivered into a standard 50 Ω coaxial cable (see Appendix D) is 23 mV. That level of voltage pulse corresponds to the typical threshold voltages, V_T, which are covered by the fast electronics which are commercially available (see Chapter 2.9).

A photograph of a photomultiplier tube and the resistive voltage divider supplying the multiple dynodes is shown in Fig. 2.12 where we get an idea of the layout of a photomultiplier tube.

An oscilloscope signal trace of the output of a photomultiplier tube driven by a scintillator is shown in Fig. 2.13. Notice that the peak output current is 50 mA and the rise time of the signal is a few ns. This plot serves to illustrate again the order of magnitude of the currents and times which were quoted in

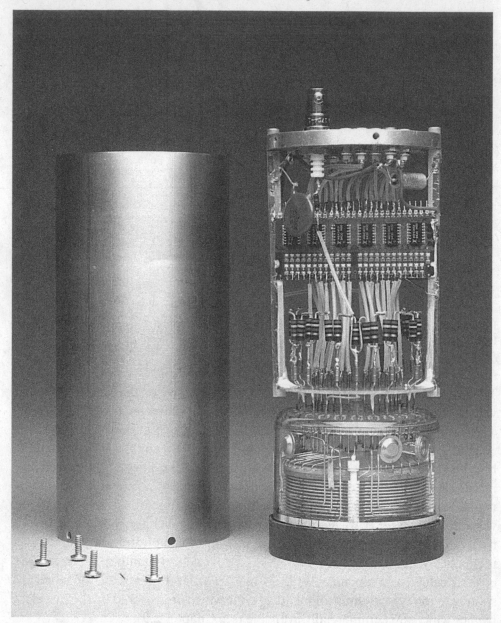

Fig. 2.12.   PMT and 'base' assembly showing the dynode structure in the PMT and the system used to supply voltage drop to the successive dynodes. (Photo – Fermilab.)

the numerical example given above. Note that there is a stochastic width to the peak height of the pulse. Its fractional spread is determined by the number of independent primary events or the number of emitted photoelectrons (see Appendix J). For low signal levels we can alternatively look at the fraction of

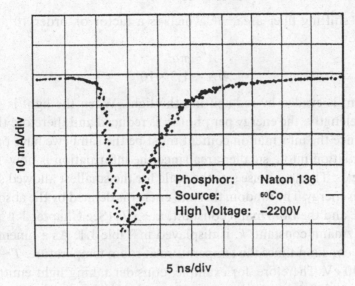

**Typical Anode Output with a Burle 8575
Photomultiplier Tube.**

Fig. 2.13. Oscilloscope trace of a PMT current pulse output. Note the rise time of ~
2 ns and the pulse width of ~ 10 ns full width at half maximum, FWHM. The peak
current is ~ 50 mA and the signal pulse height spread is ~ 10%. (From Ref. 2.10.)

time no photoelectrons are emitted and hence, see Eq. 2.12, independently esti-
mate $\langle N_{pe} \rangle$. In Fig. 2.13 the fractional spread is ~ 10% indicating that $\langle N_{pe} \rangle$ is
~ 100, which is 16 times larger than our numerical example.

2.8 Wavelength shifting

There is an interesting alternative solution to the problem of the collection of
scintillation light called 'wavelength shifting' (WLS). Typically the Liouville
theorem tells us that when we are bringing light out and routing it using total
internal reflection to a photomultiplier tube, the 'light pipe' must have an area
which does not diminish or we will lose photons. However, these light pipes are
unwieldy and ungainly and mean a loss of active detector area because they
take up a finite amount of space. The required matching photocathode surface
area is also expensive. (See Fig. 2.14a.)

There is a way to evade this difficulty by using a fiber which takes the scintillation light and shifts it to a longer wavelength, captures it by total internal
reflection and finally pipes it out to a phototransducer of some sort. Now, at
first blush this violates the Liouville theorem. For example, in a typical application one may have a ratio of the primary scintillator area, A, to the secondary

wavelength shifting fiber area, A', which is a factor of order 10^4. (See Fig. 2.14b.)

$$A' \sim \pi a^2$$
$$A/A' \sim 1.3 \times 10^4 \qquad (2.15)$$

The theorem is evaded by red shifting the light. When the light is shifted to longer wavelengths, the energy per photon is reduced and therefore the light is 'cooled'. Since the information content must be the same, we have packed the same information into a smaller area since the information is now 'stored' at lower energies. Thermal noise sets the scale for the smallest allowed storage of information energy. The random thermal energy is defined by the absolute temperature, T, and the Boltzmann constant, k, $\sim kT$. (See Chapter 1.)

The Boltzmann constant, k, is displayed in Table 1.1. As a mnemonic, the Boltzmann constant times the temperature at room temperature, $T \sim 300$ K, is roughly $1/40$ eV. Therefore, for example, consider taking light emitted in the blue at about 4500 Å and cooling it to 5000 Å in the green, wavelength shifting by about 500 Å. That 'cooling' corresponds to an energy shift of ~ 0.28 eV.

$$\lambda \sim 4500 \text{ Å} \rightarrow 5000 \text{ Å}$$
$$\Delta\varepsilon \sim 0.28 \text{ eV} \qquad (2.16)$$

The increased 'packing factor' by which we can compress the information is given by the exponential of the energy shift, $\Delta\varepsilon$, divided by kT, which defines the theoretical upper limit of information content compression. That ratio is of order 7.3×10^4 in this example, showing that we can achieve very large compression factors. That in turn allows us to make scintillation detectors which are effectively almost fully active since the wavelength shifter fiber uses up only a minuscule fraction of the area of the detector for optical readout. We will return to 'hermetic' detectors in Chapter 12.

$$e^{\Delta\varepsilon/kT} = 7.3 \times 10^4$$
$$(kT)_{300 \text{ K}} \sim 1/40 \text{ eV} \qquad (2.17)$$

A picture of the two different kinds of light pipes is given in Fig. 2.14. Figure 2.14a is a photograph of a traditional scintillation assembly with an equal area light pipe. Note how cumbersome the light piping is, even though the equal area law is not even strictly maintained. In contrast, Fig. 2.14b shows a scintillation assembly with wavelength shifter (WLS) readout. It should be clear from the photographs that the ability to cool the information allows for a much more compact and hence 'hermetic' readout.

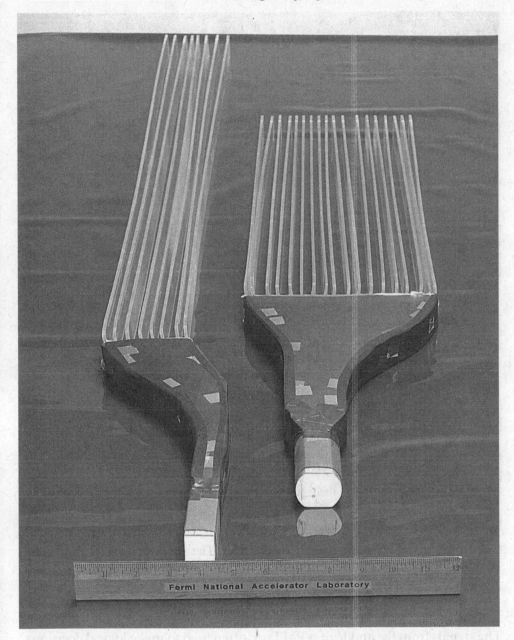

Fig. 2.14a. 'Light pipe' assembly without wavelength shifting. (Photo – Fermilab.)

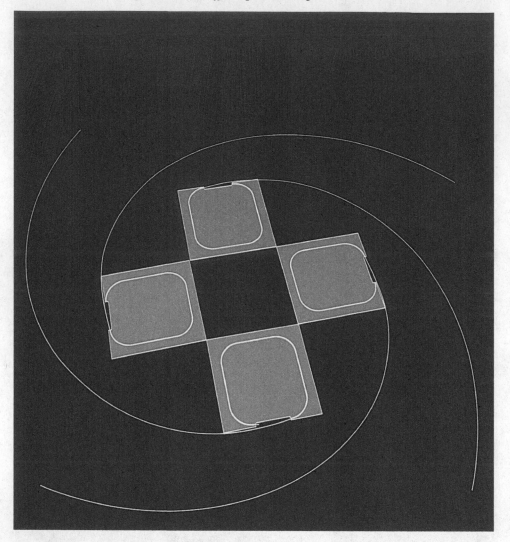

Fig. 2.14b.   Array using wavelength shifting fiber. The area ratio of scintillator plate to WLS fiber is $\sim 10^4$. (Photo – Fermilab.)

2.9 Coincidence logic and deadtime

The signals from the anode of a PMT are often used to make logical decisions, i.e. decisions based on Boolean logic. The basis of these decisions is first the initial decision between logical 'ϕ' and logical '1' and subsequent AND, OR, and NOT electronic operations on the signal. A typical two-fold AND coincidence setup is shown in Fig. 2.15. Signals on inputs 1 and 2 are first 'discriminated' by passing through circuits that give logical '1' for a fixed time (τ_1 or τ_2) if and only if $V > V_T$ ('threshold' voltage see Appendix I). These signals, C_1 and

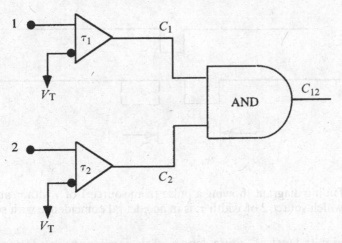

Fig. 2.15. Schematic of a typical discriminator and coincidence logic for a two input AND operation.

C_2, are input to an AND 'gate' which gives a logical '1' output, C_{12}, if and only if both of its inputs are '1', or true, at some point in time (an AND 'coincidence' gate). The method commonly used for most high speed signal connections is to use coaxial cables loaded with a dielectric medium. Although an extremely useful tool, we relegate the discussion of these cables to Appendix D.

Clearly if a particle beam passes through both scintillation counters 1 and 2, the counting rate of the AND gate, C_{12} represents the true particle beam rate. It is also possible that there are false counts due to signals accidentally coinciding in time at the inputs to the AND gate. The timing diagram shown in Fig. 2.16 makes it clear that the time interval appropriate to an accidental two-fold AND coincidence, C_{12}^A, is $(\tau_1 + \tau_2)$.

$$C_{12}^A = C_1 C_2 (\tau_1 + \tau_2) \qquad (2.18)$$

In Eq. 2.18 C_{12}^A is the accidental AND rate, C_1 and C_2 are the 'singles rates' and $\tau_1 + \tau_2$ is the 'live time'.

Clearly, a detector with a fast response time is desirable since it makes for a reduced rate of false coincidences. Typically we use logic times, τ_i, which are just sufficient to allow for the intrinsic time 'jitter' of the detector if we are in a high background rate environment. The reason is that this is the shortest live time we can use consistent with full efficiency for real coincidences. Therefore, it yields the minimum accidental rate. An example of one cause of PMT time 'jitter' was shown in Fig. 2.11. Scintillator and PMT logical decisions can be made with live times $\tau \leq 5$ ns, for spatially small scintillators. For example, if $C_1 = C_2 = 1$ MHz, then with $\tau_1 = \tau_2 = 5$ ns the accidental counting rate is $C_{12}^A = 10$ kHz. Digging low rate true signals out of high rate backgrounds often

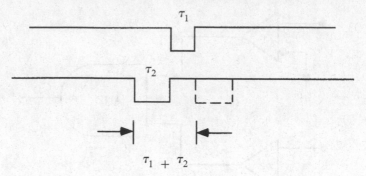

Fig. 2.16.   Timing diagram showing a pulse from source 1 of width τ_1 and the range of times for which source 2 of width τ_2 is in accidental coincidence with source 1.

requires a higher level of coincidence, with corresponding lower accidental rate.

Why were photons not seen as quanta prior to the twentieth century? An example of the numerology should help to see why. Suppose you have a 1 W light bulb and you view it with an 'eye' of area 1 mm×1 mm. Assuming a photon of 1 eV, the number of photons entering your eye at a distance of 10 m from this rather dim bulb is still 10^{10} γ/s or 10 photons/ns. Since humans function as millisecond detectors, we perceive the light as a continuous 'fluid'.

In the language of coincidence, with $\tau \sim 10^3$ s, $C\tau$ is $\sim 10^7$, and the eye integrates over millions of photons. As another example the fluorescent lights in your home are flickering on and off at $f = 60$ Hz. Since your eyes integrate due to persistence of vision you are unaware of this behavior and believe you are reading this book under a steady light. Clearly, fast devices are needed to detect the real nature of our environment, and most modern detectors react on the scale of ns.

We close this chapter with a short discussion of the concept of detector 'dead-time'. Suppose we have a detector which is dead to subsequent input signals for a time τ after producing a signal. A typical example would be a Geiger counter which must recharge after a pulse discharges it. Define C' to be the observed output counting rate and C to be the true incident rate. The fraction of time the detector is dead is simply $C'\tau$. Therefore the rate which is lost is

$$C - C' = \text{lost rate} = C(C'\tau)$$
$$C' = C/(1 + C\tau)$$
$$C' \to 0 \text{ as } \tau \to \infty \qquad (2.19)$$
$$\to C \text{ as } \tau \to 0$$

Clearly, both detectors and electronics with fast pulse recovery, or small dead time τ, are of value in that the observed rate is close to the true rate as long as

$C\tau \ll 1$. For example, if $C = 1$ MHz and $\tau = 5$ ns, then $1 + C\tau = 1.005$ or there is $\sim$ a 0.5% loss of signal. The desire to measure the true reaction rate efficiently is why we make particle detectors with short deadtimes.

Exercises

1. Consider a 1 MeV photon incident on a lead atom. Estimate the radius, a, and the ground state energy, ε, of the innermost electron. What is the momentum of the outgoing electron? What is λ_{DB}/a?

2. Estimate the photoelectric cross section using Eq. 2.7 for a 1 MeV photon on lead. Compare to Fig. 1.10 to check your answer.

3. For a particle moving with $\gamma = 5$, evaluate β, $1 - \beta$ and the approximation used in Eq. 2.8. Are these approximations good?

4. Find the time of flight difference over a 1 m flight path for pions and kaons with a 1 GeV momentum.

5. How many average detected photons are needed to make a measurement with a 1% inefficiency?

6. You might wonder how to get 100 ps time of flight resolution using PMT with 2 ns rise times. The answer is to have lots of photoelectrons and to have the time resolution defined by the fluctuation on the rise time. How many photoelectrons do you need?

7. Compare the gain of a 6 and a 10 stage tube when the gain per stage is 3.

8. Estimate the 'transit time' of a PMT, the time it takes to go from the photocathode to the anode. Treat the motion as a single drop through a fixed potential, starting at zero kinetic energy. Take a tube distance of 20 cm and a 2000 V potential. Why can we use Newtonian mechanics?

9. Suppose you have two signals which are 5 ns out of time. You want to insert a cable (see Appendices) to put them in time. How many cm of cable do you need to add to the faster signal?

10. Suppose you have two counters with random rates of 100 kHz. The resolving time is 10 ns and the real counting rate is 1000 Hz. What is the accidental rate? Suppose you want an accidental reduced rate. Show that with a third counter, $C_{123} \sim C_1 C_2 C_3 \tau^2$. How much has the accidental rate been reduced?

References

[1] *Quantum Physics*, S. Gasiorowicz, John Wiley & Sons, Inc. (1996).
[2] *Theory and Practice of Scintillation Counting*, J.B. Birks, Pergamon Press (1964).
[3] *Photomultiplier Tubes, Principles and Applications*, Philips Photonics (1995).

[4] *The Art of Electronics*, P. Horowitz and W. Hill, Cambridge University Press (1980).

[5] *Pulse Electronics*, R. Littauer, McGraw-Hill (1965).

[6] R.L. Garwin, *Rev. Sci. Inst.* **31** 1010 (1960).

[7] *Techniques of High Energy Physics*, D.M. Ritson, Wiley Interscience (1961).

[8] *X-Rays in Atomic and Nuclear Physics*, N.A. Dyson, Longmans (1973).

[9] *SCIFI93, Workshop on Scintillating Fiber Detectors*, A.D. Bross, R.C. Ruchti, and M.R. Wayne, World Scientific (1995).

[10] *Photomultipliers and Accessories*, Thorn EMI Electron Tubes (1994).

3
Cerenkov radiation

Vision – the art of seeing things invisible.
Jonathan Swift
And God said, Let there be light: and there was light.
Genesis 1:3

Cerenkov radiation is emitted when a particle moving in a medium has a velocity exceeding the velocity of light in that medium. Since the angle of emission and the intensity of radiation depends on the velocity of the particle, Cerenkov radiation can be used to determine the velocity of a charged particle. The emitted Cerenkov light can be converted into a fast electrical pulse using a PMT (see Chapter 2). For a particle with energy above 1 GeV, only a negligible fraction of its energy is radiated as photons, making the process 'non-destructive'. The energy of the incident particle is assumed to be the same in vacuum and in the medium.

3.1 Units

We begin with Maxwell's equations and introduce the key concepts of index of refraction, skin depth, and plasma frequency. These concepts find immediate application in 'deriving' the results for Cerenkov radiation, and will also appear in the next chapter 'Transition radiation'. The basic relations are given in Table 3.1 in both CGS and MKS units. As stated previously, CGS units will normally be used in symbolic calculations. However, it is easy, using the relationships given in Table 3.1, to convert back and forth from CGS to MKS units. In fact, numerical calculations will always be quoted in MKS units which are more widely used in engineering applications. It is important for us to develop a facility in both systems of units.

The basic difference between the two systems of units, as shown in Table 3.1, is in the relationship between the sources and the potentials. Clearly, there is an arbitrariness in the definition of charge in terms of physical quantities such as forces or the work done on particles. In the Gaussian, CGS, system the static electric potential V is just the charge divided by the distance from the charge to the observation point. By contrast, in MKS units, the electric potential has

Table 3.1. *Electromagnetic relations*

Quantity	Gaussian CGS	SI
Charge	$2.997\,924\,58\times10^9$ esu	$=1\,\mathrm{C}=1\,\mathrm{A\,s}$
Electron charge e	$4.803\,206\,8\times10^{-10}$ esu	$=1.602\,177\,33\times10^{-19}$ C
Potential	$(1/299.792\,458)$ statvolt	$=1\,\mathrm{V}=1\,\mathrm{J\,C^{-1}}$
Magnetic field	10^4 gauss $=10^4$ dyne/esu (ergs/esu)	$=1\,\mathrm{T}=1\,\mathrm{N\,A^{-1}\,m^{-1}}$
Lorentz force	$\mathbf{F}=q\left(\mathbf{E}+\dfrac{\mathbf{v}}{c}\times\mathbf{B}\right)$	$\mathbf{F}=q(\mathbf{E}+\mathbf{v}\times\mathbf{B})$
Maxwell's equations	$\nabla\cdot\mathbf{D}=4\pi\rho$	$\nabla\cdot\mathbf{D}=\rho$
	$\nabla\times\mathbf{H}-\dfrac{1}{c}\dfrac{\partial\mathbf{D}}{\partial t}=\dfrac{4\pi}{c}\mathbf{J}$	$\nabla\times\mathbf{H}-\dfrac{\partial\mathbf{D}}{\partial t}=\mathbf{J}$
	$\nabla\cdot\mathbf{B}=0$	$\nabla\cdot\mathbf{B}=0$
	$\nabla\times\mathbf{E}+\dfrac{1}{c}\dfrac{\partial\mathbf{B}}{\partial t}=0$	$\nabla\times\mathbf{E}+\dfrac{\partial\mathbf{B}}{\partial t}=0$
Materials	$\mathbf{D}=\varepsilon\mathbf{E},\ \mathbf{H}=\mathbf{B}/\mu$	$\mathbf{D}=\varepsilon\mathbf{E},\ \mathbf{H}=\mathbf{B}/\mu$
Permittivity of free space	1	$\varepsilon_0=8.854\,187\ldots\times10^{-12}\,\mathrm{F\,m^{-1}}$
Permeability of free space	1	$\mu_0=4\pi\times10^{-7}\,\mathrm{N\,A^{-2}}$
Fields from potentials	$\mathbf{E}=-\nabla V-\dfrac{1}{c}\dfrac{\partial\mathbf{A}}{\partial t}$	$\mathbf{E}=-\nabla V-\dfrac{1}{c}\dfrac{\partial\mathbf{A}}{\partial t}$
	$\mathbf{B}=\nabla\times\mathbf{A}$	$\mathbf{B}=\nabla\times\mathbf{A}$

Static potentials (Coulomb gauge)

$$V = \sum_{\text{charges}}\frac{q_i}{r_i} = \int\frac{\rho(\mathbf{r}')}{|\mathbf{r}-\mathbf{r}'|}d^3x' \qquad\qquad V = \frac{1}{4\pi\varepsilon_0}\sum_{\text{charges}}\frac{q_i}{r_i} = \frac{1}{4\pi\varepsilon_0}\int\frac{\rho(\mathbf{r}')}{|\mathbf{r}-\mathbf{r}'|}d^3x'$$

$$\mathbf{A} = \frac{1}{c}\sum_{\text{currents}}\frac{I_i}{r_i} = \frac{1}{c}\int\frac{\mathbf{J}(\mathbf{r}')}{|\mathbf{r}-\mathbf{r}'|}d^3x' \qquad\qquad \mathbf{A} = \frac{\mu_0}{4\pi}\sum_{\text{currents}}\frac{I_i A}{r_i}\,\frac{\mu_0}{4\pi}\int\frac{\mathbf{J}(\mathbf{r}')}{|\mathbf{r}-\mathbf{r}'|}d^3x'$$

Relativistic transformations
(v is the velocity of the primed frame as seen in the unprimed frame)

$$\mathbf{E}'_\parallel = \mathbf{E}_\parallel \qquad\qquad\qquad \mathbf{E}'_\parallel = \mathbf{E}_\parallel$$

$$\mathbf{E}'_\perp = \gamma\!\left(\mathbf{E}_\perp + \frac{1}{c}\mathbf{v}\times\mathbf{B}\right) \qquad\qquad \mathbf{E}'_\perp = \gamma\!\left(\mathbf{E}_\perp + \mathbf{v}\times\mathbf{B}\right)$$

$$\mathbf{B}'_\parallel = \mathbf{B}_\parallel \qquad\qquad\qquad \mathbf{B}'_\parallel = \mathbf{B}_\parallel$$

$$\mathbf{B}'_\perp = \gamma\!\left(\mathbf{B}_\perp - \frac{1}{c}\mathbf{v}\times\mathbf{B}\right) \qquad\qquad \mathbf{B}'_\perp = \gamma\!\left(\mathbf{B}_\perp - \frac{1}{c^2}\mathbf{v}\times\mathbf{E}\right)$$

$$\frac{\mu_0}{4} = 10^{-7}\text{ N A}^{-2}; \qquad\qquad c = \frac{1}{\sqrt{\mu_0\varepsilon_0}} = 2.997\,924\,58\times10^8\text{ ms}^{-1}$$

$$\frac{1}{4\pi\varepsilon_0} = c^2\times10^{-7}\text{ N A}^{-2} = 8.987\,55\ldots\times10^9\text{ m F}^{-1}; \quad \frac{\mu_0}{4} = 10^{-7}\text{ N A}^{-2}$$

an additional factor $1/4\,\pi\varepsilon_0$. That added factor obviously changes the relationship between the charge and the potential, and hence the charge and the physical forces.

Similarly, the relationship between the current and the vector potential **A** differs. In CGS units the potential **A** is simply the current divided by the distance and the speed of light, whereas in MKS units the permeability of free space appears as an additional factor, $\mu_0/4\pi$. Going from potential V, **A** to field **E**, **B** to Lorentz force **F**, Table 3.1 shows that there is another factor of c for CGS units. Since $(4\pi\varepsilon_0)(\mu_0/4\pi) = 1/c^2$ the conversion becomes clear. Additional discussion is available in Ref. 3.1. We will freely convert back and forth in what follows as an exercise in gaining facility with electromagnetic units.

3.2 Index of refraction

Maxwell's equations in the sourceless case, but within a medium defined by dielectric and dimagnetic constants ε and μ, are given below.

$$\nabla\cdot\mathbf{E}=0$$

$$\nabla\times\mathbf{E}+\frac{1}{c}\frac{\partial\mathbf{B}}{\partial t}=0, \quad D=\varepsilon\mathbf{E}$$

$$\nabla\cdot\mathbf{B}=0 \tag{3.1}$$

$$\nabla\times\mathbf{B}-\frac{\mu\varepsilon}{c}\frac{\partial\mathbf{E}}{\partial t}=0, \quad \mathbf{B}=\mu\mathbf{H}$$

The content of Maxwell's equations in the absence of sources is that the electromagnetic fields are divergenceless and that the 'rotation' or curl of the electric field is due to the time variation of the magnetic field while the curl of the magnetic field is due to the time variation of the electric field. In a medium the two curl equations lead to a second order wave equation with a wave propagation speed v that is not the vacuum velocity c but is given by c divided by the index of refraction, n. Assuming a plane wave for both **E**, **B** with exp $[i(\mathbf{k}\cdot\mathbf{x}-\omega t)]$ behavior and substituting into Eq. 3.1:

$$v=c/\sqrt{\mu\varepsilon}$$

$$k=\sqrt{\mu\varepsilon}(\omega/c)$$

$$[k^2-\mu\varepsilon(\omega/c)^2]\mathbf{E}=0 \tag{3.2}$$

$$n=\sqrt{\mu\varepsilon}$$

$$k=\omega/(c/n)=\omega/v$$

Representative values of n are given in Table 1.2 for different materials. For gases $n - 1 \sim 10^{-4}$, while for solids and liquids, $n - 1 \sim 0.3$. Both the permitivity and permeability of the medium contribute to the index of refraction, although most often $\mu = 1$.

3.3 Optical theorem

We can connect the macroscopic propagation of light in the medium, characterized by n, to the microscopic atomic process of scattering characterized by the total cross section (see Chapter 1), through the forward scattering amplitude, $A(0)$, Eq. 1.8. We baldly assert that if incoherent scattering off atomic electrons with number density $n_e = N_o\rho(Z/A)$ is the cause of n, then;

$$n - 1 = 2\pi N_o\rho(Z/A)[A(0)]/k^3 \tag{3.3}$$

The departure of the index of refraction from one is due to scattering. An imaginary part to $A(0)$ indicates absorption. In Appendix B we argue that the bound atomic electrons are confined to a region of volume $\underline{V} \sim \frac{4}{3}\pi a^3$. Hence the index is roughly

$$n - 1 \sim \frac{3}{2}\left[\frac{1}{ka}\right]^3 [A(0)]$$

$$\tag{3.4}$$

$$\mathrm{Im}(n) \sim \frac{3}{2}\left[\frac{1}{ka}\right]^3 \left[\frac{\sigma_\mathrm{I} k^2}{4\pi}\right]$$

Thus, for example, for an electromagnetic cross section of size $\sigma_\mathrm{EM} \sim \alpha^2(\pi a^2)$, see Chapter 10, the index n is $\sim \frac{3}{18\pi}\alpha^2(\lambda/a)$. For optical photons, $\lambda \sim 5000$ Å, $a \sim 1$ Å, we find $\mathrm{Im}(n) \sim 0.016$, which is of the right order of magnitude, as seen in Table 1.2.

3.4 Conducting medium and skin depth

Let us consider now the slightly more complicated case where there are currents in a neutral medium that are the sources. For the case of a medium with conductivity σ (not to be confused with cross section) with a current density $\mathbf{j}$, the modified Maxwell's equation are;

$$\nabla \cdot (\mu\mathbf{H}) = 0$$

$$\nabla \cdot (\varepsilon\mathbf{E}) = 0$$

$$\nabla \times \mathbf{E} + \frac{\mu}{c}\frac{\partial \mathbf{H}}{\partial t} = 0 \tag{3.5}$$

$$\nabla \times \mathbf{H} - \frac{\varepsilon}{c}\frac{\partial \mathbf{E}}{\partial t}\frac{4\pi\sigma}{c}\mathbf{E} = 0$$

$$(\mathbf{j} = \sigma\mathbf{E})$$

If we again assume a plane wave form for the electric and magnetic fields, the two curl equations again lead to a wave equation which takes the form;

$$\mathbf{H} = (\mathbf{k} \times \mathbf{E})c/\mu\omega$$

$$i(\mathbf{k} \times \mathbf{H}) + i\varepsilon\frac{\omega}{c}\mathbf{E} - \frac{4\pi\sigma}{c}\mathbf{E} = 0$$

(3.6)

Clearly, Eq. 3.6 implies that $\mathbf{H}$, $\mathbf{k}$ and $\mathbf{E}$ are again mutually orthogonal. Eliminating the magnetic field $\mathbf{H}$ we again find a second order wave equation. Note that in Eq. 3.7 if we let the conductivity of the medium go to zero we recover Eq. 3.2.

$$\left[k^2 - (\mu\varepsilon\frac{\omega^2}{c^2} + 4\pi i\frac{\mu\omega\sigma}{c^2}) \right]\mathbf{E} = 0$$

(3.7)

$$k^2 = \mu\varepsilon\left(\frac{\omega}{c}\right)^2\left[1 + i\left(\frac{4\pi\sigma}{\omega\varepsilon}\right) \right]$$

The complex nature of the square of the wave number implies that the assumed plane wave solution has an exponentially damped behavior. Therefore, the electric field inside a conducting medium is not rigorously zero, as in the static case with infinite conductivity, but penetrates a finite amount. The characteristic penetration distance, δ, which we define to be the 'skin depth' is inversely proportional to the imaginary part of the wave number.

$$|\mathbf{E}| \sim e^{-x/\delta} \sim e^{-\mathrm{Im}(k)x}$$

$$\delta = 1/\mathrm{Im}(k)$$

(3.8)

$$\delta/c \sim 1/\sqrt{\sigma\omega\mu}$$

The skin depth approaches 0 as $\sigma \to \infty$, since in that case the charges are infinitely mobile and hence free to move around and reorient themselves so as to completely exclude the electric field. For finite conductivity the skin depth is finite and depends on frequency as $1/\sqrt{\omega}$. This implies that high frequency waves stay on the surface of a conductor.

For example, in copper, using the appropriate values for the conductivity and permeability, we find that the skin depth is 6.6 cm/$\sqrt{f}$ when the frequency f is expressed in hertz. In copper at 60 Hz the skin depth is roughly 0.85 cm. At

100 MHz the skin depth is roughly 7.1 μm. Therefore, in your home wiring efforts you can choose to simply crimp the wires since the electric fields will penetrate through the oxide insulator. On the other hand, it is a good approximation to say that radio frequency fields stay totally on the surface of any conductor and require elaborate waveguide 'plumbing' structures.

3.5 Plasma frequency

Now let's consider a special case where the conductivity, σ, is due to the existence of a number density, n_e, of electrons per unit volume which are free to move. This is a rough approximation to the behavior of electrons in the medium. Normally in this text we use n to denote number density and N denotes number, while ρ is used to denote either charge or mass density or resistivity (see the Glossary at the end of this volume).

In the presence of an electric field the free charges are accelerated. The non-relativistic expression for their acceleration **a** is $e\mathbf{E}/m$, (see Table 3.1 for the Lorentz force equation).

$$\mathbf{a} = \frac{e}{m}\mathbf{E} \tag{3.9}$$

For high frequency electric fields the displacement of the electron is small because the motion is limited by the short period of the harmonic driving force. Therefore, we obtain an approximate expression for the velocity by ignoring those small displacements and assuming $\mathbf{x}=0$ for the spatial part of **E**, which goes as $e^{i\mathbf{k}\cdot\mathbf{x}}$, $v=a/\omega$. We write the current density **j** in terms of the electron number density n_e and the velocity **v** of the electrons. We can then write down the conductivity due to the existence of a free electron density. In this particular case we have a purely imaginary conductivity which decreases with frequency as $1/\omega$.

$$\mathbf{v} \sim \frac{ie\mathbf{E}_o}{m\omega}e^{-i\omega t}, \quad \mathbf{E} \sim \mathbf{E}_o e^{-i\omega t}$$
$$\mathbf{j} = n_e e\mathbf{v} = \sigma\mathbf{E} \tag{3.10}$$
$$\sigma \sim i[e^2 n_e/m\omega]$$

Looking back to Eq. 3.7 we see that if there is a purely imaginary conductivity then the relationship between k and ω for the plane wave solution is real, implying oscillatory propagation as long as $k^2>0$. We define a 'plasma frequency' ω_p.

$$k^2 = \mu\left(\frac{\omega}{c}\right)^2\left[\varepsilon - \left(\frac{4\pi e^2 n_e/m}{\omega^2}\right)\right] \quad \text{(Eq. 3.7, Eq. 3.10)}$$

$$\varepsilon \sim \sqrt{1 - \left(\frac{\omega_p}{\omega}\right)^2}, \quad \omega > \omega_p \text{ free propagation}$$

(3.11)

The plasma frequency, ω_p, is dependent only on the fine structure constant, α, the density of free electrons and the mass of the electron. For ω greater than ω_p the electric permitivity is real and there is free propagation. For $\omega < \omega_p$ the wave is reflected.

Assigning a number density for electrons, which assumes all electrons in the atom are free, and recalling the relationship between the Bohr radius, a_o, and the electron Compton wavelength, λ_e derived in Chapter 1, we have an alternative expression for ω_p in terms of the binding energy of the hydrogen atom, E_o. Note that, since the factor $N_o\rho(Z/A)$ is the number of electrons per mole, and πa_o^3 is roughly the volume of an atom, we expect $\hbar\omega_p$ to be equal to the binding energy of the hydrogen atom times a number of order 1.

$$\hbar\omega_p = \sqrt{4\pi\alpha n_e \lambda_e} \quad \text{(Eq. 3.11)}$$

$$n_e \equiv N_o\rho(Z/A)$$

$$a_o = \lambda/\alpha$$

(3.12)

$$\hbar\omega_p \sim \left[\sqrt{\pi\left(\frac{N_o\rho Z}{A}\right)a_o^3}\right]E_o, \quad E_o = -\frac{mc^2}{2}\alpha^2$$

For a typical density of 1 gm/cm^3 and for $Z/A = 1/2$, $\hbar\omega_p$ is $\sim$18.8 eV. Specifically, in lithium, which is used in foils of transition radiation detectors, the plasma energy is 14 eV. We will return to ω_p in our discussion of transition radiation in the next chapter.

It is interesting to recall that the Earth's ionosphere is a dilute plasma with an electron density $n_e \sim 10^{-5}$ e/cm^3. Hence, $\omega_p \sim 6 \times 10^7$ Hz, and radio waves $\omega < \omega_p$ are reflected from the ionosphere, making short wave radio 'bounces' possible. Higher frequency radio, e.g. FM at $\omega \sim 100$ MHz, is not reflected as you can infer by noting the line of sight microwave relay towers that dot our countryside. Note that numerically, $\hbar\omega_p \cong 29\,\text{eV}\sqrt{\rho_e Z/A}$, ρ in g/cm^3.

3.6 Two 'derivations' of the Cerenkov angle

We now turn specifically to Cerenkov radiation, having quickly reviewed Maxwell's equations. A charged particle in uniform unaccelerated motion in

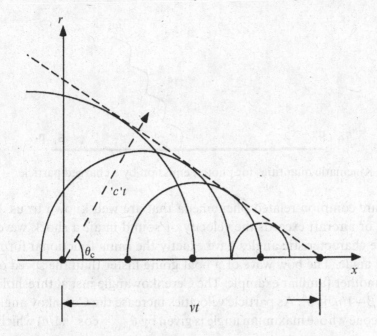

Fig. 3.1. Cerenkov cone construction using Huygens' principle.

vacuum does not radiate, see Chapter 10. However, if a particle is moving with a uniform constant velocity in a medium, its electric field will interact with the medium. This interaction can cause the emission of real photons. Energy and momentum are balanced by the medium. This concept was discovered theoretically by a Russian physicist, Cerenkov, in the early part of the twentieth century. The effect is now in common use in detectors used in high energy physics.

A construction for the geometry of the Cerenkov emission angle is given in Fig. 3.1. Particles move with constant velocity in a medium whose index of refraction, n, is greater than 1. Hence, the effective velocity of light in the medium is less than the vacuum velocity of light. If the particle velocity exceeds the velocity of light in the medium, a shock wave or 'Mach cone' is set up in the medium. The construction shows that the Huygens wavelets emitted from each point in time add up constructively along a line defined by the Cerenkov angle θ_c. Since the particle energy is constant (non-destructive detection), $c\beta$ is the constant particle velocity.

$$'c' = c/n$$
$$\cos\theta_c = {}'c'/v = {}'c't/vt \qquad (3.13)$$
$$= c/nv = 1/\beta n$$
$$v > {}'c' \text{ and } \cos\theta_c < 1 \text{ if } \beta > 1/n$$

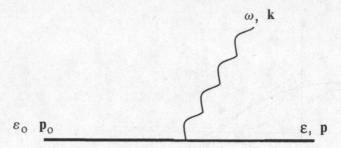

Fig. 3.2.   Kinematic quantities for photon emission by a charged particle.

There are common related phenomena that are well known to us. When a projectile or aircraft exceeds the velocity of sound in air a shock wave is sent out whose characteristic angles have exactly the same functional form as the Cerenkov angle. The bow wave of a boat going faster than the speed of water waves is another familiar example. The Cerenkov angle just at threshold velocity, when $\beta = 1/n$, is $0°$. As particle velocities increase the Cerenkov angle opens up into a cone whose maximum angle is given by $\theta_{max} = \cos^{-1}(1/n)$ which occurs as $\beta \rightarrow 1$.

An example is light emission in water, with $n = 1.33$. The threshold velocity is $\beta > 0.752$, which means $T_e > 0.26$ MeV and $T_\mu > 54$ MeV. Therefore, the emission of Cerenkov light in water can be used to distinguish between electrons and muons with kinetic energies in the few MeV range. This technique is used in neutrino experiments at Kamioka and elsewhere.

For an alternate 'derivation' of the Cerenkov angle consider Fig. 3.2 where we show single photon emission by a charged particle. Energy and momentum conservation (see Appendix A) require that

$$\varepsilon_0 = \varepsilon + \omega$$
$$\mathbf{p}_0 = \mathbf{p} + \mathbf{k} \tag{3.14}$$

Squaring both equations, and assuming $\omega/c = k$, we find that for free photons;

$$\omega = k_\parallel \beta_0 c \tag{3.15}$$

Therefore, applying this derivation to emission with an index of refraction, since $\omega = ck/n$, Eq. 3.2,

$$k/n = \beta k \cos \theta_c$$
$$n\beta \cos \theta_c = 1 \tag{3.16}$$

The relationship should hold in general given that the derivation basically used only energy and momentum conservation. Note that in the case of vacuum, $n = 1$, since β is always less than 1, $\cos \theta$ is greater than 1. We thus regain the statement that free particles in uniform motion do not emit photons.

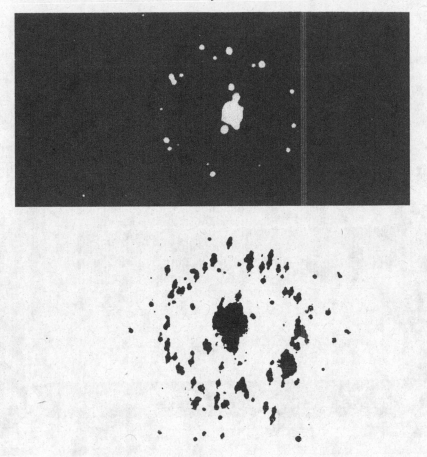

Fig. 3.3. Cerenkov cone for detected photons and ionization products in a gas filled detector for ~ 10 individual particle passages. (From Ref. B.1 and Ref. B.2, with permission.)

An example of the reality of the Cerenkov emission cone is given by data shown in Fig. 3.3. In this case the photons emitted at the Cerenkov angle are focused onto a plane and are detected in a gas filled device (see Chapter 8). In these particular plots there are several individual particle passages which are added together. We can see the cone built up by the individual emitted photons. The 'hits' at the center of the circle are due to ionization energy (see Chapter 6) deposited by the incoming particle itself.

A photograph of an actual 'ring imaging' device showing the gaseous detector used to detect the ultraviolet photons by passage through a UV transmitting window and subsequent photoelectric detection is shown in Fig. 3.4. This is a functioning device which has been used in a high energy physics

Fig. 3.4.   Photograph of a ring imaging detector chamber showing the thin UV transmitting window. (Photo – Fermilab.)

experiment. The transmission of a CaF_2 window 3 mm thick is compared to
quartz and LiF in Fig. 3.5 as a function of photon wavelength and/or photon
energy. Clearly these windows pass light down into the far UV, $\lambda \sim 1500$ Å,
which is well matched to the useful absorption bands of the detecting vapors,
TMAE and TEA.

The use of such a 'ring imaging Cerenkov counter', or RICH, as a particle
identification device is illustrated in Fig. 3.6. Charged particles are momentum
analyzed by measuring their trajectories in a magnetic field (Chapter 7). For a
given momentum the observed Cerenkov cone angle, θ_c, is then measured.

The data show three distinct bands of correlation between the Cerenkov
angle and the momentum, which correspond to particles of three distinct
masses, pions π (mass ~ 0.14 GeV), kaons K (mass ~ 0.5 GeV), and protons p
(mass ~ 0.94 GeV). In Fig. 3.6 we see the angle $|\theta_c|$ beginning at threshold at 0°

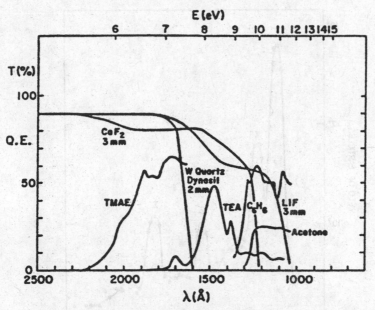

Fig. 3.5.   Transmission of different window materials as a function of photon wave-length compared to absorption in different vapors. (From Ref. B.2, with permission.)

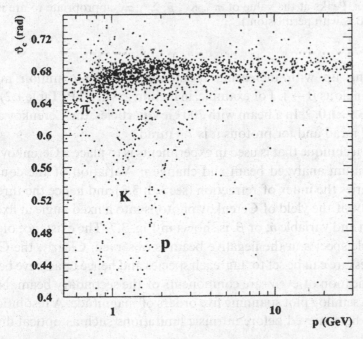

Fig. 3.6.   Plot of observed Cerenkov cone size as a function of momentum. The bands for π, K, and p are evident indicating the utility of a Cerenkov detector for particle identification. (From Ref. 4.9, with permission.)

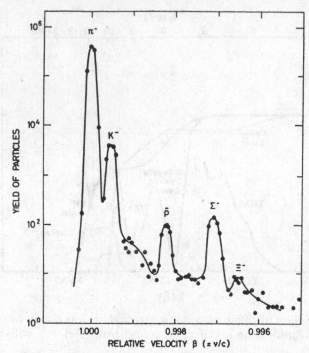

Fig. 3.7.   Yield of Cerenkov photons in a negative beam of fixed momentum as a function of n. Peaks at the value of π^-, K^-, $\bar{p}$, Σ^-, Ξ^- appropriate to are indicated. (From Ref. B.1, with permission.)

and opening up with increasing momentum to the common maximum Cerenkov angle as $\beta \rightarrow 1$. For example, in helium gas at STP (Table 1.2) we have $1 - n \equiv \delta = 3.9 \times 10^{-5}$. In a beam with 200 GeV particles, the Cerenkov angle for pions is 8.7 mrad and for protons it is 7.4 mrad.

Another technique that is used in experiments is to place a Cerenkov counter in a momentum analyzed beam and change n. Variation of the counter gas pressure varies the index of refraction (see Eq. 3.3) and hence the threshold β value. A plot of the yield of Cerenkov photons into a fixed angle at fixed beam momentum and variable n, or β, is shown in Fig. 3.7. The existence of five distinct particle species in the negative beam is observed. Clearly, the Cerenkov counter pressure can be set to 'tag' each species and hence to achieve beam particle identification of even rare components of the secondary beam. Note that Fig. 3.7 is a semilog plot spanning five orders of magnitude. A resolution of $d\beta \sim 10^{-4}$ can be achieved before intrinsic limitations such as optical dispersion are reached.

3.7 A 'derivation' of the frequency spectrum

The emission of Cerenkov radiation is due to the interaction of the field of a charged particle with a medium. Hence, it is truly a form of energy loss, dE/dx (see Chapter 6). A rigorous derivation is not attempted here. If that is desired, the reader should consult a standard text, e.g. Ref. 3.1, on electromagnetism or consult Appendix E for an outline of the derivation. The photons in the case of ionization energy loss are 'virtual', transmitting the forces to the atomic electrons of the medium. In the Cerenkov effect real photons are emitted.

Energy loss by ionization will be derived in Chapter 6. For our purposes now we simply quote the result for the energy loss per unit path length by ionization, dE_I/dx up to slowly varying logarithmic factors.

$$dE_I/dx = \left(\frac{N_o \rho Z}{A}\right) \frac{4\pi\alpha^2}{\beta^2 c^2}(\chi_e) \tag{3.17}$$

Inserting Eq. 3.12 for the plasma frequency, $(\hbar\omega_p)^2 = 4\pi\alpha[N_o\rho(Z/A)]\chi e$, Eq. 3.17 can be written in the suggestive form:

$$\frac{dE}{dx} = \frac{\alpha}{\beta^2}\left[\frac{\hbar\omega_p}{c/\omega_p}\right] \tag{3.18}$$

We now assert that the results for ionization can be carried over to also describe the related process of Cerenkov radiation and ascribe the energy loss, dE, to the emission of N photons, $dE = (\hbar\omega)dN$ where dN is the number of emitted photons of frequency ω causing energy loss dE. Since $\varepsilon^2 = 1 - (\omega_p/\omega)^2$, Eq. 3.11,

$$\frac{dE}{dx} \sim \frac{\alpha}{\beta^2}[(1-\varepsilon^2)\omega^2]\left(\frac{\hbar}{c}\right)$$

$$\frac{dN}{dx} \sim \frac{\alpha}{c}\frac{(1-\varepsilon^2)}{\beta^2}\omega \tag{3.19}$$

or

$$\frac{d^2N}{d\omega\,dx} \sim \frac{\alpha}{c}\frac{(1-\varepsilon^2)}{\beta^2}$$

This is in no sense a derivation. However, it does manage to show the deep connection between ionization and Cerenkov radiation. The fact that $d^2N/d\omega dx$ is independent of ω and is proportional to α/c times a function of ε and β is 'demonstrated'. In vacuum $\varepsilon^2 = 1$ and there is no radiation. The fact that the function is really $\sin^2\theta = 1 - 1/\beta^2\varepsilon$ and not $(1-\varepsilon^2)/\beta^2$ can only appear in a full and rigorous derivation. Assuming such a derivation, and using $\varepsilon = n^2$, we then attain the Cerenkov result.

$$\frac{d^2 N_c}{d\omega \, dx} = \left(\frac{\alpha}{c}\right) \sin^2\theta_c = \frac{\alpha}{c}\left(1 - \frac{1}{\beta^2\varepsilon}\right)$$

(3.20)

$$N_c = \frac{\alpha}{c}\sin^2\theta_c \Delta x \, \Delta\omega$$

Note that in Eq. 3.20 all energies for the photon have equal probability. That fact puts a premium on full detection efficiency over as wide a photon energy range as possible. Indeed, that is the reason for using UV transmitting 'windows' which are an important feature of Cerenkov devices.

A finite length, L, of radiator leads to a more complex emission distribution. Note that the existence of a diffraction pattern, shown in Eq. 3.21, is due to the fact that there is a finite coherence length for the radiation. We cannot have an infinitely precise emission angle because of the uncertainty relationship inherent in wave phenomena. This fact will be of critical importance in the next chapter. Note that diffraction is important, $\delta \sim 1$, only if $L < \lambda$. In the case of optical photons any radiator longer than 1 cm is 'long'. We will revisit this issue in Chapter 4, where the radiating foils can be 'short'.

$$\frac{d^2 N_c}{d\omega d\cos\theta} = \frac{\alpha}{c}\left(\frac{\sin\delta}{\delta}\right)^2 \sin^2\theta (L^2/\lambda)$$

(3.21)

$$\delta = \left(\frac{1}{n\beta} - \cos\theta\right)\left(\frac{\pi L}{\lambda}\right)$$

In the limit that the radiator becomes very long, the diffraction pattern, $(\sin\delta/\delta)^2$, becomes extremely sharp and approaches a 'delta function' which can then be integrated over. In this approximation there are $\alpha\sin^2\theta_c/c$ photons emitted at the Cerenkov angle, per unit path length and per unit frequency interval. Numerically, we can evaluate the factors in Eq. 3.20 to find out that the yield is 365 photons times $\sin^2\theta_c$ per eV of frequency range per centimeter of radiator length.

$$\hbar c = 2 \times 10^{-5}\text{eV cm}$$

(3.22)

$$\frac{d^2 N_c}{d(\hbar\omega) \, dx} \sim 365\sin^2\theta_c/(\text{eV cm})$$

3.8 Examples and numerical values

For example, take a 10 meter long gas radiator filled with nitrogen. The index $(n-1)$, is about 3×10^{-4} per atmosphere (see Table 1.2). Visible photons cover about 1 eV of frequency range, $\hbar c \sim 2000$ eV Å. We then expect, for fully efficient light collection, 220 photons collected. A PMT with a 25% quantum efficiency would yield a healthy current pulse suitable for high speed operation (see Chapter 2).

A photograph of a typical large area Cerenkov detector is given in Fig. 3.8. We see the reflecting mirrors which bounce the emitted Cerenkov light into the light collecting cones on the sides. By looking into the mirrors we can see the circular PMT windows which are used for detecting the light.

The relationship between the index of refraction, the Cerenkov angle and the γ of the particle traversing the detector is shown below. Note that δ is used in this chapter to denote both the photon optical phase difference in emission, Eq. 3.21, and the value of $n-1$. In subsequent chapters δ will refer only to the optical phase difference. Taking the small angle, high energy, and $n \sim 1$ limit:

$$n \equiv 1 + \delta, \quad 1 - \beta \sim 1/2\gamma^2, \quad \gamma = 1/\sqrt{1 - \beta^2}$$

$$\theta_c^2 \sim \frac{1}{\gamma_{TH}^2} - \frac{1}{\gamma^2} \tag{3.23}$$

$$\gamma_{TH}^2 \sim 1/2\delta$$

The threshold gamma factor, γ_{TH}, is proportional to $\sqrt{\delta}$. The maximum Cerenkov angle, θ_c^{max}, occurs when the particle has extremely large γ. The maximum number of photons, N_c^{max}, thus goes as $1/\gamma_{TH}^2$. In terms of index of refraction, N_c^{max} is proportional to δ.

$$\theta_c^{max} = 1/\gamma_{TH}$$

$$N_c^{max} \sim \delta \sim 1/2\gamma_{TH}^2 \tag{3.24}$$

For example, in nitrogen δ is 3×10^{-4}, which means that γ_{TH} is 40.8 or that the threshold for pions is 5.6 GeV, for kaons 20.2 GeV and for protons 38.2 GeV. Below those momenta the incident particles will not radiate photons. The maximum Cerenkov angle in this gas is only 1.4° so the light is thrown very forward. That is the reason for the forward mirrors (the hole for incident beam passage is in the center) seen in Fig. 3.8.

The number of photons is proportional to Δx and $\Delta \omega$. Increasing the counter length, L, is expensive. The other way to increase the number of photons is to extend the frequency range $\Delta \omega$ over which we can capture photons. As an example, calcium fluoride windows have transmission above 50% for photon energies of about 9 eV and below (See Fig. 3.5.) or wavelengths above about 1380 Å. That gives a rather larger frequency range than the ~ 1 eV visible range we have previously assumed.

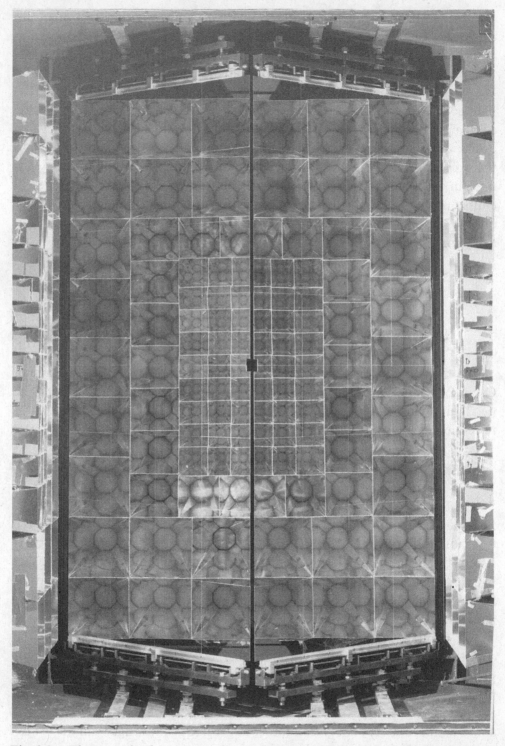

Fig. 3.8.   Photograph of a gaseous Cerenkov detector showing the mirrors, light collecting cones and the circular windows of the photomultipliers (FNAL E687). (Photo – Fermilab.)

$$\lambda = \frac{\hbar c(2\pi)}{\hbar\omega} = \frac{12\,400\,(eV\text{Å})}{E_\gamma(eV)} \tag{3.25}$$

for CaF_2, $E_\gamma < 9$ eV, $\lambda > 1380$ Å

Particle identification for particles with high momenta requires an attempt to go to lower indices of refraction. Lower indices of refraction, such as helium, mean fewer photons produced, as we can see looking at Eq. 3.24. A way to recoup photons is to use a larger frequency range $\Delta\omega$. However, at very high momentum the required resolution, $d\beta$, cannot be obtained. For example, the limitation due to dispersion, n a function of ω, or optical aberration limits the achievable resolution in β (see Fig. 3.7). Particle identification at very high energies requires a technique which does not depend on β, since at high energy $\beta \rightarrow 1$ independent of particle type. Both time of flight (Chapter 2) and Cerenkov radiation (Chapter 3) become increasingly difficult and the use of 'transition radiation' (Chapter 4) is called for.

Exercises

1. Explicitly work out the intermediate steps from Eq. 3.1 to Eq. 3.2.
2. Show that, if the cross section is geometric, the index of refraction is expected to go as $1/k$ or increase linearly with wavelength (decrease inversely with frequency).
3. Consider a transmission line cable carrying 1 ns pulses. For 1000 Mz frequencies, show that 3 skin depths of attenuation (95%) requires 0.0006 cm of copper thickness of cable. The cable has to be at least this thick not to 'leak' out the signal.
4. What is the plasma frequency appropriate to a plasma of say one proton per cubic meter as is found in interstellar space?
5. Fill in the details of the derivation of Eq. 3.15 from Eq. 3.14. Show that for real photons energy and momentum cannot be simultaneously conserved.
6. Work out the emission angles for a 2 GeV momentum pion, kaon and proton in a water filled Cerenkov counter.
7. What are the energy thresholds for Cerenkov light production for pions, kaons, and protons in a helium counter?
8. What are the emission angles in a helium counter for a 200 GeV beam of pions, kaons, and protons? What are the radial separations at a distance of 10 m from the radiator?
9. Find the number of photons emitted by a 1 cm long water Cerenkov counter if a 1 eV band of photon energies is accepted by the photon transducer.

10. Assuming optical photons of 5000 Å wavelength and a 1 m radiator with a 1 mrad emission angle, evaluate the diffraction broadening of the emission angles. Compare that to the separation found for particles in Exercise 8.

11. For $n = 1.0003$ compare the Cerenkov angle for a 50 GeV momentum kaon calculated exactly and by means of the approximations used in the text. Is this a good approximation?

References

[1] *Classical Electrodynamics*, J.D. Jackson, John Wiley & Sons, Inc. (1962).
[2] P.A. Cherenkov, *Phys. Rev.* **52** 378 (1937).
[3] *Nobel Lectures in Physics*, P.A. Cherenkov, I.M. Frank, and I.E. Tamm, Elsevier Publishing (1964).
[4] *Sub-atomic Physics*, H. Fraugnfelder and E.M. Henley, Prentice-Hall, Inc. (1974).
[5] P. Lecompze, *et al.*, *Phys. Scr.* **23** 377 (1981).
[6] J. Litt and R. Meunier, *Annu. Rev. Nucl. Sci.* **23** 1 (1973).

4

Transition radiation

The light of the spirit is invisible, concealed in all beings.
The Upanishads

In the previous chapter we explored the relationship of Cerenkov emission to ionization loss. The wave vector, k, was shown in Chapter 3 to be related to the dielectric constant ε as $k \sim \sqrt{\varepsilon}\,(\omega/c)$. If ε is real, there is free photon propagation. If ε is >1, the threshold velocity defined in Chapter 3 is $<c$. If ε is <1, the threshold velocity is $>c$ and no real photon emission is possible in an infinitely long radiator. However, for finite length radiators the emission angle is not unique, as discussed in Chapter 3. Because of the diffraction broadening of the distribution of emission angles, a particle with velocity $<c$ can still emit photons with $\cos \theta < 1$. We call this phenomenon the 'sub-threshold' emission of Cerenkov light.

The sub-threshold emission of photons when particles traverse a thin segment of material is called transition radiation because it is caused by a transition in the index of refraction between the vacuum and the material. We will find that the emitted photons are in the x-ray range of energies and that the number of emitted photons depends on γ and not β. Therefore, the technique is most useful at very high energies when the previously discussed methods of velocity measurement become very difficult.

4.1 Cerenkov radiation for a finite length radiator

Transition radiation occurs when we are below threshold for Cerenkov radiation, $\beta_{\mathrm{TH}} > 1$, but in a finite length, L, medium. Let us consider the radiation pattern from a thin film due to the passage of a charged particle. The distance traveled in the film is L, the wavelength in the medium is λ', and the angle of emission of the radiation in the medium is θ'. The geometric definitions of the quantities used are given in Fig. 4.1.

Recall in the previous chapter on Cerenkov radiation that for a finite wave train there is a diffraction pattern due to the fact that the Cerenkov radiation

75

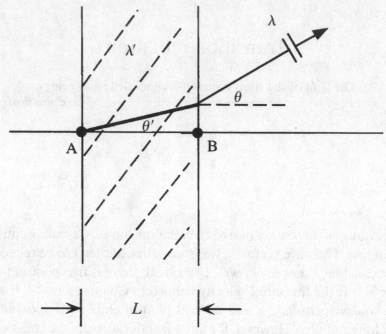

Fig. 4.1.   Definition of quantities used in deriving the transition radiation effect. This effect can be thought of as the pattern due to a finite length Cerenkov radiator.

was only emitted over a finite distance. The diffraction pattern due to the phase difference δ was also indicated (Eq. 3.21).

$$\frac{d^2N}{d\omega\,d\Omega} = \frac{\alpha}{c}\sin^2\theta'\left(\frac{1}{2\pi}\right)\left(\frac{\sin\delta}{\delta}\right)^2\left(\frac{L^2}{\lambda'}\right)$$

(4.1)

$$\delta = \frac{\pi L}{\lambda'}\left(\cos\theta' - \frac{1}{n\beta}\right)$$

A very sharp emission angle in the medium, labeled as θ', is an approximation to the diffraction pattern of a very long radiator, given as $(\sin\delta/\delta)$ in Eq. 4.1. For short radiators the problem can be treated as slit diffraction from the entrance and exit points labeled as A and B in Fig. 4.1. The wavelength in the radiating medium, λ', is related to the index of refraction $n = \sqrt{\varepsilon}$, Eq. 3.1, and the photon frequency as discussed in Chapter 3.

$$\lambda' = 2\pi c/\omega\sqrt{\varepsilon}$$

$$\delta = \frac{\pi L}{\lambda'}\left(\cos\theta' - \frac{1}{\beta\sqrt{\varepsilon}}\right)$$

(4.2)

$$= \frac{\omega L}{2c}\left(\sqrt{\varepsilon}\cos\theta' - \frac{1}{\beta}\right)$$

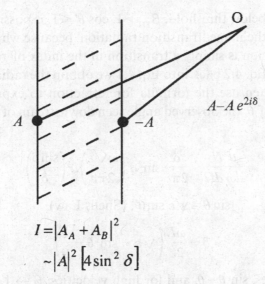

$$I = |A_A + A_B|^2$$
$$\sim |A|^2 \left[4 \sin^2 \delta\right]$$

Fig. 4.2. Interference pattern at observation point O due to amplitudes A and $-A$ for emission from the front and back of the film with optical path length difference equivalent to the phase difference between the amplitudes.

The optical path difference to the observation point, O, shown in Fig. 4.2, leads to a phase difference, 2δ, between waves produced from points A and B. This phase difference is related to the geometric path difference. If $\mathbf{x}$ is the observation point, then $2\delta = (\mathbf{k'}_A - \mathbf{k'}_B) \cdot \mathbf{x}$.

Thus the optical phase difference, which is the argument in the $(\sin \delta/\delta)^2$ diffraction pattern, is given in Eq. 4.2 in terms of the radiator length, L, the emitted wavelength, λ' and of the emitted angle, θ'. Clearly, the diffraction peak is centered on $\cos \theta' = 1/\beta\sqrt{\varepsilon}$, the 'Cerenkov angle', with an angular width of $\Delta\theta' \sim \lambda'/L$. Hence for a very thin radiator, the chance for sub-threshold radiation is not necessarily negligible.

4.2 Interference effects

The interference of the amplitudes A and $-A$ from points A and B is controlled by the phase difference δ, as seen in Fig. 4.2. The intensity, I, is $4|A|^2 \sin^2 \delta$. Note that if $\delta \to \pi/2$ the intensity $\to 4|A|^2$ which is the maximum constructive interference value, while if $\delta \to 0$ the interference is destructive and the intensity $\to 0$. For random phases $I \sim 2|A|^2$. Given that the phase at the entrance and exit transition differs by 180°, we need to have a sufficient optical path length difference to avoid destructive interference. This point becomes important, as we will see later, in the practical design of detectors.

The diffraction broadening of the angular distribution, $\delta \sim [\cos\theta' - \beta_{TH}/\beta]$, implies that even below threshold, $\beta_{TH} > 1$, $\cos \theta' < 1$, is possible. This observation gives rise to the name 'transition radiation' because what is necessary for real photon emission is simply a transition in the index of refraction. Putting the expression in Eq. 4.2 back into Eq. 4.1 we obtain the radiation pattern with diffraction. We then use the formula for refraction to express the radiation pattern in terms of θ, the observed angle seen downstream of the foil as defined in Fig. 4.1.

$$\frac{d^2N}{d\omega \, d\Omega} \sim \frac{\alpha}{2\pi c} \sin^2\theta' \left(\frac{\omega\sqrt{\varepsilon}}{2\pi c}\right) L^2 \left(\frac{\sin\delta}{\delta}\right)^2$$

$$\sin\theta = \sqrt{\varepsilon} \sin\theta', \text{ (Snells Law)} \tag{4.3}$$

$$\delta \sim \frac{\omega L}{2c} \left(\sqrt{\varepsilon - \sin^2\theta} - \frac{1}{\beta}\right)$$

For small angles, $\sin\theta \sim \theta$, and for high velocities, $\beta \to 1$, $1 - \beta \sim 1/2\gamma^2$. We can expand the expression for δ and obtain the approximate optical phase difference.

$$\delta \sim \frac{\omega L}{4c} \left(\left(\frac{\omega_p}{\omega}\right)^2 + \theta^2 + \frac{1}{\gamma^2}\right)$$

$$\tag{4.4}$$

$$\frac{d^2N}{d\omega \, d\Omega} \sim \frac{\alpha}{\pi^2 \omega} \theta^2 \left[\frac{\sin\delta}{\delta}\right]^2 \left[\frac{\omega L}{2c}\right]^2, \quad \varepsilon \sim 1$$

The factor $(\theta^2 + 1/\gamma^2)$ will be motivated as the phase difference between waves in vacuum in Chapter 10. The extra term results from the fact that this is propagation in a medium characterized by a plasma frequency, ω_p, which leads to the additional term $(\omega_p/\omega)^2$ beyond what will be found for the vacuum case. The doubly differential cross section is proportional to α and has a typical dipole pattern, θ^2. If phase factors are ignored, it goes roughly as $1/\omega$.

This present heuristic treatment is not sufficient, although it brings us quite close to the true result. The correct expression for the doubly differential cross section for transition radiation is given below.

$$\frac{d^2N_{TR}}{d\omega \, d\Omega} \sim \frac{\alpha}{\pi^2 \omega} \theta^2 \sin^2\delta \left[\left(\frac{1}{\delta}\right)^2 - \left(\frac{1}{\delta_v}\right)^2\right] \left(\frac{\omega L}{2c}\right)^2,$$

$$\tag{4.5}$$

$$\delta_v = \frac{\omega L}{4c} \left(\theta^2 + \frac{1}{\gamma^2}\right) \text{ (Chapter 10)}$$

4.3 The vacuum phase shift

Comparing to Eq. 4.4, we see that the exact result, Eq. 4.5, can be thought of as arising from having adjusted Eq. 4.4 by the subtraction of the vacuum phase shift, δ_v, that we would get for propagation if the thin film were a vacuum, i.e. no 'transition' in the index of refraction. The factor in the phase difference is proportional to the vacuum factor, which we will derive in Chapter 10, $(\theta^2 + 1/\gamma^2)$. The insertion of the vacuum phase factor insures that there is no radiation without a transition in the index of refraction; i.e. if $\delta \to \delta_v$ then $d^2N_{TR}/d\omega\, d\Omega \to 0$.

4.4 The frequency spectrum

The frequency spectrum of transition radiation, in comparison to Cerenkov emission, is not flat. The emission spectrum has high frequency components because its very existence depends on the finite width diffraction pattern, $\delta \sim \omega L/c$, which is inherent in the Cerenkov process for a finite length radiator. Typically, Cerenkov light is in the optical region, while transition radiation is in the x-ray region. This characteristic high frequency emission is a consequence of needing enough diffraction broadening to enable sub-threshold Cerenkov emission to occur at a real angle.

We can obtain an informative expression by integrating Eq. 4.5 over all angles while ignoring interference by setting $\sin^2 \delta = 1$. The result for the photon frequency spectrum is

$$\frac{dN_{TR}}{d\omega} = \frac{2\alpha}{\pi\omega}[-1 + [\ln(1 + 1/y^2)](y^2 + 1/2)]$$

(4.6)

$$y = \omega/\gamma\omega_p$$

We define the frequency scale in terms of the characteristic frequency $\gamma\omega_p$. Note that $dN_{TR}/d\omega$ falls off rapidly for $y > 1$ or $\omega > \gamma\omega_p$. The expression for $dN_{TR}/d\omega$ also diverges at low frequency, but this has no practical impact, and is an artifact of the approximations used. A plot of $dN_{TR}/d\omega$ in arbitrary units is shown as a function of y in Fig. 4.3. The $1/y$ behavior at small y is evident as is the steeper fall for $y \gg 1$ indicating the high frequency cutoff. For $y \ll 1$, a limiting case is $dN_{TR}/d\omega \to 2\alpha/\pi\omega[\ln(1/y)]$. Looking at Eq. 4.6, we see that the integral over a practical range of frequency $\Delta y \sim 1$, near $y \sim 1$ is $N_{TR} \sim \alpha$. Thus we argue that in transition radiation $\sim \alpha$ photons per interface are emitted with energies typically $y \sim 1/3$ or $\gamma(\hbar\omega_p)/3$. These approximate considerations lead to some 'rules of thumb' for transition radiation detectors. We can basically read

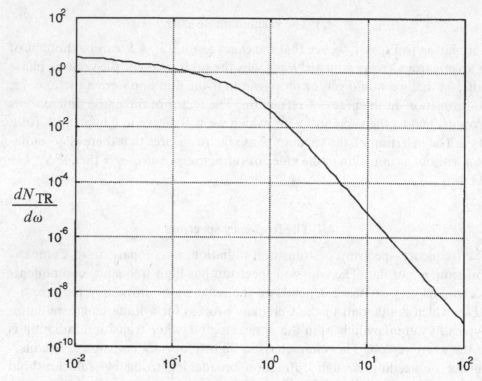

Fig. 4.3.   Frequency spectrum of TR photons, where frequency is defined in units of $\gamma\omega_p$, $y = \omega/\gamma\omega_p$.

them off from the phase factor and the coupling given in Eq. 4.4. As will be seen in the searchlight effect (Chapter 10), the typical emission angle is $1/\gamma$ which is a consequence of special relativity.

$$N_{TR} \sim \Delta E/\varepsilon_\gamma \sim \alpha$$

$$\varepsilon_\gamma \sim \hbar\omega \sim \frac{\gamma}{3}(\hbar\omega_p) \qquad (4.7)$$

$$\langle\theta\rangle \sim \frac{1}{\gamma}$$

The contour of $y^2 d^2 N_{TR}/d\Omega\, d\omega$ is shown in Fig. 4.4. We have ignored the $\sin^2\delta$ factor in Eq. 4.5 in making this plot because L is another free variable, and we have assumed that L is adjusted such that $\sin\delta \sim 1$. The peaking of the distribution at $\theta = 1/\lambda$ and $\omega \sim \gamma\omega_p$ is quite evident. If we accept that $y \sim 1$, then the phase difference is $\delta \sim (\omega L/4c)(1/\gamma^2)$.

We can heuristically think of the transition radiation as being due to the

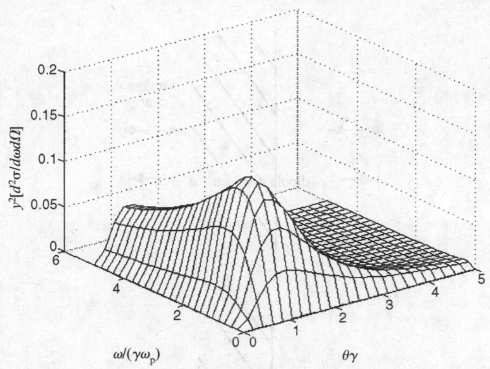

Fig. 4.4. Contour plot of $y^2 d^2 N_{TR}/d\Omega\,d\omega$ as a function of $y = \omega/(\gamma\omega_p)$ and $z = \gamma\theta$. As expected, the contours peak at $y \sim z \sim 1$, in agreement with Eq. 4.7.

changing dipole moment formed by the charge and its 'image' charge on the opposite side of the interface. The image charge is a way of solving the boundary conditions which is equivalent in effect to the induced charge density at the interface (see also Chapter 8) needed to match the boundary conditions. The charge and image charge pair makes a dipole which changes with time, and changes direction at the point of crossing the interface thus causing radiation (see Chapter 10). The interface is the point where the pointing of the dipole vector flips direction indicative of maximum acceleration. A schematic view of the geometry is shown in Fig. 4.5.

4.5 Dependence on γ and saturation

Transition radiation is a useful tool in high energy physics because it has an emitted energy, $\varepsilon_\gamma \sim \gamma$, which increases with γ as opposed to Cerenkov radiation, where all the radiation patterns collapse together as β approaches 1. That behavior is what makes transition radiation attractive for particle identification at very high energies where the Cerenkov technique and time of flight become

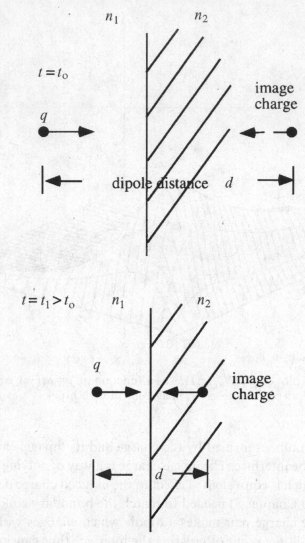

Fig. 4.5. Schematic representation of the dipole image charge – charge pair for a charged particle approaching a transition in indices of refraction $n_1 \neq n_2$.

problematic. However, this increase in photon energy with γ does not continue without limit. As with all detectors, there are intrinsic limitations.

There is a characteristic length that we define to be the distance, ξ, over which phase changes are substantial for fixed physical distance L. These phase differences must be large since the amplitudes, see Fig. 4.2, are out of phase for the front and back transitions and destructive interference must be avoided. Therefore, there is a maximum value of γ, called the saturation value, which depends on foil length, L, and radiation wavelength, λ, above which destructive interference begins to occur. That value of γ, γ_{SAT}, is inversely proportional to $\sqrt{\lambda}$. There is therefore a minimum foil thickness which is needed to avoid saturation. Note that the phase δ, which contributes a factor $\sin^2\delta$ to $d^2N_{TR}/d\omega\, d\Omega$, is simply L/ξ.

$$\xi = \frac{\lambda}{\pi} \bigg/ \left[\left(\frac{\omega_p}{\omega}\right)^2 + \theta^2 + \frac{1}{\gamma^2}\right], \quad \delta \sim L/\xi \quad \text{(Eq. 4.4)}$$

$$L \sim \xi \sim (\lambda\gamma^2) \tag{4.8}$$

$$\gamma_{SAT} \sim \sqrt{\pi L/\lambda}$$

Looking at Eq. 4.8 we might argue that we could go to thicker foils to increase the optical path difference and thus raise the γ range over which transition radiation can be used. However, that increases the self absorption of the medium which is not a winning strategy. Typically we must compromise between attenuation in the foil and losses due to destructive interference. Note that the Z^5 behavior of the photoelectric cross section (Chapter 2) argues for the use of low Z foils.

4.6 TRD foil number and thickness

As a numerical example, consider the detection of 200 GeV pions. The γ factor for pions is 1428 which means that a typical emission angle is about 0.7 mrad. Using the rule of thumb, the energy of the emitted photons ε_γ (Eq. 4.7) is about 8.9 keV which means we are in the x-ray region. Using polyethylene radiators we have $\hbar\omega_p$ of about 21 eV. We can use other foils, such as Li with $\hbar\omega_p = 14$ eV, to try to tune γ_{SAT}. Since $\omega \sim \gamma\omega_p$, we have some freedom to change the $1/\sqrt{\lambda}$ factor in γ_{SAT}. However, the Li foils are hard to work with. If we want to keep from saturating at those frequencies, the polyethylene foil thickness should be more than 14 μm. Since each foil only emits roughly α photons we need at least $1/\alpha$ or about 100 foils of material, with 100% detection efficiency, in order to observe just one x-ray per incident pion. The thickness of polyethylene for the emission of one photon is 0.19 cm while at 10 keV the mean free path in carbon is about 0.5 g/cm^2 (see Chapter 2) or about 0.5 cm indicating the importance of absorption since the 'stack' of foils is about $\langle L \rangle$ thick.

A photograph of an actual transition radiation detector (TRD) assembly used in an experiment is shown in Fig. 4.6. This consists of a radiator stack of many foils of polyethylene followed by a xenon filled proportional wire chamber (PWC). These chambers are devices which we will discuss in a later chapter. The reason for a detector consisting of a xenon gas fill is pretty clear from our discussion of the photoelectric effect. We want the high Z gas in order

Fig. 4.6.   Photograph of a TRD assembly. Each module, like the one being held, contains $\sim 1/\alpha$ foils followed by a Xe filled chamber. Note the gas flow indicators showing Xe for the detector (Fermilab E769). (Photo – Fermilab.)

to have a large absorption cross section for keV photons which means a large photoelectric cross section, and $\sigma_{\mathrm{PE}} \sim Z^5$ (Chapter 2).

Now, as mentioned before, the TRD is not transparent to its own emitted photons as was the case for the Cerenkov counter. An approximate expression for the number of 'useful' foils, N_{EF}, in a stack of N foils is given below.

$$N_{\mathrm{EF}} = (1 - e^{-N\sigma_{\mathrm{PE}}L})/(1 - e^{-\sigma_{\mathrm{PE}}L})$$

$$\rightarrow 1, \quad \sigma_{\mathrm{PE}} \rightarrow \infty \qquad (4.9)$$

$$\rightarrow N, \quad \sigma_{\mathrm{PE}} \rightarrow 0$$

For Cerenkov counters we could simply increase the radiator length to get more photons. In the present case that simple ploy does not work due to self absorption.

There is a cutoff at low energy due to the very large photoelectric cross section, $\sigma_{PE} \sim 1/\omega^{7/2}$, which we mentioned in Chapter 2. There is also a cutoff at high energies; see Fig. 4.3. Hence the effective number of foils is a strong function of the energy of the emitted x-ray photons, and there is a finite frequency range for the useful detection of photons. We had previously simply assumed this limited detector bandwidth and had used the estimate $\Delta y \sim 1$ around $\langle y \rangle \sim 1/3$. This discussion provides, after the fact, the justification for these assumptions.

4.7 TRD data

Some data taken with a transition radiation detector exposed to both pions and electrons of the same energy is shown in Fig. 4.7. The energy spectrum of both x-rays and ionization is shown. Clearly there is an exponential falloff with energy in both cases. The electrons are the relativistic objects with a large γ factor and they produce photons in the 10 to 50 keV range. The emission probability for pions, which have a much lower γ factor due to their increased mass, is much reduced. However, they do deposit ionization energy (Chapter 6) with the occasional large deposition due to delta ray fluctuations.

Also shown in Fig. 4.7 is the distribution of the statistical likelihood for electrons and pions showing that there is reasonable separation between the two. The likelihood (see Appendix J) is made up using multiple samples of the pattern of detected energy in distinct, independent, transition radiation detector modules. This technique is in the spirit of multiple ionization samples to accurately estimate dE/dx which is discussed in Chapter 6. One such module in a stack of TRD modules is shown in Fig. 4.6. The stack itself is visible in the background. It is clear that TRDs are useful for particle identification in the very high energy regime, even within the limitations due to 'saturation' phenomena which we have mentioned.

Exercises

1. Explicitly work through the algebra going from an amplitude $A - Ae^{2i\delta}$ to the form for I quoted in Fig. 4.2.
2. Evaluate the phase shift in Eq. 4.5 at zero emission angle. For $\delta = 1$ and a plasma frequency of 10 eV what minimum thickness L is needed for particles with $\gamma = 1000$?
3. Suppose we have a 10 GeV beam composed of electrons and pions impinging on a stack of polyethylene foils. Evaluate the γ factors and then the typical photon energy of transition radiation. Assuming a stack of 1000 foils how many photons are emitted? What is the total emitted energy for electrons and pions?

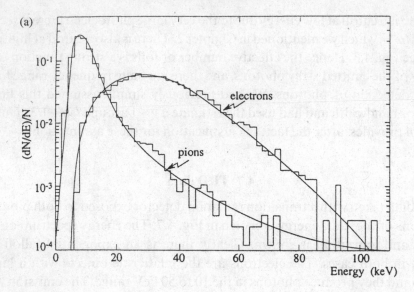

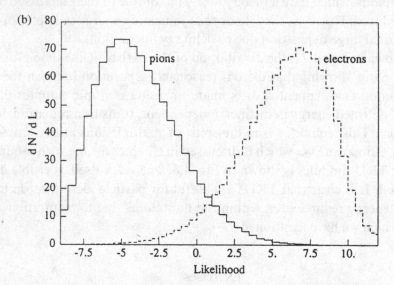

Fig. 4.7. Data taken with a TRD detector showing: (a) the mixture of x-ray and ionization energy for incident pions and electrons. Note the peak for electrons at ∼ 30 keV. (b) Likelihood function for pions and electrons using several samplings of the detected energy. (From Ref. 4.8, with permission.)

4. For the situation described in Exercise 3, what is the phase δ for polyethylene foils 0.001 cm thick for electrons and pions? Are there saturation effects?

5. Consider a stack of 1000 foils, each 0.001 cm thick. Consider 10 GeV pions and electrons. Find the mean emitted photon energy and the mean free

path for the emitted photons. Are photoelectric absorption effects important?

References

[1] W.M. Allison and P.R.S. Wright, Oxford, Dunp, 35/83 (1983).

[2] A. Bodek, *et al.*, *Z. Physik* **C18** 289 (1983).

[3] V.L. Ginzburg and I.M. Frank, *JETP* **16** 15 (1946).

[4] X. Artru, *et al.*, *Phys. Rev. D* **12** 1289 (1975).

[5] C.W. Fabjan and W. Struczinski, *Phys. Lett.* **57B** 484 (1976).

[6] T. Ludlam, *et al.*, *Nucl. Instrum. Methods* **180** 413 (1981).

[7] C.W. Fabjan, *et al.*, *Nucl. Instrum. Methods* **185** 119 (1981).

[8] S. Abachi, *et al.*, 'The D0 Detector,' *Nucl. Instrum. & Methods in Phys. Res. A* 388 (1994).

Part IIB

Scattering and ionization

In Chapters 2, 3, and 4 we examined non-destructive velocity measurements with detectors which convert the emitted light (scintillator, Cerenkov, and TR respectively) into an electrical signal. We now move to the second phase of the examination of non-destructive readout, where a charged particle transfers energy to the nuclei (Chapter 5) or the electrons (Chapter 6) of the detecting medium. In the case of transfer to nuclei, the direction of the incident particle is altered, which limits how 'non-destructive' the measurement of the properties of the particle we wish to determine is. In the case of transfer to the electrons, the resulting ionization energy loss is detected. Again, the fractional energy transfer to the detecting medium limits just how 'non-destructive' the readout will be. In the extreme case the incident particle may be slowed and halted in the detecting medium. However, for particles with energy above 1 GeV passage through a few cm of gaseous detector deposits only keV of energy, while 1 cm of solid plastic causes only of order MeV energy loss (0.1% for a 1 GeV particle).

5

Elastic electromagnetic scattering

Shoot pool Fast Eddy.
Minnesota Fats (The Hustler)

In this chapter we look at the scattering of an incident charged particle by the detecting medium. This knowledge is needed if we are to understand the physics of charged particle detectors which are based on detecting the energy transferred in the collision. We initially look at Coulomb collisions. Single collisions are termed Rutherford scatters caused by collisions with atomic nuclei. Multiple Coulomb collisions are then examined leading to important insights about the angular loss of information which occurs in scattering. Finally, we consider not angular deflection but energy transfer in Coulomb collisions. In that case, the scattering off atomic electrons is relevant. In particular, ionization of the atom with the ejection of freed electrons, 'delta rays', is an important topic, which is discussed in this chapter. The transfer of some of the energy of a charged particle to the medium is the basis of many of the devices discussed in Chapters 6 through 9.

5.1 Single scattering off a nucleus

We begin by considering single scattering off the Coulomb field of the nucleus by incoming charged particles. In performing scattering experiments it is important to realize that the atomic electrons are diffused over a characteristic size of ~ 1 Å as we said in Chapter 1. The nucleus is concentrated in a region a factor of 10^5 smaller. As first noted by Rutherford, the fact that we observe large angle scatters is a statement that there are small point-like scattering centers within the atom. It is the nucleus that is capable of causing these large angle scatters because all of the charge is concentrated at a point.

The observation of point-like structures is a recurring theme in physics. It has been often repeated; in molecules (atoms), in atoms (nuclei), in nuclei (nucleons) and in nucleons themselves (quarks). At each stage so far in our exploration of the microworld, objects have been discovered to be composites of yet smaller objects.

91

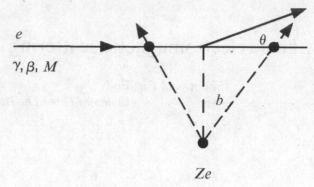

Fig. 5.1. Kinematic definitions for a Coulomb collision of a charged particle by a nucleus of atomic number Z at impact parameter b causing a scattering angle θ.

The kinematic definitions which we use are shown in Fig. 5.1 where a singly charged particle with velocity βc and mass M is incident on the field of a nucleus at rest with atomic number Z at an impact parameter, or transverse distance, b. The incoming particle suffers a scattering by angle θ. Note that for small angle scattering the incoming to outgoing symmetry of the problem tells us that for central forces the impulse is transverse. The longitudinal impulse integrates to zero, as indicated in Fig. 5.1.

The force, F, is electric, so it goes as $1/r^2$. The non-relativistic characteristic time for the interaction, Δt, is defined by the fact that the force falls off rapidly with distance and thus is only active on time scales where the incident particle is near to the scattering center.

$$F(b) = Ze^2/b^2$$

$$\Delta t = 2b/v$$

(5.1)

The net momentum impulse, Δp_T, is in the transverse direction and can be approximated as the force at the point of closest approach, $F(b)$, times the characteristic time over which that force is active. The characteristic transverse momentum transfer divided by the incident momentum gives the scattering angle. The Rutherford form for the angle, θ_R, in the small angle approximation is given in Eq. 5.2. Clearly for high Z targets the scattering angle is larger because the electric field is stronger. For small impact parameters we see a stronger field and therefore the scattering angle is larger. For low velocities we spend a longer time near the nucleus and therefore low velocity also means large scattering angles.

$$\Delta p_T \sim F(b)\Delta t, \quad \mathbf{F} \equiv d\mathbf{p}/dt$$

$$\theta \sim \Delta p_T/p, \quad \Delta p_T = 2Z\alpha/bv$$

$$\theta_R \sim 2Z\alpha/pvb$$

(5.2)

Another form for θ_R is $\theta_R \sim (Z\alpha/b)/T$ where T is the projectile kinetic energy. We recognize θ_R to be the ratio of the potential energy at $r = b$ to the total kinetic energy, $\theta_R = U(b)/T$. For example, consider 1 MeV kinetic energy, T, protons, $T = \varepsilon - M$, incident on lead (Pb) nuclei at an impact parameter of 100 fm. The Rutherford scattering angle, Eq. 5.2, is radians or $\theta_R \sim 68°$ so that the scattering angles can, indeed, be large.

5.2 The scattering cross section

What has just been 'derived' is the relationship between the scattering angle and the impact parameter. Clearly, in an actual experiment, we cannot aim an incident particle at a particular nucleus so that b is not measured. Since we cannot aim, all areas transverse to the incident beam are taken to be equally probable and the scattering probability, dP, is proportional to the transverse area element $d\mathbf{b}$. Since there is a one to one relationship between impact parameter, b, and scattering angle, θ, we can simply relabel the scattering. Note that the exact form of the relationship depends on the form of the forces, so that we can, in principle, infer the functional form of the force law by observing the scattering angle distribution (see Chapter 5.11).

$$dP \sim d\mathbf{b} = d\sigma$$

$$d\sigma \sim b \, db \, d\phi = \frac{d\sigma}{d\Omega} d\Omega, \quad d\Omega = \sin\theta \, d\theta \, d\phi$$

(5.3)

$$\frac{d\sigma}{d\Omega} = \frac{b}{\sin\theta}\left(\frac{db}{d\theta}\right)$$

Equation 5.3 says that the number incident at a given impact parameter b is equal to the number scattered into the corresponding angle θ. Referring back to Eq. 5.2, we can then immediately write down the small angle approximation for the Rutherford differential scattering cross section, $d\sigma_R/d\Omega$.

$$\frac{d\sigma_R}{d\Omega} \sim \left(\frac{Z\alpha}{Mv^2}\right)^2/\theta^4$$

(5.4)

The differential cross section has the power law behavior, $\sim 1/\theta^4$, which is characteristic of the Coulomb scattering of point particles. It is a general characteristic of point-like processes that they obey power law scattering distributions and have power law energy dependence. Collective phenomena, or distributed

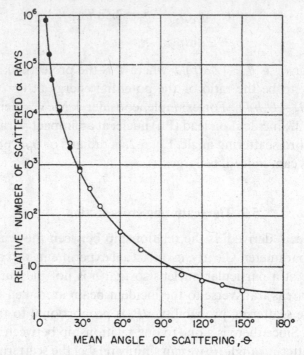

Fig. 5.2.   Data taken by Rutherford on the scattering angular distribution for incident alpha particles on silver. (From Ref. 5.6, with permission.)

sources, have more complicated and, typically, much more rapidly falling distributions as we saw in Chapter 2 for the photoelectric effect. Some data actually taken by Rutherford and collaborators are shown in Fig. 5.2. Note that measurements extending to 150° scattering angle are shown indicating that large angle scatters were seen.

5.3 Feynman diagrams

It turns out that the same result obtains in quantum mechanics. The 'Feynman diagram' for Rutherford scattering is shown in Fig. 5.3. The exchanged 'virtual photon' is responsible for the deflection of the incident charge. A word is in order here on 'Feynman diagrams'. We will use them simply as a tool to make conclusions about the powers of α (see Chapter 1) which the rate for a process contains, and occasionally to make dynamical inferences. We recall that an isolated electron cannot emit a photon and still conserve energy and momentum (see Chapter 3). The amplitude for that emission is proportional to the total charge e. The uncertainty principle allows for emission of a 'virtual' photon if subsequently that photon is reabsorbed by another charge in a time Δt which

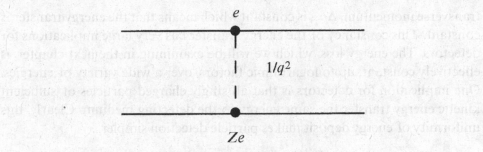

Fig. 5.3. Feynman digram for Rutherford scattering.

is consistent with a limited energy transfer ΔE, of size $\Delta E \leq \hbar/\Delta t$. The amplitude for emission and absorption then goes as $e^2 = \alpha$, so that the rate goes as α^2. The amplitude also goes as $1/\Delta E$ in second order perturbation theory where ΔE is the energy difference between initial and final unperturbed states. The amplitude is large when ΔE is small. This in turn means that any 'virtual' particle would like to be as 'real' as possible which for photons means $\Delta E \rightarrow 0$, or as 'soft' as possible. This is the main dynamical inference.

The scattering amplitude is, in the Born approximation, the matrix element of the interaction potential between the free, non-interacting, initial and final states, $A \sim \langle f | H_\mathrm{I} | i \rangle$. If we take those states to be plane waves, the scattering amplitude is the Fourier transform of the interaction potential, $A \sim \int e^{k_\mathrm{f} Z} H_\mathrm{I} e^{-k_\mathrm{i} Z} d\mathbf{r} = \int e^{iqZ} H_\mathrm{I} d\mathbf{r}$. The Fourier transform of a $1/r$ Coulomb interaction potential is, in momentum transfer space, proportional to $1/q^2$. Therefore, another way to look at the dynamics shown in Fig. 5.3 is first to recall that $\mathbf{q} = \mathbf{k}_\mathrm{i} - \mathbf{k}_\mathrm{f} = \Delta \mathbf{p}_\mathrm{T}$. The square of the amplitude, $A \sim 1/q^2$, which is proportional to the cross section, $d\sigma/d\Omega \sim |A|^2$, then goes like $1/\theta^4$ since $q \sim p\theta$ which agrees with our classical 'derivation' of Rutherford scattering. The virtual photon 'wants' to be as real as possible, so that $A(q)$ is large when q is small.

$$q \sim \Delta p_\mathrm{T} \sim p\theta$$

$$A(q) \sim (Ze)e/q^2 \tag{5.5}$$

$$|A(q)|^2 \sim (Z\alpha)^2/\theta^4 \quad \text{(Eq. 5.4)}$$

5.4 Relativistic considerations

The transverse electric field at a given impact parameter due to the relativistic field transformation increases as γ, see Table 3.1. However, the collision time decreases. For uniform linear incident motion, the time dilation factor, γ, means that the laboratory collision time Δt goes as $1/\gamma$. Therefore, the net impulse in transverse momentum, Δp_T, is constant which means that the energy transfer is constant. This constancy of the energy transfer has very large implications for detectors. The energy loss, which we will be examining in the next chapter, is effectively constant, up to logarithmic factors, over a wide variety of energies. One implication for detectors is that all singly charged particles of sufficient kinetic energy transfer the same energy to the detecting medium. Clearly, this uniformity of energy deposit makes particle detection simpler.

$$E_T(b) \sim \gamma$$

$$\Delta t \sim 2b/\beta\gamma c \tag{5.6}$$

$$\Delta p_T \sim eE_T(b)\Delta t \sim \text{constant}$$

We will return to the residual relativistic effects in Chapter 6.

5.5 Multiple scattering

Recall that Rutherford scattering is due to the field of the nucleus. Gauss's law tells us that when a particle is incident outside the volume of the atom there is screening of the electric field of the nucleus by the intervening charges of the electrons. The net charge is zero which means that there is a minimum scattering angle θ_{min} (maximum impact parameter) when b is roughly equal to the first Bohr radius, a_o.

$$\theta_{min} \sim 2Z\alpha/pva_o \quad \text{(Eq. 5.2)} \tag{5.7}$$

Therefore, the differential cross section integrated over all angles does not diverge due to the $1/\theta^4$ behavior because there exists a minimum angle cutoff. We observe a total elastic cross section equal to the geometric cross section which we discussed in Chapter 1. Looking back at Eq. 5.3 it is intuitively clear that we will recover the geometric cross section because for every impact parameter less than the atomic radius there is a non-zero scatter. The integral of $d\sigma = 2\pi b\,db$, where b ranges from 0 to the Bohr radius, is just the geometric cross section.

$$\sigma \sim \int_{\theta_{min}}^{\infty} \left(\frac{d\sigma}{d\Omega}\right) 2\pi\theta\,d\theta = \int_{0}^{a_o} 2\pi b\,db$$

$$\sim \pi a_o^2 \sim 1/\theta_{min}^2 \sim \int_{\theta_{min}}^{\infty} d\theta/\theta^3 \tag{5.8}$$

Since we know the distribution of scattering angles, the mean scattering angle in a single collision is calculable. The mean angle is defined to be the scattering angle weighted by the angular distribution (see Appendix J), where the distribution function is $d\sigma/d\Omega$.

$$\langle\theta^2\rangle \equiv \frac{\int\theta^2\frac{d\sigma}{d\Omega}d\Omega}{\int\frac{d\sigma}{d\Omega}d\Omega} \sim \frac{\int(d\theta/\theta)}{\int(d\theta/\theta^3)} \quad (\text{Eq. 5.4})$$

(5.9)

$$\sim 2\theta_{min}^2[\ln(\theta_{max}/\theta_{min})]$$

Because Rutherford scattering is weighted so strongly towards the minimum scattering angle (maximum impact parameter), up to logarithmic factors, $\langle\theta^2\rangle$ is equal to twice the square of the minimum scattering angle.

If we now consider the passage of a charged particle through a thick block of material there are N scatters on average in that material. The scattering angle compounds as a random stochastic process, with a mean given by the central limit theorem. The mean square multiple scattering angle $\langle\theta_{MS}^2\rangle$ for N scatters is N times the mean square angle for a single scatter $\langle\theta\rangle^2$ (see Appendix J).

$$\langle\theta_{MS}^2\rangle = N\langle\theta^2\rangle$$

$$N = (N_o\rho\sigma/A)dx \quad (\text{Chapter 1})$$

(5.10)

$$= dx/\langle L\rangle$$

The number of scatters is given by the distance traveled divided by the mean free path between single scatterings. Inserting the expression for the mean square of the scattering angle from Eq. 5.9, Eq. 5.7 and the Rutherford scattering cross section from Eq. 5.8 we obtain the mean of the square of the angle for multiple Coulomb scatterings.

$$\langle\theta_{MS}^2\rangle = \left[\frac{N_o\rho\,dx}{A}\right]2\pi\left[\frac{2Z\alpha}{p\beta c}\right]^2[\ln(\,)]$$

(5.11)

5.6 The radiation length

The factor of θ_{min}^2 from Eq. 5.9 and the factor $1/\theta_{min}^2$ from σ cancel out, so that we are left with a simple expression up to logarithms. It is traditional to use X_o units (see Chapter 10) in expressions for the multiple scattering angle. However, it should always be kept in mind that this is not a radiative process. It is fortuitous that radiation length units are applicable but it is, in fact, misleading. The radiation length is derived in Chapter 10 and numerical values are provided in Table 1.2.

$$\langle \theta_{MS}^2 \rangle = \frac{dx}{X_o} m^2 / (\alpha \beta^2 p^2)$$

$$(5.12)$$

$$X_o^{-1} = \frac{16}{3} \left(\frac{N_o}{A} \right) (Z^2 \alpha)(\alpha^2/m^2)[\ln(\ )]$$

A characteristic energy, E_s, is defined, 21 MeV, which is related to the electron mass and the electromagnetic coupling constant α. In these units $\langle \theta_{MS}^2 \rangle$ is proportional to the path length in radiation length units, which we will define to be the variable t. Since this is a stochastic or random walk process, the multiple scattering angle is proportional the square root of the path length.

$$E_s \equiv \sqrt{\frac{4\pi}{\alpha}} (mc^2) = 21 \text{ MeV}$$

$$(5.13)$$

$$\sqrt{\langle \theta_{MS}^2 \rangle} = \frac{E_s}{p\beta} \sqrt{\frac{dx}{X_o}}$$

A convenient way to think of the physics is to assign a multiple scattering transverse momentum impulse, $(\Delta p_T)_{MS}$. Any incident singly charged particle suffers this same average impulse when going through a block of material and being scattered by the nuclei of the material. The transverse energy scale is set by the target physics.

$$\theta_{MS} = (\Delta p_T)_{MS}/p, \quad (\Delta p_T)_{MS} = \frac{E_s}{\beta} \sqrt{\frac{dx}{X_o}} = \frac{E_s}{\beta} \sqrt{t} \qquad (5.14)$$

For example, in traversing one radiation length of material, a 20 GeV pion will suffer a mean scattering by angle $\theta \sim 0.001$, or 1 mrad ($\Delta p_T \sim 21$ MeV).

5.7 Small angle, three dimensional multiple scattering

Let us quickly mention the actual statistical distribution of the quantities involved in multiple scattering. For small deflections the distributions in the two directions perpendicular to the direction of motion are independent, basically since $\theta^2 \sim \theta_x^2 + \theta_y^2$, $d\Omega \sim d\theta_x d\theta_y$, and θ^2 is distributed with a Gaussian dependence. Since the location of the particle and its momentum vector in a plane require two variables, a total of four are needed to specify the three dimensional problem. A possible choice for the y coordinate is shown in Fig. 5.4. Note that θ_y is normally distributed with RMS $= \sqrt{\langle \theta_y^2 \rangle} = \sqrt{\langle \theta_{MS} \rangle^2/2}$, $\langle \theta_x^2 \rangle + \langle \theta_y^2 \rangle = \langle \theta^2 \rangle$. The (y, θ_y) variables are correlated, with error matrix M_{MS}, and with zero means, $\langle y \rangle = \langle \theta_y \rangle = 0$. (See Appendix J.)

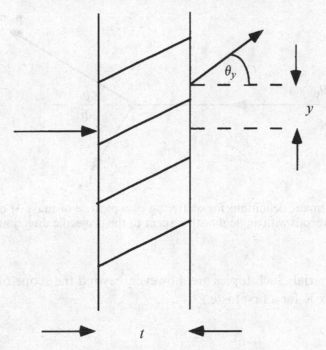

Fig. 5.4. Definition of multiple scattering quantities in a plane. The exit position, y, and exit angle, θ_y, upon traversing a depth t in X_0 units is shown.

$$M_{MS} = \begin{bmatrix} \langle \theta_y^2 \rangle & \langle \theta_y y \rangle \\ \langle \theta_y y \rangle & \langle y^2 \rangle \end{bmatrix}, \begin{bmatrix} \theta_y \\ y \end{bmatrix}$$

$$= \begin{bmatrix} 1 & t/2 \\ t/2 & t^2/3 \end{bmatrix} \langle \theta_y^2 \rangle$$

(5.15)

It is interesting to note that we can find uncorrelated variables. We assert that θ_y and $y_s = y - \theta_y t/2$ are uncorrelated. This can be established by diagonalizing the matrix, M_{MS}, and finding the eigenvectors and is left as an exercise to the reader. The diagonalized matrix is M'_{MS}.

$$M'_{MS} = \begin{bmatrix} 1 & 0 \\ 0 & t^2/12 \end{bmatrix} \langle \theta_y^2 \rangle, \begin{bmatrix} \theta_y \\ y_s \end{bmatrix}$$

(5.16)

The physical reason for the decoupling of y_s and θ_y is that the coupling comes about because large scattering angles, θ_y, necessarily mean large deflections, y. The variable y_s removes this effect by subtracting from y the deflection due to scattering by an angle θ_y at the center of the material. Note that the error about the mean is then reduced by a factor 2 ($\sigma_{ys} \sim \sigma_y/2$). These error matrices are very useful in making mathematical models for particles moving in the presence of

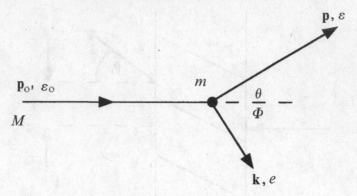

Fig. 5.5.   Kinematic definitions for scattering of a particle of mass M off a target of mass m which recoils with angle Φ with respect to the projectile direction.

scattering material. Such topics are, however, beyond the scope of this book. (See Appendix K for a first taste.)

5.8 Maximum momentum transfer

Looking at Eq. 5.1 it is clear why so far we have been considering scattering off the nucleus. The factor Z and the point-like nature of the nucleus make it important for angular deflections. What we wish to do now is to consider not the angular deflection of the projectile but the energy loss which it suffers in transferring energy to the target. The kinematic definitions for the energy transfer are given in Fig. 5.5 where an incident particle of defined momentum, energy and mass, (p_0, ε_0 and M) scatters off a target particle of mass m which recoils with angle Φ and momentum $\mathbf{k}$. We label the projectile scattering angle by θ as before. The collision kinematics is dictated, as usual, by the conservation of energy and momentum.

$$\mathbf{p}_0 = \mathbf{p} + \mathbf{k}$$

$$\varepsilon_0 + m = \varepsilon + e \tag{5.17}$$

We define the kinetic energy, T, of the target in the final state by subtracting the rest mass from the recoil energy, e. We can, with considerable tedious algebra, extract an expression for Q as a function of the recoil scattering angle, Φ. (See Appendix A.)

$$T \equiv e - m$$

$$Q \equiv T/m \tag{5.18}$$

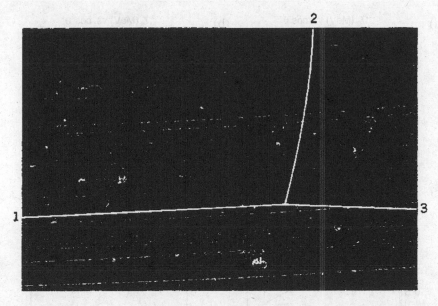

Fig. 5.6.   Bubble chamber photograph (enhanced) showing an elastic scattering with recoiling target at an angle $\Phi \sim 90°$. (From Ref. 5.2, with permission.)

The maximum value of the energy transfer to the target, Q_{max}, comes at zero degree recoil angle when the target is boosted straight ahead. The minimum value, $Q = 0$, looking at Eq. 5.19, clearly comes at 90° recoil angle when we have a 'grazing' collision. Anyone who has shot pool knows intuitively that this is correct.

$$Q = 2 \left\{ \frac{(\beta\gamma M\cos\Phi)^2}{(m + \gamma M)^2 - (\beta\gamma M\cos\Phi)^2} \right\}$$

$$Q_{max} = \frac{2(\beta\gamma)^2}{\left[1 + \left(\frac{m}{M} \right)^2 + \left(\frac{2\gamma m}{M} \right) \right]} \rightarrow 2(\beta\gamma)^2 = 2\left(\frac{p_0}{M} \right)^2, \quad \Phi = 0° \quad (5.19)$$

$$Q/Q_{max} \sim \cos^2\Phi$$

A bubble chamber photograph, in which the trajectories are enhanced, showing an elastic scatter where the target recoils at effectively 90° is shown in Fig. 5.6. (As an aside, we will not further discuss the 'bubble chamber' since it is no longer an active research tool. Suffice it to say that the ionization released by a charged particle heats the bubble chamber liquid, e.g. liquid hydrogen,

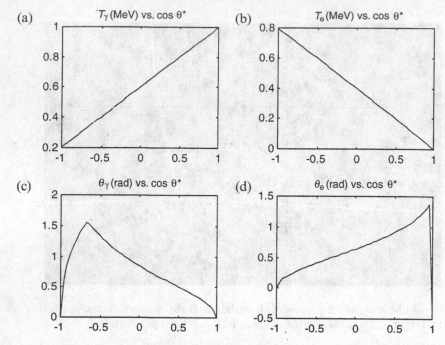

Fig. 5.7. $\gamma+e\rightarrow\gamma+e$ elastic scattering with incident photons of 1 MeV. (a) Scattered γ kinetic energy, (b) recoil e kinetic energy, (c) scattered γ angle, (d) recoil e angle, as a function of cosine of the center of momentum scattering angle.

which then locally boils. The small bubbles are illuminated and then photographed.) The dynamics in this case also limit recoils to small momentum transfer, just as Rutherford scattering peaks at small q. Therefore, the projectile has a small deflection in these grazing collisions while the target recoils with low velocity at wide angles, $Q \sim 0$.

The maximum energy transfer Q_{max} simplifies if $\gamma m/M \rightarrow 0$. In that case, the scaled maximum energy transfer, Q_{max}, is just twice the momentum of the projectile divided by its mass squared. The ratio of recoil energy to the maximum possible, Q_{max}, is approximately the square of the cosine of the recoil angle, Φ.

For elastic scattering at fixed incident energy there is only one free variable which can be taken either to be the scattering angle or the impact parameter or the recoil momentum. It is clear from Eq. 5.19 that there is a functional relationship between the recoil energy and the scattering angle. In Fig. 5.7 we show, for example, the scattered energy and angle and the recoil energy and angle, all as a function of the CM scattering angle for $\gamma+e$ elastic scattering with incident 1 MeV photons. For grazing collisions, $\cos\theta^* \sim 1$, $T_\gamma \sim 1$ MeV, $\theta \sim 0°$ and $T_e \sim 0$ MeV $\theta_e \sim 90°$ as expected.

5.9 Energy transfer

In Eq. 5.2 we've already derived the momentum transfer in an elastic collision. Looking now at the energy transfer, and assuming non-relativistic recoil, we conclude that it is the light targets to which it is easy to transfer energy.

$$\Delta p_T \sim 2\alpha/bv. \quad Z = 1$$

$$\Delta\varepsilon \sim \Delta p_T^2/2m \quad\quad\quad (5.20)$$

$$\Delta\varepsilon \sim 2\alpha^2/b^2v^2m$$

Because of the existence of the factor $1/m$ in the expression for the recoil energy, $\Delta\varepsilon$, light particles are the ones to which energy can be transferred. This is quite familiar in everyday life. If you throw a Ping-Pong ball at another Ping-Pong ball there can be a large energy transfer. If you throw a Ping-Pong ball at a Mack truck, the truck will not recoil and the incident ball will retain its full initial energy. Therefore, we expect that incident charged particles will preferentially transfer energy to the atomic electrons, which are ~ 2000 times lighter than protons. The electrons will not, however, be involved in large angle scatterings, as we said, because large scattering angles come from interactions with a localized charge.

5.10 Delta rays

We expect that occasionally the recoil electrons will gain enough energy to be removed from their bound state and be kicked into the ionization continuum (see Chapter 1). For historical reasons these freed electrons are called delta rays. An example of delta rays being ejected by a charged particle traversing material is seen in Fig. 6.6 in the next chapter and in Fig. 5.6 of this chapter. They are seen to recoil at large angles and at low energies. Instead of impact parameter b, we choose the kinetic energy of the recoil, T, to relabel the scattering process. The relationship of $T(b) = \Delta\varepsilon$ in Eq. 5.20 is used.

$$d\sigma = d\mathbf{b} = b\ db\ d\phi = (d\sigma/dT)dT$$

$$\frac{d\sigma_\delta}{dT} = 2\pi b \left(\frac{db}{dT}\right) = \frac{db}{dT} = (b/2T) \quad\quad (5.21)$$

$$\frac{d\sigma_s}{dT} = \pi\ b^2/T = \left[\frac{2\pi a^2}{\beta^2 C^2 T^2 m}\right]$$

The characteristic $1/\theta^4$ projectile angular behavior translates into a $1/T^2$ recoil energy behavior. The projectile scattering angles are dynamically constrained to be small by the $1/\theta^4$ factor and similarly the recoil energies are constrained to be small by the $1/T^2$ factor.

We now convert the cross section to the mean free path for delta ejection. There are Z atomic electrons which are incoherently added. The source is spread over a size $\sim a_0 \sim 1$Å, and the wavelength of the projectile is assumed to be less than that. Hence there is no phase coherence over the size of source as there is, for example, in Bremsstrahlung in the Coulomb field of the nucleus (Chapter 10). The $1/T^2$ divergence is still just a mathematical artifact as there is a cutoff at the maximum impact parameter, $b \sim a_0$. For fast incident particles, $\beta \to 1$.

$$\frac{dN_\delta}{dT d(\rho x)} = 2\pi \left(\frac{N_0 Z}{A}\right) \frac{\alpha^2 \lambdabar_e}{T^2} \tag{5.22}$$

Numerically the factor $2\pi(N_0 Z/A)\alpha^2 \lambdabar_e$ is $0.078/(\text{MeV} \cdot \text{g/cm}^2)$, if Z/A is $\sim 1/2$. If we express the minimum recoil kinetic energy in MeV, the number of collisions giving a recoil electron greater than 1 MeV in g/cm^2 units is approximately 7.8%. The number goes with the cutoff energy T_0 as $1/T_0$.

$$dN_\delta/d(\rho x)|_{T>T_0} \sim \frac{7.8\%}{\text{g/cm}^2}, \quad T_0 = 1\text{MeV} \tag{5.23}$$

These 'delta rays', which we will see in examining bubble chamber pictures later (see Chapter 6), confuse the measurement of points along the trajectory of the particle. We will also see later that they are responsible for some of the irreducible fluctuations in the energy deposited by an incident charged particle in traversing a detecting medium.

5.11 Other force laws

It is amusing to explore the relationship between other force laws and the scattering angular distribution. Suppose the power law were different. How would that change the $d\sigma/d\Omega \sim 1/\theta^4$ behavior (Eq. 5.4)? Suppose the force law is not $F(b) = Ze^2/b^2$, but rather

$$F(b) = a/b^N \tag{5.24}$$

Assuming that the time interval is unchanged, the relationship between scattering angle and impact parameter is

$$\theta = (a/pv)b^{1-N} \sim U(b)/T \tag{5.25}$$

Using Eq. 5.3, we find the generalized scattering angular distribution to be

$$d\sigma/d\Omega = \frac{1}{\theta} \frac{(2T/a)}{(N-1)} [b]^{N+1} \tag{5.26}$$

Using Eq. 5.25 to eliminate b in favor of θ

$$d\sigma/d\Omega = \left(\frac{1}{N-1}\right)\left(\frac{2T}{a}\right)^{1/(1-N)} \theta^{(2N/(1-N))} \qquad (5.27)$$

For $N=2$ we recover $d\sigma/d\Omega(N=2) = (2T/a)^{-1}\theta^{-4}$ with $a = Z\alpha$. Note that more localized forces, e.g. $N=3$, give more localized impact parameters or more isotropic angular distributions. For $N=3$, $d\sigma/d\Omega$ goes as $1/\theta^3$.

Exercises

1. Consider a 10 MeV alpha particle incident on a lead nucleus. At what impact parameter, b, is there a 5° deflection?
2. What is the momentum transfer, q, for the parameters of Exercise 1?
3. What is the minimum scattering angle for a 10 MeV alpha particle if the atomic radius of lead is taken to be 10 Å?
4. A calorimeter is constructed with 25 radiation lengths of lead. What deflection will a 1 GeV muon suffer in traversing this device? What is the transverse momentum impulse due to multiple scattering?
5. Diagonalize the multiple scattering matrix given in Eq. 5.15. Show that the result is as given in Eq. 5.16 with eigenvectors as quoted in the text.
6. Show that if $M \gg m$, then $Q \to 2(\beta\cos\Phi)^2/[1-(\beta\cos\Phi)^2]$. Plot $Q(\Phi)$ for $\beta = 0.1$ and 0.99.
7. Evaluate the energy transfer for $v=c$ and a 100 fermi impact parameter. Is this value above the ionization potential?
8. Show explicitly that, if there is an inverse cube force law, the deflection angle $\sim 1/b^2$ and the differential cross section goes as $\sim 1/\theta^3$.
9. For an inverse cube force law show explicitly that the energy transfer T goes as $\sim 1/b^4$ and that the differential energy loss distribution goes as $1/T^{3/2}$.
10. Find the energy transfer to an electron for collisions with $b=10$ fermi and $b=1$ Å for $v=c$ incident charged particles.

References

[1] *Quantum Physics of Atoms, Molecules, Solids, Nuclei and Particles*, R. Eisberg and R. Resnick, John Wiley and Sons (1974).
[2] *An Introduction to Quantum Physics*, A. P. French and E. F. Taylor, W. W. Norton & Co., Inc. (1978).
[3] *Radiations from Radioactive Substances*, E. Rutherford, J. Chadwick and C.D. Ellis, Cambridge University Press (1930).
[4] R. Sternheimer, *Phys. Rev.* **88** 857 (1952).
[5] Allison, Warshaw, *Rev. Mod. Phys.* **25** 779 (1953).
[6] *Elementary Modern Physics*, A.P. Arya, Addison-Wesley Publishing Co. (1974).

6

Ionization

Men love to wonder, and that is the seed of science
Ralph Waldo Emerson

The detection of the charge freed in a medium by Coulomb collisions with an incident charged particle, and subsequently collected on an electrode, is the basis of the detection of charged particles. The groundwork for discussing the physics of these processes was laid in Chapter 5. In this chapter we work through the implications. Some comments on the fluctuations in the energy deposit are also made. Finally, a comparison of the relative strength of ionization processes and radiative processes (see Chapter 10) is made leading to the concept of the critical energy which is basic to EM calorimetry (see Chapter 11) and the detection of high energy muons (see Chapter 13).

6.1 Energy loss

In Chapter 5, while examining the elastic scattering of an incident charged particle off the atomic electrons, we derived the distribution of delta rays and the energy transfer, $d\varepsilon$, at a given impact parameter b. The notation $d\varepsilon$ for a single scatter and dE for the total loss of energy is adopted in this section.

$$d\varepsilon/d\mathbf{b} \sim 2\alpha^2/b^2v^2m \text{ (Eq. 5.20)} \tag{6.1}$$

We know we cannot aim the projectile which implies that all areas transverse to the incident particle have equal probability, $d\sigma = d\mathbf{b}$. We then integrate the energy loss at a given impact parameter over all impact parameters to get the mean energy loss in a single collision, . Assuming that there are minimum and maximum impact parameters we find

$$d\varepsilon \sim \int 2\pi b\, db (d\varepsilon/d\mathbf{b})$$

$$\sim \frac{4\pi\alpha^2}{mv^2}[\ln(b_{max}/b_{min})] \tag{6.2}$$

Note that the ionization loss depends only on the charge and velocity of the projectile, not the mass. We now sum over all collisions to get the ionization energy loss dE_1.

$$\frac{dE_1}{d(\rho x)} \sim \left(\frac{N_o Z}{A}\right) d\varepsilon$$

$$\frac{dE_1}{d(\rho x)} \sim 4\pi \left(\frac{N_o Z}{A}\right) (\alpha^2 \lambda_e) \left(\frac{1}{\beta^2}\right) [\ln(\ )]$$

(6.3)

As we saw in Chapter 5, there is also a distribution of energetic delta rays. They in turn can make secondary ionization which can be several times larger than the primary energy loss which is represented by Eq. 6.3. We ignore secondary processes in what follows but caution the reader of their practical importance.

6.2 Minimum ionizing particle

The expression for $d\varepsilon$, Eq. 6.2, is the energy lost in a single collision. The ionization loss, dE_1, is the energy lost in traversing a block of material expressed in g/cm^2 units. Equation 6.3 evaluated at $\beta = 1$ can be found for the typical value of $Z/A \sim 1/2$. In that case the factor in front of the logarithm is 0.16 MeV/(g/cm^2) (see Eq. 5.22). As a rule of thumb and as a mnemonic, a correct representation of the logarithmic factors leads to a minimum ionizing particle losing about 1.5 MeV for every g/cm^2 traversed. Numerical estimates are given in Chapter 6.6 below. Note that as $\beta \rightarrow 1$ the energy loss per unit path length goes to a minimum. The jargon is that these are 'minimum ionizing particles' or MIPs.

$$MIP \sim 1.5 \text{ MeV/(g/cm}^2) \sim [dE_1/d(\rho x)]_{min}$$

(6.4)

Table 1.2 lists the minimum energy loss for an incident singly charged particle. For the purposes of visualization the minimum (dE/dx) from Table 1.2 is plotted in Fig. 6.1 as a function of atomic number Z. Note that the variation in value, scaled to Z/A, is less than 60% across the full periodic table. Clearly the MIP is a useful general concept for detector designers.

For example, it is easy to see why vacuum is needed in, say, TV picture tubes. The TV tube acceleration voltage is about 10 keV. In air, Table 1.2, a minimum ionizing particle loses ~ 1.8 keV/cm. Thus the 10 eV electron in an air filled picture tube would not even reach the screen, let alone be well imaged.

Note that the near constancy of the energy loss by ionization expressed in g/cm^2 means that, if space is limited, there is a premium on using the highest

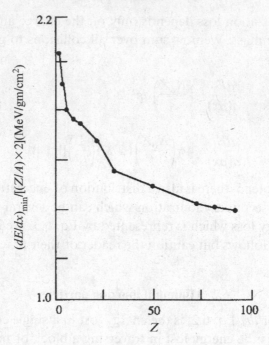

Fig. 6.1.   Minimum value of dE/dx scaled by Z/A as a function of Z. The value of 1.5 MeV/(g/cm^2) is valid to ~20% for all but the lightest elements.

density materials. Lead and steel are commonly used materials for reactor shielding, for example. Cheaper but less effective materials, include concrete and dirt.

6.3 Velocity dependence

Notice that the energy loss per unit path length goes like $1/\beta^2$. Hence, as a particle slows down and begins to stop it starts to lose energy very rapidly. It will preferentially deposit more energy per unit path length when it is going slowly. The jargon is 'Bragg peak' in the deposited ionization energy.

$$\frac{dE_I}{dx} \sim \frac{1}{\beta^2} \sim \frac{M^2}{p^2} \tag{6.5}$$

Fundamentally the $1/\beta^2$ behavior reflects the length of time that the particle is near the scattering center. This behavior has strong practical implications. For example, protons can be cunningly used in cancer therapy. Their initial energy is chosen so that most of their energy can be deposited at the end of their range where they stop deep in the human body at a tumor site and not in the intervening healthy tissue. For that reason a variable energy proton accelerator is of

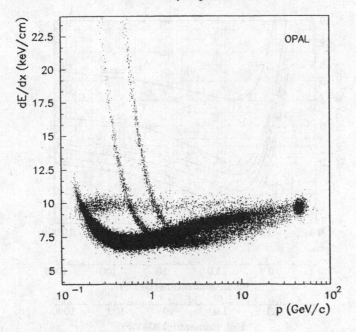

Fig. 6.2.   Data showing the ionization dE_I/dx as a function of particle momentum showing e, π, K, and p bands. The M^2 band separation and $1/p^2$ behavior is evident. (From Ref. 6.9, with permission.)

great usefulness in cancer therapy and some have been built specifically for that purpose. Since people are largely made of water, a 400 MeV momentum or 85 MeV energy proton goes about 10 cm before stopping (see Fig. 6.5 below) in a patient, which sets the energy scale of medical proton synchrotrons.

A plot of data on dE/dx as a function of momentum is shown in Fig. 6.2 for a detector exposed to several different particles. Looking at Eq. 6.5 we see that dE/dx for fixed momentum goes like the square of the mass for low velocities. What we see in Fig. 6.2 are bands going as $1/\beta^2$ for the pions, mass 140 MeV, kaons, mass 494 MeV, and protons, mass 938 MeV. The horizontal 'MIP' band is due to the relativistic electrons. All the bands converge to the same minimum value of dE/dx, as expected. Note that the vertical scale is $\sim$ keV/cm so that we know the detector is gas filled. The momentum is measured by measuring the particle path in a magnetic field as will be discussed in Chapter 7. Precise measurement of the momentum and ionization loss in a detector, Eq. 6.5, allows us to do 'particle identification' in the sense of 'measuring' the mass of the particle.

Theoretical curves using the full and exact theory show dE_I/dx in g/cm^2 units as a function of $\beta\gamma$ in Fig. 6.3. We see that when $\beta\gamma$ is of order 1 to 10, the particles are minimum ionizing. The minimum $dE_I/d(\rho x)$ value for different

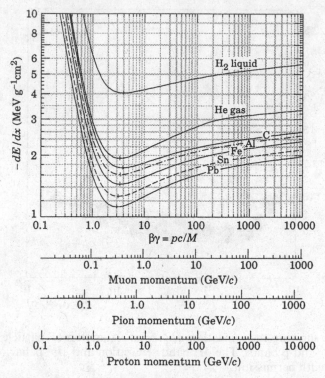

Fig. 6.3.   Theoretical curves for the ionization $dE_1/d(\rho x)$ as a function of $\beta\gamma$. The near universality of the curves validates Eq. 6.3 which is universal if $Z/A \sim 1/2$, up to logarithmic factors. (From Ref. 1.1.)

materials from carbon to lead is roughly 1.5 MeV g/cm². At lower values of $\beta\gamma$ we see the $1/\beta^2$ dependence. At very high values of $\beta\gamma$ we see a slow rise of the ionization energy loss per unit path length. This 'relativistic rise' will be discussed in Section 6.6.

The fact that the ionization is essentially a common constant (minimum ionizing) is also extremely useful for other kinds of particle identification. In Fig. 6.4 is plotted the measured energy deposition in an ionization tracking detector with many samples. There is clearly a component due to a single minimum ionizing particle and a second component due to what appear to be single tracks depositing the energy appropriate to two minimum ionizing particles. As we will see in Chapter 10, the angle between the electron and positron (the 'opening angle') in the pair conversion of photons is very small for high energy photons. In fact, it may be so small that the detector cannot spatially resolve the two tracks. By measuring the ionization of these unresolved double tracks, photon conversions may be isolated. Note that a doubly charged track, e.g. a He nucleus, would leave a four times MIP signal since energy loss goes as the

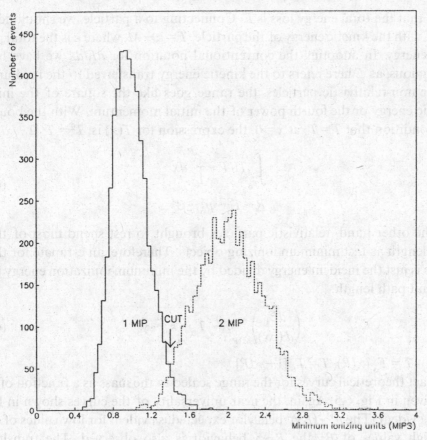

Fig. 6.4.   Data from an ionization detector with many samples showing the distribution of ionization energy for one and two tracks (γ conversions). (From Ref. 6.10, with permission.)

square of the incident charge, as we discuss later, and would be off scale in Fig. 6.4.

6.4 Range

Particles lose energy in the material and after traversing sufficient material they will have deposited all of their kinetic energy. The jargon for this is 'coming to the end of their range' or 'ranging out'. A crude estimate for the range, $\underline{R}$, is given below.

$$T = \frac{p^2}{2M}$$

(6.6)

$$\frac{dE_1}{dx} \sim \frac{1}{p^2} \sim \frac{1}{T} \frac{dT}{dx}$$

Note that the total energy loss is E. Connecting to a particle, we should identify E with the kinetic energy of the particle, $T = \varepsilon - M$, where ε is the total particle energy. In adopting the conventional notation for dE/dx we have been ambiguous, as E here refers to the kinetic energy transferred to the medium.

For non-relativistic particles, the range goes like the square of the initial kinetic energy or the fourth power of the initial momentum. With the boundary condition that $T = T_0$ at $x = 0$, the expression for $T(x)$ is, $T^2 = T_0^2 [1 - x/\underline{R}]$.

$$\int_{T_0}^0 T dT \sim \int_0^R dx \tag{6.7}$$

$$\underline{R} \sim T_0^2 \sim p_0^4 \sim \beta_0^4$$

On the other hand, relativistic particles brought to rest spend most of their path length as fast minimum ionizing objects. Therefore, an estimate for their range is just the incident energy divided by the minimum ionization energy loss per unit path length,

$$\left(\frac{dE_I}{d(\rho x)} \right)_{\text{MIP}} \underline{R} \sim T_0 \sim \varepsilon_0 \sim \gamma_0 \tag{6.8}$$

or, $T_0 - T = T_0 (x/R)$, $T = T_0 [1 - x/R]$.

Exact theoretical curves for the range scaled to the mass as a function of $\beta\gamma$ are given in Fig. 6.5. Again, the near universality of the curves shown in Fig. 6.5 is evident. The $p_0^4 \sim (\beta\gamma)^4$ behavior expected is evident for low values of $\beta\gamma$. At high values of $\beta\gamma$ the $\underline{R} \sim \gamma$ behavior is also observed. The transition between non-relativistic and relativistic behavior occurs for $\beta\gamma$ in the range 1–10.

Some examples of ionization and stopping are shown in Fig. 6.6a which is a 'bubble chamber' photograph of a $\pi^- - p$ interaction. The pion beam enters from below, center-right. Two neutral particles are produced, each of which decays into two charged particles. In particular, one of the charged particles in the right most decay stops in the liquid. We can see the increase in the bubble density or ionization near the end of the range. We also see several delta rays (see Chapter 5) knocked out by the incident beam which enters at the bottom of this figure. They 'range out', losing energy and therefore going in ever tightening spirals in the magnetic field in which the bubble chamber is immersed (see Chapter 7). In Fig. 6.6b we see nuclear target fragments stopping in a bubble chamber filled with a heavy liquid rather than hydrogen. From our derivation in Chapter 5, it is clear that an incident particle of charge ze would have $\Delta p_T \sim ze$ and therefore $d\varepsilon \sim (ze)^2$. Therefore, the ionization of a track goes as the square of its charge. Clearly, the nuclear fragments seen in Fig. 6.6b are multi-

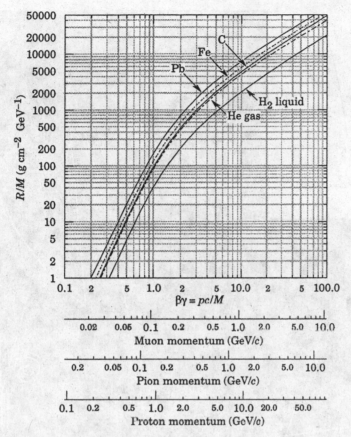

Fig. 6.5. Theoretical curves for range, $\underline{R}/M$ in g/(cm² GeV) as a function of $\beta\gamma$. (From Ref. 1.1.) For $\beta\gamma < 1$ the β^4 behavior is evident, while the γ behavior for $\beta\gamma > 10$ is also observed.

ply charged. In fact, a careful measurement of the ionization loss and momentum also yields a measure of z.

Another interesting visualization of particles coming to the end of range is shown in Fig. 6.7 where a proton beam extracted from a synchrotron stops in the air. We see that as the beam particles slow down and lose energy the transverse impulse due to multiple scattering spreads the beam out transversely. Figure 6.7 thus illustrates both range and multiple scattering. In Chapter 8 we will discuss transverse diffusion in gases. We will find that diffusion is a fundamental limitation to the accuracy of position measurements for both wire chambers (Chapter 8) and solid state detectors (Chapter 9).

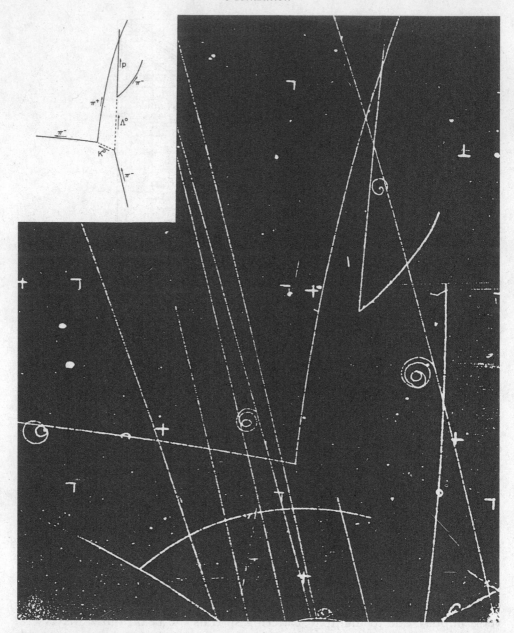

Fig. 6.6a.   Bubble chamber picture showing a decay pion stopping in the liquid. Note the increased ionization near the end of the range. Note also the 'knock on' or 'delta ray' electrons along the path of several tracks, which are low momentum, which lose energy (reduced radius of curvature) and which eventually stop. (From Ref. 10.9, with permission.)

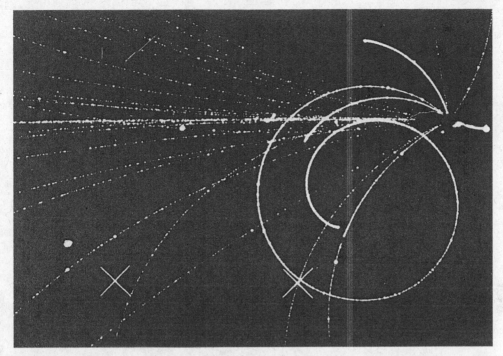

Fig. 6.6b. Photograph of a proton interaction with a Xe nucleus. Note the heavily ionizing nuclear fragments which lose energy and eventually stop in the chamber. (From Ref. 6.12, with permission.)

6.5 Radioactive sources

The use of radioactive sources in detector R&D is quite common. We need to gain a familiarity with the potential hazards of their use in order to insure safe research. The unit for radioactive dose is the rad. In MKS units it is the energy deposited per unit weight, 1 rad $= 10^{-2}$ joule/kg. To set a scale, a human yearly dosage of 1 rad (compared to the irreducible backgrounds of 0.2 rad/year due to cosmic rays) should be avoided. A third of a kilorad instantaneous dose is lethal. By comparison, detectors at high luminosity accelerators being built today need to withstand lifetime doses up to hundreds of Mrad.

Some commonly used radioactive sources are listed in Table 6.1. The type and the energy of the radioactive decay products are also shown. Note that 'β' sources have a spectrum of emitted electron energies due to the underlying three body, $n \rightarrow p \; e^{-} \nu$, decay kinematics. The quoted 'end point' is the maximum electron energy.

These sources, along with cosmic ray muons, $\langle E_\mu \rangle \sim 2$ GeV, are commonly used to exercise detectors in the laboratory without the need to expose them to beams from accelerators. Thus, these sources are extremely useful tools. For

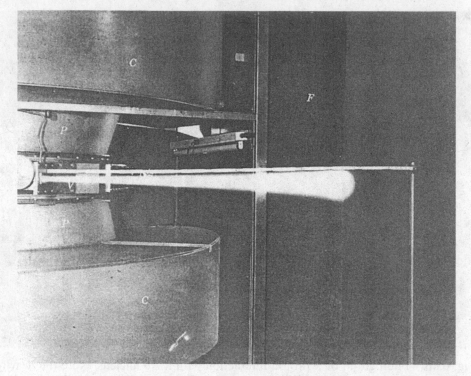

Fig. 6.7.   Proton beam extracted from a synchrotron. Note the beam stops in the air. Note also the multiple scattering spread of the transverse beam size as the momentum decreases toward the end of the range. (From Ref. 6.11, with permission.)

example, a ruthenium source with e^- near the 'end point' or maximum energy of a continuous β decay distribution provides electrons with $\beta\gamma\sim6$ and thus provides a simple source of 'almost minimum ionizing' particles. Radioactive decay is also used in dating of artifacts. For example, ^{14}C has a lifetime for β decay of 8270 years. Hence, detection of the 0.156 MeV 'end point' electrons allows us to date recent artifacts which contain carbon.

For example, an electron near the ruthenium end point with $\beta\gamma=6$ has a range of about 1.5 g/cm^2. Therefore a ruthenium source can be used to create coincidences in two or three thin plastic scintillators (Chapter 2). This energy scale for sources also explains why the protective source holders we encounter in the laboratory are a few cm of high density material thick.

Photon, alpha particle, and neutron sources are also available, as are cosmic ray muons. We will return to cosmic ray muons in Chapter 12. Typically, sources supply particles with energies of a few MeV because they arise from nuclear transitions and the binding energy per nucleon is ~8 MeV (see Chapter 12).

Table 6.1. *Cosmic rays and radioactive sources*

Type	Source	Energy/Flux (MeV)
μ	Cosmic rays $\pi \rightarrow \mu\nu$ decays	$dN/dE \sim 1/E^2$ Flux $\sim 0.013/(cm^2 \cdot s)$ ($E \geq 1$ GeV)
γ	^{55}Fe ^{137}Cs ^{60}Co	0.055 0.66 1.17
$e^-(\beta)$	^{90}Sr ^{106}Ru	2.28 End point 3.54 End point
α	^{241}Am	5.44
n	^{252}Cf	$\langle T_n \rangle = 2.14$

6.6 The logarithmic dependence and relativistic rise

We must also make some mention of the 'relativistic rise' which is evident at high energies. For example, the plot shown in Fig. 6.3 shows a rise in dE/dx for $\beta\gamma > 10$. The slow rise of ionization with $\beta\gamma$ comes from the so far neglected logarithmic factor. As indicated in Eq. 6.2, the argument of the logarithm is the ratio of the maximum and the minimum impact parameters.

We found in Chapter 5 an expression for the maximum momentum transfer, corresponding to the minimum impact parameter. At the minimum ionizing point $\beta\gamma$ is ~ 1. At that point the maximum energy transfer, T_{max}, is simply twice the target electron mass or about 1 MeV. The maximum impact parameter comes when passing near the outer part of the atom. In energy terms it corresponds to the ionization potential, $\langle I \rangle$, for the outer electrons. We take $\langle I \rangle$ to be of order 10 eV, see Fig. 1.3.

$$\ln(b_{max}/b_{min}) \sim [\ln(T_{max}/\langle I \rangle)]$$

$$T_{max} = 2m(\beta\gamma)^2$$

$$\sim 2m \sim 1 \text{ MeV(MIP)} \tag{6.9}$$

$$\langle I \rangle \sim 10 \text{ eV}$$

$$\ln(T_{max}/\langle I \rangle) = 11.5$$

Our estimate for Eq. 6.3 is then $0.16(11.5) = 1.84$ MeV/(g/cm^2) which is much closer to the exact calculation (Eq. 6.4).

The impulse, $\Delta p_T \sim e E_T(b) \Delta t$, at fixed impact parameter, is constant, which leads to the concept of minimizing ionizing particles.

$$|\mathbf{E}_T| \sim \gamma$$

$$\Delta t \sim 1/\gamma \tag{6.10}$$

$$\Delta p_T \sim \text{const}$$

Therefore, the arguments outside the logarithm are independent of γ. What about the logarithm itself? The maximum energy transfer when $\gamma \to \infty$ is, see Chapter 5, $Q_{max} \to \gamma M/m$ or $T_{max} \to \gamma M$ and scales as γ. The maximum impact parameter comes, in the theory of Bethe, when the collision time Δt is equal to the atomic orbital time $\sim 1/\langle \omega_o \rangle$. We know that the collision time also suffers a time dilation. Therefore the maximum impact parameter goes as γ and as $\langle \omega_o \rangle^{-1}$.

$$b_{min}^{-1} \sim T_{max} \sim \gamma M$$

$$b_{max}^{-1} \sim \hbar \langle \omega_o \rangle / \gamma \tag{6.11}$$

$$(b_{max}/b_{min})_{Bethe} \sim \gamma^2 M c^2/(\hbar \langle \omega_o \rangle)$$

This is not yet the whole story because there is also a density effect. The maximum impact parameter has been assumed to be cut off by the ~ 1 Å atomic radius. However, for very large γ the electric field may interact with several atoms as it stretches out transversely across the medium. How many atoms it interacts with depends on the density of the medium. For these large values of γ the medium will, in response, polarize. Everything that we did so far has assumed that only one atom is encountered during each collision and we simply summed over single collisions.

In the interest of space, we simply state that the density effect, due to $E_T \sim \gamma$, lowers the dimensionality of the dependence on γ by one power so that finally the relativistic rise goes like $\ln \gamma$.

A measured relative ionization is shown in Fig. 6.8. Relative to the minimum ionization at $\beta\gamma \sim 1$ there is a 50% increase by the time we reach $\beta\gamma \sim 1000$ at 1 atmosphere pressure. The logarithmic dependence of the ionization on $\beta\gamma$ is quite evident in this semilog plot. Note that the ionization loss is also seen to depend on the density. The linear ionization rise with $\ln(\beta\gamma)$ is reduced at high density because of the blocking polarization of the medium. The exact theory of energy loss can be found in the references given at the end of this chapter.

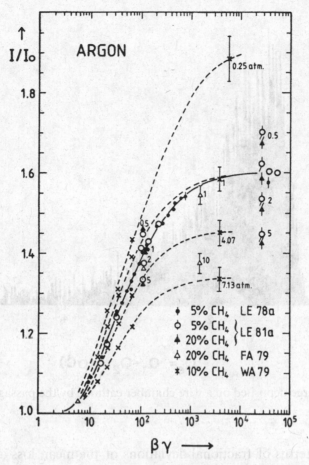

Fig. 6.8.   Relative ionization as a function of $\beta\gamma$ for argon–CH_4 gas at several pressures (densities). (From Ref. B.1, with permission.)

6.7 Fluctuations

The ionization loss, dE/dx, as derived above should be identified with the mean energy loss. We might expect that there is a long 'tail' at high losses due to delta ray ejection or other processes. We can make a large number of samples of dE/dx and reject any large fluctuations from the mean. This 'truncated mean' method is used in determining the mean ionization deposit in many samples, for example Fig. 6.2, Fig. 6.4 and Fig 6.8. In addition, in a dilute medium, for example a gaseous detector, there can be a statistical fluctuation in the number of produced primary ion–electron pairs.

In passing through a dilute material of depth L, the most probable energy loss is $\langle \Delta E \rangle \sim \langle dE/dx \rangle L$. The 'Landau distribution' of deviations, $\Delta E - \langle \Delta E \rangle$,

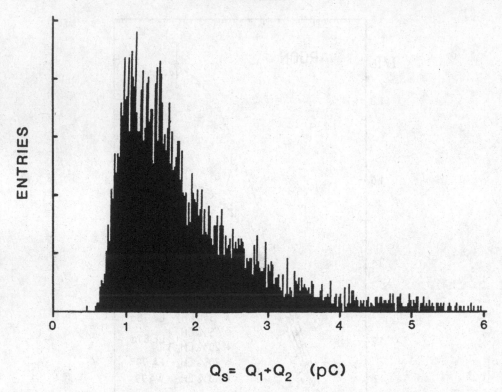

Fig. 6.9. Charge deposited on a wire chamber cathode by the passage of a MIP.

expressed in terms of fractional deviations of the mean loss about the most probable energy loss is

$$P(\lambda) = \frac{1}{\sqrt{2\pi}} e^{-1/2(\lambda + e^{-\lambda})} \tag{6.12}$$

$$\lambda \equiv \frac{(\Delta E - \langle \Delta E \rangle)}{\langle \Delta E \rangle}$$

This distribution is asymmetric about the mean and is skewed to large values of λ as expected since it represents the effects of energetic delta rays and other processes yielding a large amount of charge.

A plot of the charge deposited in a gas filled proportional chamber (Chapter 8) is shown in Fig. 6.9 for a MIP incident on the detector. The existence of a minimum ionizing peak and a long 'Landau tail' is evident.

There is also the phenomenon known as 'straggling'. For example, particles with the same initial energy do not all stop in the same distance, x. There are fluctuations in the energy loss, and also multiple scattering fluctuations in the

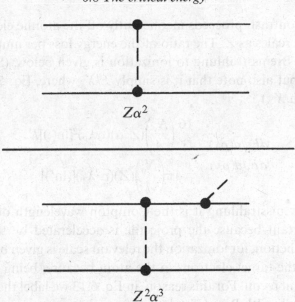

$$Z\alpha^2$$

$$Z^2\alpha^3$$

Fig. 6.10. Schematic representation of the critical energy where ionization losses (numerator) become equal to radiation losses (denominator).

path length. The details of straggling can be found in the references given at the end of this chapter.

6.8 The critical energy

We now compare two distinct mechanisms for the energy loss caused by charged particles. In Chapter 10 we will derive the radiative Bremsstrahlung mechanism. The question obviously arises – which mechanism dominates and over what kinematic regime does it do so? This question can be addressed most simply and intuitively by drawing the appropriate Feynman diagrams as shown in Fig. 6.10. We only use them to count interaction vertices. Lines are used to represent charged particles, while dashed lines represent photons.

The ionization energy loss is effectively constant with energy while the Bremsstrahlung energy loss increases linearly with energy. Radiation (see Chapter 10) is enhanced by relativistic effects. Therefore, we expect that at high energies (see Chapter 1) the radiative process will dominate. On the other hand, looking at Fig. 6.10 and counting the number of vertices, the ionization goes as $Z\alpha^2$, Eq. 6.3, while Bremsstrahlung has another vertex and goes as α^3.

However, we can overcome the extra factor of α both with the energy rise and by noting that the nucleus acts as a geometric point on these energy scales so that the Bremsstrahlung process is coherent over the size of the nucleus.

Ionization, in contrast, proceeds incoherently off the atomic electrons over an atomic size and scales as Z. The ratio of the energy loss per unit path length in g/cm^2 units for Bremsstrahlung to ionization is given below. (See Chapter 10 for $dE_B/d(\rho x)$, but also note that it is simply E/X_o where, Eq. 5.12 defines the radiation length X_o.)

$$\frac{dE_B/d(\rho x)}{dE_I/d(\rho x)} \sim \frac{\frac{16}{3}\left(\frac{N_o}{A}\right)(Z^2\alpha)(\alpha\lambda_P)^2[\ln(\ )]E}{4\pi\left(\frac{N_o}{A}\right)(Z)(\alpha^2\lambda_T)[(\ln')]} \tag{6.13}$$

Note that in Bremsstrahlung it is the Compton wavelength of the projectile which is important because the projectile is accelerated by the field of the nucleus. In distinction, for ionization the relevant scale is given by the Compton wavelength of the target electrons in the atom because, being light, they can soak up energy in recoil. For this reason, in Eq. 6.13 we label the target and the projectile carefully with P and T subscripts.

We now define an energy at which the loss due to ionization is equal to the loss due to radiation. This is called the critical energy, E_c. As seen from Eq. 6.13, up to ratios of logarithmic factors E_c goes like $1/Z$ and is proportional to the ratio of projectile, m_P to target, m_T, masses.

$$E_c \sim \frac{3\pi}{4}\left(\frac{m_P}{m_T}\right)\left[\frac{m_P c^2}{Z\alpha}\right] \tag{6.14}$$

$$\sim 166 \text{ MeV}/Z$$

The critical energy goes like the square of the projectile mass since heavy particles radiate less. For example, muons at low energies ($E < 300$ GeV) are observed to ionize and not to radiate. Incident electrons have equal projectile and target masses. A plot of the energy loss for electrons on lead as a function of the electron energy is given in Fig. 6.11. The ionization loss, which falls as $1/\beta^2$, crosses over the Bremsstrahlung energy loss which is rising with energy. The crossover occurs at about 8 MeV, which is then the critical energy for electrons on lead. Note also that when Bremsstrahlung dominates the fractional energy loss per unit length is defined by the radiation length, since X_o, by definition, is the radiative mean free path $\langle L \rangle$; $1/E(dE/d\rho x) \sim 1/X_o$, $E > E_c$.

A plot of the critical energy for electrons as a function of the atomic number is given in Fig. 6.12. Given that we have ignored logarithmic ratios in Eq. 6.14, the agreement with the exact calculations is adequate. In both cases the basic $1/Z$ behavior of the critical energy is evident, although the coefficient, $E_c \sim 600$ MeV/Z, is only in rough agreement with Eq. 6.14.

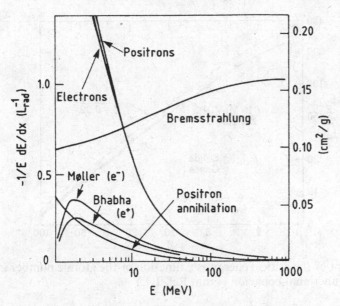

Fig. 6.11. Energy loss for electrons in Pb as a function of electron energy. At the critical energy, the Bremsstrahlung curve crosses the ionization curve, $E_c \sim 8$ MeV. (From Ref. B.2, with permission.)

As a preview of calorimetry, Chapter 11, we comment that an electromagnetic shower is a cascade alternating between Bremsstrahlung by electrons and pair production by photons. At some point, as the energy per particle in the cascade reduces below the critical energy, the radiative processes are suppressed. Therefore, the particles cease to multiply and begin to die out. The charged particles stop and are absorbed by ionization. The photons are removed via the photoelectric effect.

The dependence on m_p implied by Eq. 6.14 means that since we have ~ 8 MeV for E_c for electrons on lead, we have ~ 300 GeV for the critical energy of muons. That is why we normally think of muons as being particles that only ionize (see Chapters 12 and 13). It is only very recently at the high energies soon to be available in the design of multi-TeV colliders that the radiative energy losses of muons in muon detection systems have become important. For example the LHC, or Large Hadron Collider, which will be built at CERN will have detectors which will attempt to measure many muons which have energies well above the critical energy in lead and iron. Thus, the detector designers will have to accommodate the possibility of radiating muons.

A plot of the energy loss per unit length, in g/cm² units, as a function of the incident muon energy in iron is shown in Fig. 6.13. The crossover between ionization and Bremsstrahlung occurs at several hundred GeV as expected. Note

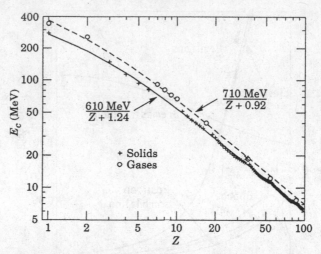

Fig. 6.12.   Plot of the critical energy as a function of the atomic number Z indicating the basic $1/Z$ functional behavior. (From Ref. 1.1.)

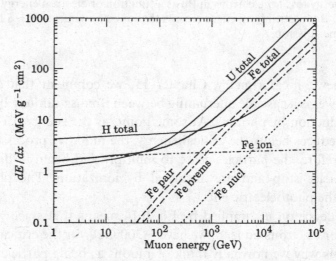

Fig. 6.13.   Energy loss, $dE/d(\rho x)$, as a function of muon energy in iron. The crossover of ionization loss and radiative losses at ~ 500 GeV is evident. (From Ref. 1.1.)

that ionization, Bremsstrahlung and 'pair creation' are the important processes. Pair creation diagrammatically looks like Bremsstrahlung (Fig. 6.10) but the photon is 'virtual' and 'decays' into an e^+e^- pair. The pair creation probability is comparable to that for Bremsstrahlung, to which it is closely related. Pair creation and Bremsstrahlung also contribute to the 'Landau tail' of fluctuations for high incident energies.

Exercises

1. Redo the calculation leading to Eq. 6.1 for an incident particle with charge ze. From the fact that $F(b) = ze^2/b$, show that $dE/dx \sim z^2$.

2. The Fermilab Tevatron beam is 1000 GeV protons. Show that this 6 km circumference machine would, if filled with air, cause the protons to lose 1.1 GeV/turn. At a velocity $= c$, estimate how long it would take to lose the total beam energy by ionization.

3. Show, starting from Eq. 6.5 that $\beta^3 d\beta \sim dx$. Therefore, establish that $\beta \sim x^{1/4}$.

4. Assume that $dT/dx \sim (dE/dx)_{min} [1/\beta^2]$. Integrate this equation with boundary conditions to show that $T_o^2 - T^2 = (dE/dx)_{min} (mc^2)x$.

5. Use the result of Exercise 4 to estimate the range of a 1 GeV/c momentum proton, using Eq. 6.4 to define minimum ionizing. Compare your approximate result to the exact results shown in Fig. 6.5.

6. Use Fig. 6.5 to estimate the thickness of solid material required to stop a 10 GeV muon. What thickness of concrete and iron is needed (Table 1.2)?

7. Compare the ~ 5250 g/cm^2 estimated in Exercise 6 to the relativistic approximation given in Eq. 6.8. Which is larger and why?

8. Take the cosmic ray rate from Table 6.1. Make a model of a person as a cube of water, 40 cm on a side (weight ~ 141 pounds). A year is $\pi \times 10^7$ s. Find the number of cosmic rays passing through this person per year.

9. The cosmic rays are mostly minimum ionizing muons. Find the energy deposited per year by the 6.6×10^8 muons passing through this person. Convert this to rads. Compare to the quoted annual background dose of 0.2 rad.

10. If the radiative energy loss $dE/d\rho x$ is E/X_o and the ionization loss is at the minimum, then E_c is simply the MIP value times X_o. Evaluate that assertion for Pb using Table 1.2. Does this agree with the correct result given in Fig. 6.11?

References

[1] N. Bohr, 'Penetration of atomic particles through matter', *K. Dan. Vidensk. Selsk. Mat.-Fys. Medd.* **XVIII** 8 (1948).
[2] 'Ionization Chambers' in Nuclear Physics *Handbook of Physics*, H.W. Fulbright, Springer Verlag (1958).
[3] *Nuclear Radiation Detection*, W. Price, McGraw-Hill (1958).
[4] H.A. Bethe, *Z. Physik* **76** 293 (1932).
[5] L.D. Landau, *J. Exp. Phys. (USSR)* **8** 201 (1944).
[6] I. Lehraus, *et al.*, *Nucl. Instrum. Methods* **153** 347 (1978).
[7] J.N. Marx and D.R. Nygren, *Physics Today*, 46 (Oct. 1978).

[8] *Experimental Nuclear Physics*, E. Segre, Wiley (1953).
[9] H. Burkhardt and J. Steinberger, 'Tests of electroweak theory at the Z resonance', *Annu. Rev. Nucl. Part. Sci.* **41** 55 (1991).
[10] S. Abachi, *et al.*, 'The D0 Detector', *Nucl. Instrum. & Methods. In Phys. Res. A* **338** 185 (1994).
[11] *Physics, Part II*, D. Halliday and R. Resnick, John Wiley and Sons, Inc. (1960).
[12] *Introduction to High Energy Physics*, D.H. Perkins, Addison-Wesley Publishing Company, Inc. (1987).

Part IIC

Position and momentum

We now begin our exploration of the non-destructive measurement of position and momentum, using the deposited ionization energy (Chapter 6) and limited in accuracy by the multiple scattering in the medium (Chapter 5). The goal is to measure the vector momentum which we indicated in Fig. I.1 as part of the general task of the detector designer.

To that end we first look at particle motion in magnetic fields in Chapter 7. Since the helical trajectory depends on the magnitude of the momentum, we deploy gaseous detectors in Chapter 8, which use localized ionization deposition to locate points along the trajectory. Since the charge is collected on electrodes, we also must explore motion in electric fields. Finally, if greater spatial accuracy is needed, silicon detectors are employed, as explicated in Chapter 9. The points are fit to a hypothesis regarding the trajectory. The simplest example of a straight line is treated in the Appendices. A more complex helical path hypothesis yields best-fit estimates for the vector momentum and their errors.

7

Magnetic fields

Of Newton with his prism . . . Voyaging through strange seas of
Thought alone.
William Wordsworth
The wheel is come full circle.
William Shakespeare, King Lear

The use of magnetic fields in particle detectors has several aspects. As we will discuss, magnetic fields are used to prepare the 'beams' that impinge on experiments. In addition, the creation of a large volume with a magnetic field and associated detectors allows us to measure the trajectory of charged particles in the field (the jargon is 'tracking') by detecting the ionization caused by the particle passage. Tracking allows us to measure the position and momentum of charged particles (see Chapter 13 for some examples of tracking in large detectors).

7.1 Solenoidal fields

We begin by considering electromagnets in 'colliding beam' applications, i.e. where two oppositely directed beams collide head on. The most common topology for such an experimental electromagnet is the solenoid. Consider a solenoid coil of N total turns, each carrying a current I, with full length L and radius a. The magnetic field, in CGS units, at a point along the axis where the ends subtend angles θ_1 and θ_2, defining n as the turns/length $= N/L$, is

$$B = \frac{2\pi nI}{c} (\cos \theta_1 + \cos \theta_2) \tag{7.1}$$

The result at the centerline of a very long solenoid approaches the elementary result which follows from Ampère's law.

$$B(\text{centerline}) = \frac{4\pi nI}{c} \left[\frac{(L/2)}{\sqrt{(L/2)^2 + a^2}} \right] \equiv B_0 \tag{7.2}$$

$$B_0 \rightarrow 4\pi nI/c \text{ as } L \rightarrow \infty, \text{ CGS}$$

In MKS units the result for the magnet centerline is (see Table 3.1, $1/c \rightarrow \mu_0/4\pi$);

$$B \rightarrow \mu_0 nI, \text{ MKS} \tag{7.3}$$

129

In the center of an infinitely long solenoid the field depends only on the current carried by each turn I and the turns per unit length n. For reference purposes, 1 tesla is 10 kG or 1 N/(Am^2). The electromagnetic constants that are needed for all our calculations can be found either in Table 1.1 or in Table 3.1.

For example, to produce a 20 kG field with 1 turn per centimeter, n, requires 16 kA of current I. Thus a copper bus (water-cooled) of 1 centimeter diameter at 16 kA produces a 2 tesla solenoidal field. The energies stored in such magnets may be enormous, $U_B \sim B^2 \underline{V}$. Typical experiments in high energy physics or magnetic fusion devices now utilize magnets with stored energies in the 10–100 MJ range.

7.2 Dipole fields – fringe fields

We can also look at a typical 'fixed target' topology for the field, i.e. where a beam is directed to a target at rest in the laboratory. A 'dipole' magnet is illustrated schematically in Fig.7.1a. The iron 'return yoke' shapes the field produced by the coils and produces a roughly constant field in the air gap. Given that the physical magnet length is $\sim L$, the field has a finite spatial extent. There is a 'fringe field' due to the constraints imposed by Maxwell's equations. For example, in the absence of sources, if the field ramps down (from magnitude B_0) in the z direction over a characteristic distance, ℓ, then a component B_z is induced. Since all fields are finite in extent, a 'fringe field' is guaranteed to exist.

$$\int \mathbf{B} \cdot d\mathbf{x} \sim B_0 (L + \ell), \quad \text{magnetic length} = L = \ell \tag{7.4}$$

Outside the sources, the curl of the magnetic field is zero. Looking at one component of that vector equation, we find that the $\partial_y B_z$ is equal to $\partial_z B_y$. Therefore, B_z arises because the change of the y component of the field is non-zero. An estimate of the size of B_z and its sign is given in Eq. 7.5 and is shown schematically in Fig. 7.1c. Note that what started as a simple uniform field has now grown a lens, i.e. B field proportional to deviation from magnet centerline, at both its exit and entrance. We discuss magnetic lenses further in Chapter 7.6 and 7.7 below.

$$\nabla \times \mathbf{B} = 0$$

$$\partial_y B_z = \partial_z B_y \sim \pm B_0 / \ell \tag{7.5}$$

$$B_z \sim (\partial_z B_y) y \sim \pm B_0 (y / \ell)$$

The divergence of the magnetic field is also zero. The $\partial_z B_z$ must be compensated by, for example, $\partial_x B_x$. That implies the existence of an x component of the field proportional to the second z partial derivative of B_y. (See Fig. 7.1d.)

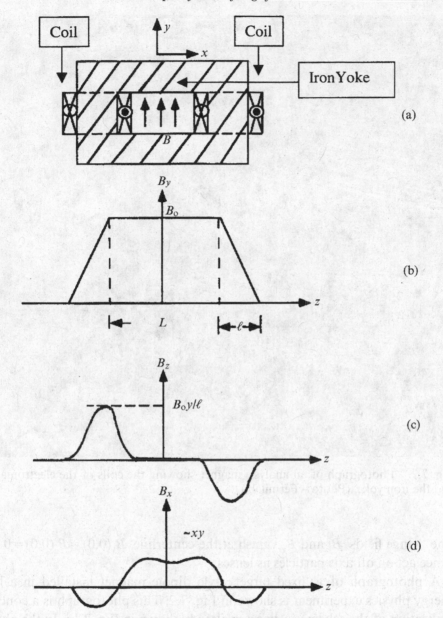

Fig. 7.1. (a) Layout of a dipole electromagnet, (b) B_y as a function of z, (c) fringe field B_z as a function of z, and (d) fringe field B_x as a function of z.

$$\mathbf{\nabla} \cdot \mathbf{B} = 0$$

$$\partial_x B_x \sim -\partial_z B_z \sim -y(\partial_z^2 B_y) \qquad (7.6)$$

$$B_x \sim -xy(\partial_z^2 B_y)$$

Fig. 7.2.   Photograph of an analysis magnet showing the coils of the electromagnet and the iron yoke. (Photo – Fermilab.)

The 'fringe fields' B_z and B_x vanish at the centerline, $B_x(0,0) = B_z(0,0) = 0$ and hence act on off axis particles as lenses.

A photograph of a 'fixed target' style dipole magnet installed in a high energy physics experiment is shown in Fig. 7.2. This photograph is a concrete realization of the schematic diagram already shown in Fig. 7.1a. In the photograph the coils of the electromagnet and the iron return yoke which shapes the field can be readily identified. A fixed target 'bubble chamber', such as that used to make the picture shown in Fig. 6.6, is also immersed in a reasonably uniform dipole field. A prepared 'beam' of particles of roughly the same momentum and all moving with about the same angle impinges on the 'bubble chamber' from below in Fig. 6.6. Assuming that momentum goes inversely with radius of curvature (Eq. 7.8 below), the momentum of the beam particles is clearly larger than that of the decay products since the radius of curvature of the beam is seen to be large.

7.3 Particle motion in a uniform field

Let us now look at the trajectory of a particle, of charge q, in a uniform solenoidal magnetic field oriented along the z axis. Since we are doing numerical estimates, we will use MKS units here. The force $\mathbf{F}$ due to the magnetic field is given by the Lorentz force equation, Table 3.1. Since this force is perpendicular to the velocity, the power dissipated is zero. Hence, the momentum vector is of constant magnitude.

$$\mathbf{F} = m\mathbf{a} = q(\mathbf{v} \times \mathbf{B}) = d\mathbf{p}/dt$$

$$\mathbf{F} \cdot \mathbf{v} = 0 \Rightarrow |\mathbf{p}| = \text{const} \tag{7.7}$$

Since there is no force along the field direction, the path in z, as a function of arc length, is a straight line with a constant direction cosine, $\alpha_z = dz/ds$. In the transverse plane, (x, y), the equations of motion imply a solution where the momentum vector simply rotates in that plane by an azimuthal 'bend' angle ϕ_B.

The centrifugal force in the non-relativistic case can be equated to the Lorentz force to yield an expression for the radius of curvature of the momentum vector in the x–y plane.

$$F_{\text{cent}} = mv^2/r = qvB$$

$$p = mv, \quad r = a \tag{7.8}$$

$$a = p/qB$$

In MKS units, for unit charge $q = e$, $1/q$ is 33.3 when the momentum, p, is given in units of GeV (10^9 eV), the magnetic field, B, is given in kG, and the radius of curvature, a, is given in meters. For example, a 100 GeV proton in a 10 kG field has a radius of a curvature of 336 m.

Since, in the non-relativistic case, $\mathbf{p} = m\mathbf{v}$ Eq. 7.7 becomes

$$d\mathbf{p}/dt = \frac{q}{m}(\mathbf{p} \times \mathbf{B}) \tag{7.9}$$

The form of Eq. 7.9 implies circular motion which we could prove by assuming a circular trajectory and substituting into the equations of motion. Since the force is transverse to momentum it always acts so as to bend the path into a circular arc just as in the case of a satellite in earth orbit. The frequency can be read off directly, $\omega p = qpB/m$, and is given by the 'cyclotron frequency', ω_c, quoted in Table 1.1.

$$\omega_c = eB/m \tag{7.10}$$

For constant field we have a constant rotation frequency which is important in the design of cyclotrons because then simple radio frequency (r.f.) drivers of fixed frequency can be used. For reference $\omega_p = 9.6 \times 10^6$ rad/(s kG) and $\omega_e = 1.7 \times 10^{10}$ rad/(s kG) for protons and electrons respectively (see Table 1.1.) For example, a 10 MeV proton cyclotron has a final $T = 10$ MeV or $p = 141$ MeV. If the bending field is 10 kG then the protons at a full energy have a 0.47 m orbital radius. The r.f. system works at $\omega = 9.6 \times 10^7$ rad/s or $f \sim 15$ MHz. These values are typical of cyclotron parameters.

The situation is modified in considering relativistic motion. Non-relativistic motion is circular with a fixed frequency, whereas in relativistic motion the inertial mass increases by a factor of γ as is evident in Eq. 7.8, $p = \gamma m v$. Thus, ω goes from ω_c to $eB/\gamma m$ which means that the radio frequency power must be applied at variable frequency. For a practical high energy circular accelerator, we need to increase the magnetic field B as the particles gain energy, or γ, in order to keep the beam within a fixed radius vacuum pipe. For example, the Fermilab Tevatron magnets are 'ramped' up to full field of ~ 30 kG during the acceleration cycle which takes ~ 10 s to go up to ~ 1 TeV.

It is this effect which defines the ultimate size of a circular accelerator. The physics of magnetic solids implies that there is a maximum field which we can get from iron magnets of about 2.0 tesla due to domain saturation; all the domains are aligned at that point. Therefore, the radius of an accelerator must grow proportionally to the maximum momentum $a \sim p_{max}/eB_{max}$, $B_{max} \sim 20$ kG. For example, a 1 TeV (10^{12} eV $= 1000$ GeV) accelerator would have a 1.65 km radius. The use of higher field 'superconducting' electromagnets allows us to somewhat relax this limit but only by a factor of two or three.

A concrete visualization of the circular trajectories is given in Fig. 6.6. This is a 'bubble chamber' picture of an event featuring the production of two neutral particles, which leave no ionization in the chamber, each decaying into two charged particles. We can tell that the charged particle pairs are positive and negative since the sense of rotation in the magnetic field in which the chamber is immersed is opposite for the elements of the pair. Looking at Eq. 7.8 the sense of rotation in a magnetic field depends on the charge. In the lower pair it is also clear that one of the decay products is much more energetic, 'stiffer', than the other one. The higher momentum track has a much larger radius of curvature. The lowermost charged particle in the pair has a lower momentum.

We can also observe that momentum conservation (Appendix A) in the decays means that the neutral parent should point back to the production vertex which is visible. In the case of the upper pair the plus and minus daughter tracks roughly share the momentum, so that the charged pair bisector goes close to the production vertex. In the lower pair the neutral parent particle's momentum vector must be near to the direction of the 'stiff' track of the pair. Indeed that track by itself approximately points back to the production vertex. We can look at these bubble chamber pictures and do the vector momentum conservation kinematics with rulers and protractors which is a very instructive exercise highly recommended to the student. You really believe in energy–momentum conservation after you do the kinematics yourself by hand.

7.4 Momentum measurement and error

It's also possible to use solid electromagnets without an air gap, simply as an iron core wound with exciting coils. The deflection angle, ϕ_B, on traversing a length, L, in a direction transverse to the magnetic field, B, is given in small angle approximation as the length divided by the radius of curvature, a. (See Appendix F and Eq. 7.8.)

$$\phi_B \sim L/a = eLB/p \tag{7.11}$$

Using Eq. 7.8 for the radius of curvature, we find that the bend angle is the transverse momentum impulse divided by the momentum.

$$\phi_B = (\Delta p_T)_B/p \tag{7.12}$$

$$(\Delta p_T)_B = eLB = 0.03BL\left(\frac{\text{GeV}}{\text{kG m}}\right)$$

This result is similar to the treatment that was given for multiple scattering where the physics defined the transverse momentum impulse which was independent of the incident particle save for its charge. The bend angle goes as the inverse of the incident particle momentum. For example, 1.5 meters of iron magnetized to near saturation at 18 kG gives a momentum impulse of roughly 0.82 GeV. Thus an 8 GeV particle bends ~ 0.1 rad $\sim 6°$ in traversing the iron.

Closer to home, we all have an electron accelerator in our home – the TV set – with which we can do experiments. Inside there is an electron gun which accelerates electrons through about 10 kV, an energy needed to excite the phosphors on the screen. We can use permanent magnets to make a roughly 1 kG field extending over about 1 cm. Thus, a refrigerator magnet put in close proximity to the TV screen will cause a noticeable deformation of the TV image. Note that a 10 keV energy electron has a 100 keV momentum, to compare to the 300 keV momentum impulse due to a 1 kG field extending over 1 cm. The student is encouraged to do this experiment.

We recall that the transverse momentum impulse due to multiple scattering has a characteristic two dimensional scattering energy of $E_s/\sqrt{2}$ or 14 MeV. Traversing the 1.5 m of iron, given that the radiation length of iron is 1.76 cm (Table 1.2), a particle crosses 85 radiation lengths and therefore suffers a multiple scattering momentum impulse of 0.13 GeV.

$$(\Delta p_T)_{MS} = \frac{E_S}{\sqrt{2}} \sqrt{L/X_o} \quad \text{(Chapter 3)} \tag{7.13}$$

The ratio of the momentum impulses is about 16%. It defines the achievable momentum resolution for a system with momentum error dominated by multiple scattering. Note that the best possible resolution, $(\Delta p_T)_{MS}/(\Delta p_T)_B$ scales as $1/B$ and $1/\sqrt{L}$. Therefore, it is extremely costly to improve performance by increasing the length, and there is a natural limitation to the magnetic field when you reach saturation in the iron, $B \sim 2$ T. Thus, the example given above is fairly typical and difficult to improve on much.

A photograph of a typical solid iron magnet is shown in Fig. 7.3. The iron toroid in this device is used both for the detection and identification of muons. We can see that the steel is excited by the copper coils which are visible in the photograph. The absorber mass is useful in removing particles which interact in the steel and do not escape into the muon tracking system which is located outside the steel and thus not visible. The momentum measurement is limited, as discussed above, to about 16%. We will return to the topic of hadronic cascades, 'punch through' and muon detection in Chapters 12 and 13.

If we are operating in vacuum rather than in steel, the angle of deflection in the field goes as $1/p$, and thus the fractional error in the momentum measurement goes as $dp/p \sim p$. The achievable resolution in this case is ultimately limited by alignment and the systematic errors of the position measuring detectors limit the final accuracy of the angular measurement. The fractional momentum resolution in the case of vacuum scales as p, whereas in the case where the multiple scattering limits it is a constant.

Assume that the bend angle is measured using detectors which make measurements having an error dx_T along an arc length L. For bend angle ϕ_B and momentum impulse $(\Delta p_T)_B$:

$$1/p = \phi_B/(\Delta p_T)_B$$

$$d(1/p) = \frac{d\phi_B}{(\Delta p_T)_B} \sim \frac{(dx_T/L)}{(\Delta p_T)_B} \tag{7.14}$$

$$= dp/p^2$$

Fig. 7.3. Photograph of iron toroids for the detection and identification of muons. (Photo – Fermilab.)

We have not specified the number of detectors, nor their optimal deployment along the path length of the track. Thus Eq. 7.14 is only indicative of the order of magnitude of the momentum error.

For example, with detectors having $dx_T \sim 100$ μm (e.g. diffusion limited gas detectors, see Chapter 8) deployed over $L = 1$ m in a field with momentum impulse $\Delta p_T = 1$ GeV, a 100 GeV particle is measured with a momentum accuracy of $\sim 1\%$.

The multiple scattering fractional error is a constant, $dp/p \sim (\Delta p_T)_{MS}/(\Delta p_T)_B$. Thus, it is conventional to parameterize the momentum resolution of a tracking device as the statistical sum ('fold in quadrature', see Appendix J) of measurement and multiple scattering errors characterized by the constants a and b respectively.

$$dp/p = \sqrt{a^2 p^2 + b^2} \equiv ap \oplus b \qquad (7.15)$$

7.5 Exact solutions – Cartesian and cylindrical coordinates

Consider now the exact solution for the orbit in a perfectly uniform magnetic field, B, oriented along the z axis. We use the arc length, s, as a track parameter since the velocity, v, is constant. Equation 7.7 using $\mathbf{p}$ and s becomes

$$ds = v dt \tag{7.16}$$

$$p d\mathbf{p}/ds = q(\mathbf{p} \times \mathbf{B})$$

This equation implies that the momentum rotates with frequency, $\omega_s = qB$ or (see Eq. 7.11) through an angle $\phi_B = s/a$. For details see Appendix F. The solutions given in Appendix F for z are straight lines.

$$z = z_o + \alpha_z s, \quad s = 0 \text{ at } z = z_o, \quad \alpha_z = \text{constant} \tag{7.17}$$

The direction cosines of the momentum rotate by ϕ_B in the (x,y) plane.

$$\begin{pmatrix} \alpha_x \\ \alpha_y \end{pmatrix} = \begin{pmatrix} \cos\phi_B & \sin\phi_B \\ -\sin\phi_B & \cos\phi_B \end{pmatrix} \begin{pmatrix} \alpha_{xo} \\ \alpha_{yo} \end{pmatrix} \tag{7.18}$$

The 'bend angle', ϕ_B is the angle of azimuthal rotation of the momentum vector. The orbit is a helix.

We integrate to get the coordinates. Initial position and momentum components are indicated by the o subscript (see Appendix F). The orbit is a circle in the (x, y) plane.

$$(x - x_o) = -(a/p)(p_y - p_{yo})$$

$$(y - y_o) = (a/p)(p_x - p_{xo}) \tag{7.19}$$

$$(x - x_o - a\alpha_{yo})^2 + (y - y_o + a\alpha_{xo})^2 = a^2(\alpha_x^2 + \alpha_y^2) \equiv a_T^2$$

The solutions given in Eqs. 7.18 and 7.19 are exact. Although a completely uniform field is not possible (see Fig. 7.1), these solutions are still useful. For example, we can differentially 'track' a particle through an inhomogeneous field because we can make a series of small 'steps', treating the field as locally uniform during each step, and applying the exact solution. This is a useful strategy, for example in 'Monte Carlo' programs which are used to construct a computer model of a device. (See Appendix K.)

The forms given above are in Cartesian coordinates which are perhaps more appropriate to fixed target geometries. In collider experiments the experimenter is more likely to have built detectors around a basic solenoid in an axial geometry for which cylindrical coordinates are more appropriate. For example see Fig. 9.6 and Fig. 9.7 for two silicon detector geometries.

The angular definitions are given in Fig. 7.4. A particle emitted from the

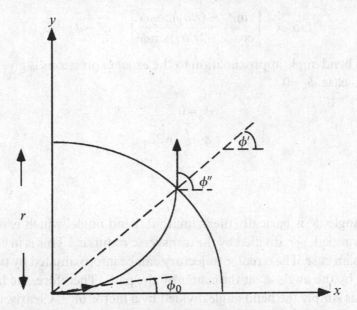

Fig. 7.4. Definitions used in constructing the path of a particle in a purely axial sole-noidal field emitted from the origin at angle ϕ_o. The particle at radius r is located at angle ϕ' and its momentum vector at that point has an azimuthal angle ϕ''.

origin with a transverse momentum vector labeled by the azimuthal angle ϕ_o describes a circle in the (x, y) plane and the trajectory is evaluated at radius r. The azimuthal angle ϕ' labels the position of the particle as it crosses radius r, whereas the azimuthal angle ϕ'' labels the angle of the momentum vector at that point. (See Appendix F.)

We define a momentum transverse to the magnetic field, p_T, along with the corresponding transverse radius of curvature, a_T. (See also Eq. 7.19.)

$$p_T = p\sin\theta \tag{7.20}$$

$$a_T = p_T/qB_o$$

The initial angle of emission at the origin is specified by the initial momentum components. The exit position, ϕ', has no solution, if $\sin(\phi' - \phi_o) > 1$, or if a_T is less than half the radius. That is because the circular path has a small enough radius that it never reaches r. These are called 'loopers' and they continue to rotate in the magnetic field until they drift out along the z direction as determined by Eq. 7.17, $z = z_o + \alpha_z s$. They are visible as circles in Fig. 6.6 where the drift direction is parallel to the line of sight and therefore, not apparent.

$$\tan\phi_o = p_{yo}/p_{xo}$$

$$\sin(\phi' - \phi_o) = -(r/2a_T) \tag{7.21}$$

$$\tan \phi'' = \left[\frac{\sin\phi_o - (r/a_T)\cos\phi'}{\cos\phi_o + (r/a_T)\sin\phi'} \right], \quad \phi'' - \phi_o = \phi_B$$

The small bend angle approximation to the exact expressions is given below, in the special case $\phi_o = 0$.

$$\phi_o = 0$$
$$\phi' \sim -r/2a_T$$
$$\phi'' \sim -r/a_T$$
$$= \phi_B$$

(7.22)

The exit angle ϕ'' is basically the azimuthal 'bend angle' which is roughly the distance traveled, $\sim r$, divided by the transverse radius a_T. This is in accord with the Cartesian case. The circular trajectory can be approximated by two straight lines bent by the angle ϕ'' at the center of the path. Therefore, the label of the exit point is simply the bend angle divided by a factor of 2. Clearly, these small angle approximations have a nice transparent geometric interpretation as illustrated in Fig. 7.5.

The circular path made by charged particles in a solenoidal field is illustrated in Fig. 7.6 which is derived from data taken at a Fermilab collider detector. Note for the various tracks that the sense of the rotation can be both positive and negative. Note also that the different radii of curvature reflect particles of different momenta emanating from the 'event vertex' (recall Fig. I.1).

In contrast to the view in the (x, y) plane, we show the trajectories in a similar detector located at CERN, the European particle physics facility, immersed in a dipole field pointing along the y (vertical) axis, in the (y, z) view, in Fig. 7.7. In this case the particle trajectories are much straighter because there is no force in the y direction. We would have precisely straight line trajectories if y were plotted vs. s. Since the (y, z) view is not exactly congruent to (y, s) coordinates, the trajectories are only approximately straight.

7.6 Particle beam and quadrupole magnets

As a final topic we consider the simplest aspects of the preparation of beams of charged particles. Of necessity, we will need to understand the next most complicated magnetic multipole; the quadrupole magnet. The simplest elements of a particle beam transport line consist of dipole magnets, quadrupole magnets and collimators. In the case of accelerators, we require an accelerating system, such as a radio frequency cavity and, since many traversals of a magnet are made in a circular machine, correction elements such as sextupoles. The study of accelerators is well outside the scope of this volume.

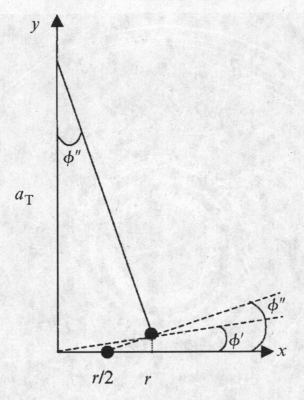

Fig. 7.5. Small bend angle approximation for motion in a uniform field. The exit angle is $\phi'' \sim r/a_T$ and the exit position is $y \cdot \phi'' r/2 \cdot r^2/2a_T$, for $\phi_o = 0$.

In traversing a dipole magnet different momenta have different bending angles. This momentum 'dispersion' effect is like the spread of colors out of a prism. Clearly, using the thick material in collimators to absorb the unwanted high and low momenta, we can 'momentum select' a beam to pass only particles with momenta within some limits.

A field free (or 'drift') space does not act on a particle. In what follows, we will represent passage through an element of the beam transport as a matrix. Multiplying the matrices of the individual elements can then represent complex systems of elements. The matrix representation, M_D, of a straight line trajectory over a distance L is

$$M_D \approx \begin{pmatrix} 1 & L \\ 0 & 1 \end{pmatrix}, \begin{pmatrix} x_o \\ x'_o \equiv (dx/ds)_o \end{pmatrix} \qquad M'_D \cong \begin{pmatrix} 1 & 1 \\ 0 & 1 \end{pmatrix}, \begin{pmatrix} x_o/\Delta s \\ (dx/ds) \end{pmatrix}$$

$$x = x_o + x'_o L \qquad\qquad x/L \sim x_o/L + x'_o \qquad\qquad (7.23)$$

$$x' = x'_o \qquad\qquad\qquad x' \sim x'_o$$

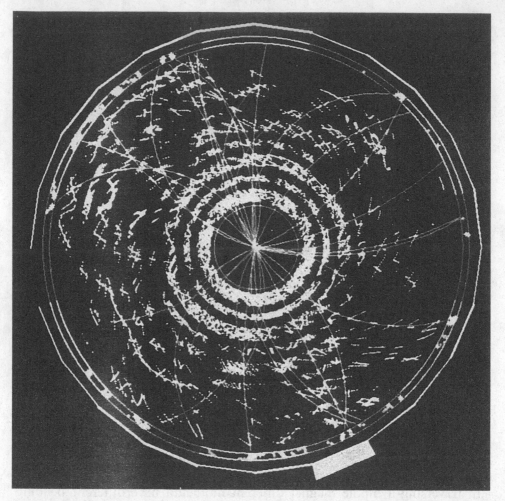

Fig. 7.6.   Path of charged particles in a solenoidal field, '(x, y) view'. Note that the sense of the radius of curvature is + or − depending on charge and that the radius of curvature is proportional to the momentum. (Photo – Fermilab.)

The arc length is s which is approximately the longitudinal distance L. The beam axis is conventionally taken to be z, and the motion is mostly along z, $ds \sim dz$. Motion in the x and y directions transverse to the beam axis is assumed to be small angle. It is traditional to use x and a 'slope' dx/ds as the two variables used to label particle position and direction. However, that leads to matrices with both dimensionless elements and elements with dimensions. An alternative formulation using $x/\Delta s$ and dx/ds is also included in what follows in the approximation that $\Delta s = L$ (the primed matrices). These objects, e.g. M'_D, have dimensionless matrix elements.

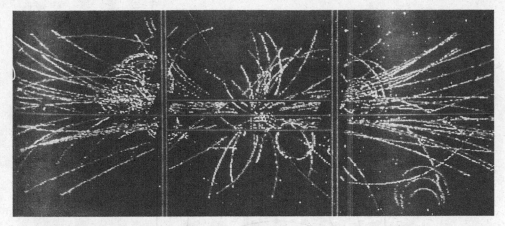

Fig. 7.7. Path of charged particles in a dipole field, (y, z) view. Note that the paths are nearly straight. (Photo – CERN Courier.)

For a dipole magnet of length L, in the approximation that it imparts a very small bend angle, the transformation is

$$x = x_0 + L^2/2a = x_0 + \phi_B(L/2)$$
$$x' = x'_0 + L/a = x'_0 + \phi_B \tag{7.24}$$

Note that the gradients in the field, Fig. 7.1, would act as thin lenses at the entrance and exit of the uniform field dipole. We ignore them here, but the quadrupole formalism we derive below can be directly applied if desired. Note that the gradient of B_z is $B' = \partial B_z/\partial y \sim B_0/\ell$. (Eq. 7.5.)

Consider next the case of a quadrupole magnet. The fields are shown in Fig. 7.8. The field $\mathbf{B}$ is zero at the magnet centerline by design. By construction there is a linear gradient, B'.

$$\nabla \times \mathbf{B} = 0 \Rightarrow \frac{\partial B_x}{\partial y} = \frac{\partial B_y}{\partial x} = B'$$

$$B_y = B'x \tag{7.25}$$

$$B_x = B'y$$

The quadrupole field, outside of the currents, is derivable from a magnetostatic potential V_B, Table 3.1, $V_B = -B'xy$, using $\mathbf{B} = -\nabla V_B$. The force equation for the motion transverse to the direction of motion, assuming $\alpha_z \cong 1$, is

$$\mathbf{F} = d\mathbf{p}/dt = \gamma m v \, d\mathbf{v}/ds = q(\mathbf{v} \times \mathbf{B}) \tag{7.26}$$

$$= \gamma m v^2 d^2\mathbf{x}/ds^2 = pv \, d^2\mathbf{x}/ds^2$$

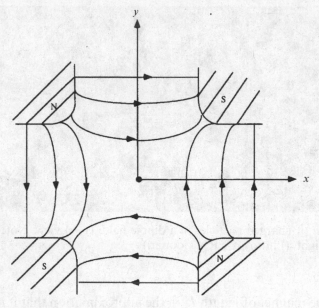

Fig. 7.8.   Field orientation for a quadrupole.

For example, in the x direction, $pv\,d^2x/ds^2 \sim -qvB_y = -qvB'x$. The equations of transverse motion are those of a simple harmonic oscillator with 'spring constant' k.

$$\frac{d^2x}{ds^2} = +kx$$

$$\frac{d^2y}{ds^2} = -ky \qquad\qquad (7.27)$$

$$k = qB'/p$$

The dimension of k is $[k] = [1/L^2]$. The equation of motion is one with solutions having harmonic oscillation in one transverse coordinate (called focusing or F type) and diverging hyperbolic trajectories in the other (call defocusing or D type). For the F type we define initial conditions x_o, $x_o' = (dx/ds)_o$ and look at the solution of the equations of motion. The first integral of Eq. 7.27 in a 'thin lens', where x is approximately constant, is $(\Delta x') = \Delta(dx/ds) = \int kx\,ds \sim k\,x_o\,L$. Thus a thin lens quadrupole imparts an angular impulse which depends on the deviation from the magnet centerline, x_o. The 'focal length', f, is then defined to be the longitudinal distance over which an incident parallel ray, with offset x_o and incident angle $x_o = 0$ is brought to a focus with displacement $x = 0$, $x' \sim kx_oL \cong x_o/f$ or

$$\frac{1}{f} = kL \qquad\qquad (7.28)$$

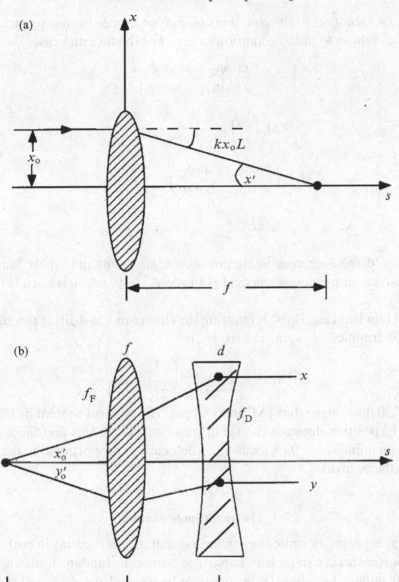

Fig. 7.9. (a) Illustration of the meaning of focal length. (b) A doublet lens system capturing a divergent beam and making it parallel, 'point to parallel'.

The geometric proof of this relationship is shown in Fig. 7.9a. All incident parallel rays are brought to the same point on the z axis, the focus. Hence the quadrupole acts as a charged particle lens. For example, a 3 m long quadrupole magnet with a 5 kG gradient over a 5 cm aperture has an 11.2 m focal length for a 100 GeV beam.

The exact solution to the equations of motion comes from integrating Eq. 7.27 twice subject to initial conditions x_o, x'_o. For the focusing case

$$\begin{pmatrix} x \\ x' \end{pmatrix} = \begin{pmatrix} \cos\phi_Q & \sin\phi_Q/\sqrt{k} \\ -\sqrt{k}\sin\phi_Q & \cos\phi_Q \end{pmatrix} \begin{pmatrix} x_o \\ x'_o \end{pmatrix}$$

$$= M_Q \begin{pmatrix} x_o \\ x'_o \end{pmatrix}$$

(7.29)

$$M_Q = \begin{pmatrix} \cos\phi_Q & \sin\phi_Q \\ -\sin\phi_Q & \cos\phi_Q \end{pmatrix}$$

$$\phi_Q = \sqrt{k}L$$

in the y or 'defocusing' coordinate, $\cos \to \cosh$, $\sin \to \sinh$ in both M_Q and M'_Q. Note that, in our previous numerical example, $\phi_Q \sim 30°$ which is not a very thin lens.

In the thin lens case, $\phi_Q \ll 1$, ignoring the change in x in drifting through the short quadrupole length, the matrix M_Q is

$$M_Q \to \begin{pmatrix} 1 & L \\ -kL & 1 \end{pmatrix} \to \begin{pmatrix} 1 & 0 \\ \pm 1/f & 1 \end{pmatrix}$$

(7.30)

In Eq. 7.30 the $\pm$ sign refers to F and D types. The physical content of Eq. 7.30 is that the position does not change in traversing a thin lens and the angle x'_o receives an impulse $\pm x_o/f$ ($-$ focusing, $+$ defocusing) proportional to the offset x_o from the beam axis.

7.7 The quadrupole doublet

Note that in geometric optics we can have simultaneous focusing in both transverse coordinates in a single lens. In particle beams, the fundamental constraint $\nabla \times \mathbf{B} = 0$ implies focusing (F) in one coordinate, and defocusing (D) in the other. Therefore, the simplest 'element' exhibiting overall beam focusing in both transverse dimensions is not the single lens, but the quadrupole doublet which we treat here in the simplest possible (thin lens) approximation. Note that this approximation is often useful as a first iteration to a complete solution, for example as 'starting values' to a complicated nonlinear computer code.

As the simplest application consider the quadrupole 'doublet' arranged as in Fig. 7.9b. We can find the conditions needed to capture a beam diverging from a point target and make it parallel. The full 'transport matrix' is the product of

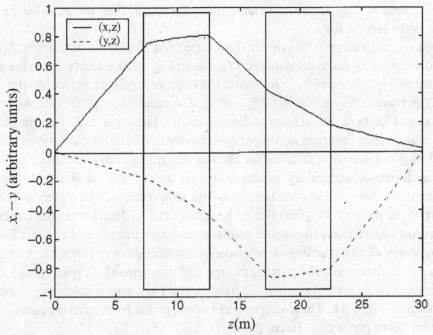

Fig. 7.10.   Plot of beam envelope for a thick lens quadrupole doublet in a 'point to point' optical configuration.

the matrices appropriate to traversing each beam element, $M_{doublet} = M_Q(F)M_D(L)M_Q(D)M_D(L_o)$. The parallel condition (for vanishing target size, $x_o = 0$, $y_o = 0$) is, $x' = M_{21}x_o + M_{22}x'_o = 0 \cong M_{22}x'_o$ or $M_{22} = 0$ where M refers to $M_{doublet}$. The final beam size is $x = M_{11}x_o + M_{12}x'_o \sim M_{12}x'_o$. The algebra in the simplifying thin lens case, which we omit as an exercise for the reader, has a solution where the quadrupole focal lengths must be

$$f_D = LL_o/f_D \tag{7.31}$$

$$f_F = L_o\sqrt{L/(L+L_o)}$$

and the final beam size is, as can be visualized in Fig. 7.9b,

$$x = x'_o\,(L_o + L + LL_o/f\,F) \tag{7.32}$$

$$y = y'_o\,(L_o + L - LL_o/f\,F)$$

This solution is found by requiring both an x and a y parallel beam. The matrix element M_{22} must vanish simultaneously for both x and y motion. Those constraints lead to a quadratic equation for f. Note that, as Fig. 7.9b indicates, a cylindrical limiting magnet aperture, see Fig. 7.11, means an asymmetric angular acceptance for the doublet. That fact is also clear from Eq. 7.32 where,

if $x = y$, then $x'_0 < y'_0$. If the first element is y focusing, then the angular acceptance is the largest for y.

As a second example, the optical envelope for a thick lens doublet is shown
in 'point to point' focus conditions, $M_{12} = 0$, in Fig. 7.10. Clearly, since the first
quadrupole is y defocusing (x focusing) the capture angle for a circular quadrupole aperture is asymmetric ($\theta_x^{max} > \theta_y^{max}$). Complete computer optimization
codes exist for both beam lines and accelerators. However, they require 'starting values' being complex nonlinear systems without analytic solutions. Thus,
good physical intuition is called for even in designing such systems.

The design of secondary beams is an art unto itself, as is the design of
accelerators. We have shown here only the simplest cases in order to give a
flavor of the physics. Magnets should be thought of as detectors in the concrete
sense that we use them to prepare beams to do experiments and to exert forces
on particles so that the use of position measuring devices (Chapter 8 and 9)
allows us to determine the particle charge and momentum. A photograph of a
beam line dipole and a quadrupole used in the Fermilab accelerator complex
is shown in Fig. 7.11. These magnets are appropriate to prepared beams, with
vacuum beam pipes of ~ 10 cm diameter.

Exercises

1. Show that Maxwell's equations are obeyed by the field, $B_y = B_0(z/\ell)$ and $B_z = B_0(y/\ell)$. If the velocity of a particle in this field has x and z components, show that the field B_z causes a vertical force.

2. Consider a 1 MeV proton cyclotron. If the guide field is 20 kG, what is the radius of the machine needed to contain the full energy protons?

3. For the cyclotron defined in Exercise 2, what is the period of rotation? The frequency?

4. How big would a new accelerator of 50 TeV = 50 000 GeV maximum energy be? Assume closely packed dipole magnets which reach 60 kG (superconducting).

5. Suppose you use silicon detectors (Chapter 9) as tracking elements. Assume you deploy them over a length of 0.5 m in a 40 kG field. Assume that, with a charge sharing measurement, we can achieve a resolution of 10 μm. What is the fractional momentum error this system obtains for a 100 GeV track?

6. Suppose you are at a radius 1 m from the interaction point and immersed in a 40 kG solenoidal field. For what values of transverse momentum can a created particle never reach your location?

7. Imagine constructing a quadrupole magnet with 20 kG pole field at 5 cm radius. For that gradient, what is k for a 100 GeV particle?

Fig. 7.11. Magnets appropriate for high energy particle beams. (a) A dipole. Note the two coils surrounded by the steel 'yoke'. (b) A quadrupole. Note the four coils surrounding the cylindrical vacuum pipe. Outside the coil is a steel yoke and water cooling plumbing. (Photos – Fermilab.)

8. Suppose the quadrupole is 10 m long. If a beam particle enters 1 cm off axis, what angular deflection does it receive? What is the focal length of the quadrupole?

9. Multiply the matrices for the thin lens doublet and derive Eq. 7.31 for the required focal lengths for making the beam parallel in both transverse directions.

10. For $L_o = 5$ m and $L = 10$ m find the focal lengths of the quadrupoles required for exiting parallel beams. What is the ratio of the collected angles, assuming equal aperture in the two transverse directions?

References

[1] R. Palmer, A.F. Tollestrup, 'Superconducting magnet technology for accelerators', *Annu. Rev. Nucl. Part. Sci.* **34** 247 (1984).

[2] R.L. Gluckstern, *Nucl. Instrum. Methods* **24** 381 (1963).

[3] R. Yamada, *et al.*, *Nucl. Instrum. Methods* **138** 567 (1976).

[4] *An Introduction to the Physics of Particle Accelerators*, D.A. Edwards and M.J. Syphers, FNAL FN (1988).

[5] *The Physics of Charged-Particle Beams*, J.D. Lawson, Clarendon Press (1988).

8

Drift and diffusion in materials, wire chambers

Time is a sort of river of passing events . . . and this too will be
swept away

Marcus Aurelius

Wire chambers are perhaps the most commonly used detection devices in present high energy physics experiments. A chamber consists, conceptually, of a gas volume in which the gas is ionized by the passage of a charged particle. Ionization has been discussed in Chapter 6. The ionization then drifts (and diffuses) in an electric (and perhaps magnetic) field toward an electrode (wire). Subsequent collection and amplification of the signal charge on the anode and the charge induced on the cathode creates a detectable signal. Since the radius of curvature of a charged particle in a magnetic field depends on momentum (Chapter 7), measurement of points on the trajectory determines **p** the vector momentum. We can then make a second redundant momentum measurement using 'destructive' measurement techniques (see Chapters 11 13).

In this chapter we develop a general treatment suitable for 'unity gain' devices, e.g. liquid argon calorimetry, or devices with gain, e.g. proportional wire chambers. We first introduce the drift velocity, mobility and the diffusion coefficient. The signal capacitively induced on the two electrodes is then considered both for a unity gain device and for a device with gain. The methods used here are later applied, in the following chapter, to silicon devices.

8.1 Thermal and drift velocity

We begin by considering the drift of the produced ionization in a uniform electric field and the accompanying diffusion of the localized ionization. First, we note that gas molecules of mass M have a thermal velocity v_T due to the thermal kinetic energy T_T, which is of order kT, where k is the Boltzman's constant and T is the absolute temperature. The numerical value of k is given in Table 1.1.

$$T_T = Mv_T^2/2 \sim \frac{3}{2}kT$$

$$v_T \sim \sqrt{\frac{3kT}{M}} \tag{8.1}$$

The thermal velocity goes like $\sqrt{T}$. Numerically, we find that the thermal velocity of N_2 at room temperature is of order 10^5 cm/s or 0.1 cm/μs. Other light molecular gas velocities at standard temperature and pressure (STP) are in the range of 0.1 to 1.0 cm/μs. In comparison, escape velocity on earth is $\sim$ 1.1 cm/μs. This fact explains why there are no light gases such as helium in our atmosphere. Fluctuations in the thermal velocity allow light gases to attain escape velocity and they are lost to the atmosphere.

For an electron, given the $1\sqrt{M}$ scaling, we expect that typical thermal velocities would be of order $\sim(4-40)$ cm/μs. Note that the thermal energy scale is well below the scale of the binding energies (see Chapter 1) which are relevant to atomic excitations, so we can consider the atoms to be inert 'billiard balls'.

Let us consider next the situation where we have an applied electric field, **E**, in addition to the random thermal motion. Between collisions, ions and electrons are accelerated by **E**. The integral of the acceleration **a** is the drift velocity v_d. The parameter τ is defined to be the mean time between collisions. Since the mean free path $\langle L \rangle$ is the mean distance between collisions, τ is approximately $\langle L \rangle / v_T$. The average drift velocity is $\langle v_d \rangle$.

$$\mathbf{a} = e\mathbf{E}/M$$

$$\langle v_d \rangle \sim a\tau \tag{8.2}$$

$$\sim \left(\frac{eE}{M}\right)\left(\frac{\langle L \rangle}{v_T}\right)$$

The drift velocity $\langle v_d \rangle$, which is imposed on top of the larger randomly oriented thermal velocity, is proportional to the applied field. We assume that the main velocity components are caused by the random thermal motion and not by the applied electric field, which is assumed to be only a small perturbation to the thermal motion.

As we discussed in Chapter 1, the inverse mean free path is proportional to the collision cross section, σ. For atomic collisions at low energy scales, the collision cross section is of the order of the geometric cross section $\cong \pi a_0^2$. The total elastic cross section for electrons on CH_4 as a function of electron kinetic energy is shown in Fig. 8.1.

The scale is set by the geometric cross section, $\sim \pi a^2$ with $a \sim 1$ Å. Since the signal current induced on an electrode is proportional to the drift velocity (see

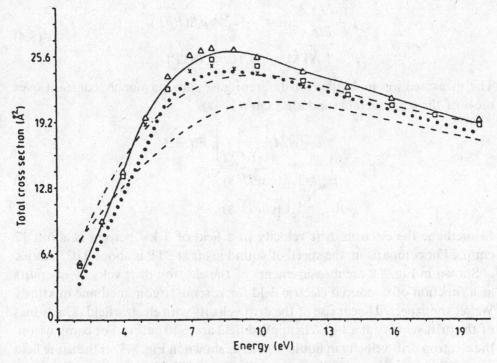

Fig. 8.1.   Cross section for electrons on CH_4 as a function of the electron energy in eV.

Eq. 8.7) and since the drift velocity is larger for electrons than ions, we concentrate mostly on electrons in what follows.

$$\langle L \rangle^{-1} \cong N_o \rho \sigma / A$$

$$\sigma \sim \pi a_o^2 \tag{8.3}$$

$$\langle v_d \rangle \sim \left(\frac{eE}{m_e} \right) \left(\frac{A/N_o \rho \sigma}{v_T} \right) = \frac{eEA}{N_o \rho \sigma \sqrt{3kT m_e}}$$

8.2 Mobility

The drift velocity is proportional to the applied electric field and inversely proportional to the collision cross section and the thermal velocity. Therefore, in this very simple-minded picture, we expect that the drift velocity per unit electric field divided by density would be a constant which is called the mobility. The electric field per unit of pressure, P, or per unit of density, ρ, with respect to STP is called the reduced electric field. The mobility, μ, is defined to be the drift velocity $\langle v_d \rangle$ per unit reduced electric field E/ρ.

$$\mu \equiv \frac{\langle v_d \rangle}{E/\rho} \sim \text{const}, \quad \langle v_d \rangle \equiv \mu E (P_o/P)$$

$$(8.4)$$

$$P_o = 1 \text{ ATM} = 760 \text{ Torr} \quad (\text{STP})$$

The measured ion mobilities for different gases are reasonably constant over most of the periodic table and are 1 cm²/($V \sim$s).

$$\mu \sim (e/M) \left[\frac{A}{N_o \sigma v_T} \right], \text{ Eq. 8.3}$$

$$\mu_{Ar} = 1.7 \text{ (cm}^2/V \text{ s)} \tag{8.5}$$

$$\mu_{CO_2} = 1.1 \text{ (cm}^2/V \text{ s)}$$

In methane the electron drift velocity in a field of 1 kV per cm is about 12 cm/μs. For comparison, the speed of sound in air at STP is about 0.033 cm/μs.

Shown in Fig. 8.2 are measurements of the electron drift velocity in cm/μs as a function of a reduced electric field for gaseous argon–methane mixtures. We see the expected linear rise of the drift velocity with electric field. The values of the drift velocity at a 1 kV/cm applied field are ~ 10 cm/μs. For comparison, the electron drift velocity in liquid argon, as shown in Fig. 8.3, at the same field is ~ 0.15 cm/μs indicating a lower mobility in the liquid.

We also see, looking at Fig. 8.2, another phenomenon. The linear rise with field no longer holds at high electric fields. At those fields the drift velocity becomes comparable to the electron thermal velocity which, as we said, is of order 4– 40 cm/μs. Our assumptions are then no longer valid, because we assumed that the drift velocity was a small perturbation on the random thermal motion.

Of some practical use is the experimental fact that a 50:50 mixture of argon:ethane has a 'saturated' drift velocity. By this we mean that the electron drift velocity $\langle v_d \rangle$ is essentially independent of the field when $E > 0.5$ kV/cm. This gas mixture is widely used in 'drift chambers' since, if E is kept high, the velocity is a constant, so that a drift time measurement is a linear measure of the distance from the point of particle passage to the anode wire. (See Fig. 8.7.)

8.3 Pulse formation in 'unity gain' detectors

Let us now turn to the signal formation in a 'unity gain' detector. The ionization produced by the passage of a charged particle is drifted to, and collected on, the electrodes which apply the electric field. There is no charge gain, as there is in a photomultiplier tube (see Chapter 2). For a parallel plate configuration with spacing d and applied voltage V_o, a field, $E = V_o/d$, exists in

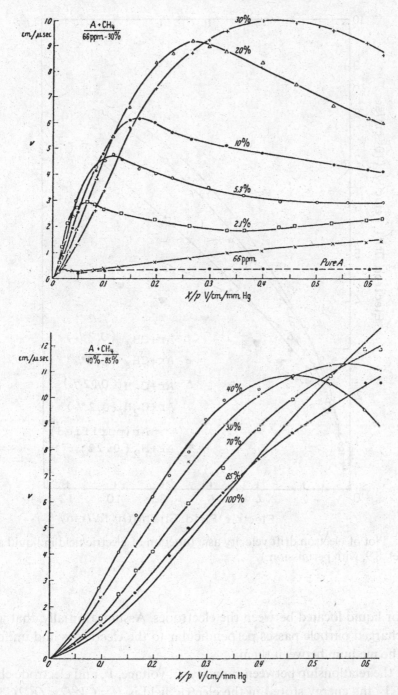

Fig. 8.2.   Drift velocity in cm/μs as a function of 'reduced' field, *E/P*. (a) Ar +CH$_4$ gas mixtures, with fraction of CH$_4$ $\leq$ 30%. (b) Ar +CH$_4$ gas mixtures, with fraction of CH$_4$ > 40%. (From Ref. 8.2, with permission.)

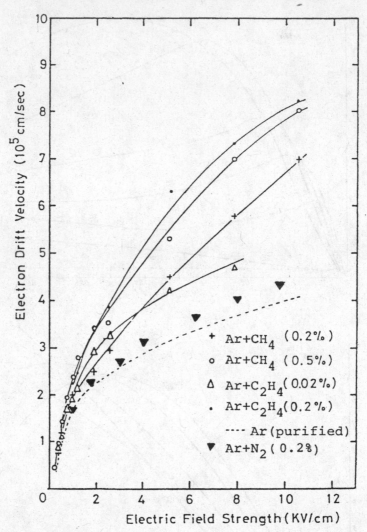

Fig. 8.3.   Plot of electron drift velocity as a function of electric field in liquid argon. (From Ref. 8.9, with permission.)

the gas or liquid located between the electrodes. Assume, initially, that a high energy charged particle passes perpendicular to the electrodes and uniformly ionizes the medium between them.

Using the relationship between capacity, C, voltage, V, and electrode charge, Q, $Q = CV$, the energy stored in the electric field is $U = CV^2/2 = Q^2/2C \sim QV$. The motion of the ionization charge in the field with time, $q(t)$, capacitively induces a current pulse on the electrodes, $I(t)$. We ignore the ions, because the size of the current pulse is proportional to the drift velocity. The change in field energy, dU, due to the drift of the ionization is equal to the work done, Fdx, by the field on the drifting charges.

$$dU = Q_o dQ(t)/C = Fdx \tag{8.6}$$

$$= [q(t)E][\langle v_d \rangle dt]$$

The charge $Q(t)$ induced on the electrodes is proportional to the ionization charge $q(t)$ in the electrode gap and the drift velocity $\langle v_d \rangle$. Note that this treatment is quite general, having used only energy conservation. Thus it can be used later in the discussion of wire chambers and silicon detectors. The expression for $I(t)$ is

$$dQ(t) = q(t)\langle v_d \rangle dt[E/V_o] = [q(t)\mu E^2/V_o]dt \tag{8.7}$$

$$\frac{dQ(t)}{dt} \equiv I(t) = q(t)\mu E^2/V_o$$

The charge/length initially in the gap is $q_s/d = Ne/d$ where q_s is the total signal charge. Note that N is the number of independent 'events', just as, for example, N is the number of photoelectrons which is statistically distributed (Chapter 2). Therefore, the fractional spread in the output signal due to the statistical variation of N, dN, is given by $dN/N \sim 1/\sqrt{N}$. The characteristic time for all the charge to be swept up by the electrodes is τ_d. For example, for a 1 cm gap filled with argon–methane and a 1 kV applied voltage, this time is 100 ns.

$$q(t) = q_s\left(1 - \frac{t}{\tau_d}\right), \quad t < \tau_d$$

$$= 0, \quad t > \tau_d \tag{8.8}$$

$$\tau_d = d/\langle v_d \rangle$$

The charge $q(t)$ in the gap is q_s at $t = 0$ and decreases linearly to zero at $t = \tau_d$. The induced current behaves similarly, $I(0) = q_s/\tau_d$, $I(\tau_d) = 0$. The total induced charge is $Q(0) = 0$, $Q(\tau_d) = q_s/2$, since half the induced charge appears on the other electrode, which we have assumed to be grounded.

$$I(t) = (q_s/\tau_d)(1 - t/\tau_d)$$

$$\int I(t)dt \equiv Q(t) \tag{8.9}$$

$$= q_s[y - y^2/2], \quad y = t/\tau_d$$

Sometimes all the charge is located at a single point, x_o, in the gap, for example, due to a short range alpha decay which ranges out almost immediately or a Compton recoil electron due to a scattering of an x-ray from ^{55}Fe decay (see

Table 6.1). In that case $Q(t)$ is different from the case where charge is spread throughout the gap.

In the 'point' ionization case, with total charge $q_s = e$.

$$dQ = \frac{q(t)E}{V_o}(\langle v_d \rangle dt)$$

$$q(t) = q_s \text{ for } t > \tau'_d \tag{8.10}$$

$$= 0 \text{ for } t > \tau'_d, \quad \tau'_d \equiv x_o/\langle v_d \rangle$$

The current in this case is constant in time, and the charge collected on the electrodes increases linearly with time up to τ'_d, i.e. $Q(0) = 0$, $Q(\tau'_d) = q_s(x_o/d)$.

$$I(t) = \frac{q_s}{\tau_d}, \quad t < \tau'_d$$

$$= 0, \quad t > \tau'_d$$

$$Q(t) = q_s t/\tau'_d \tag{8.11}$$

$$\leq q_s(x_o/d)$$

For example, a $d = 1$ mm gap in liquid argon has a collection time for a typical electric field of $\tau_d = 0.25$ μs. The collected charge, $Q(t)$, as a function of time is shown in Fig. 8.4 for point ionization, provided by short range alpha irradiation, and for line ionization provided by traversal of minimum ionizing particles. Clearly linear and quadratic behavior are seen, respectively, with a time scale set by $\sim \tau_d$. The signal noise, which is seen on the traces of Fig. 8.4, will be discussed in Chapter 9.

8.4 Diffusion and the diffusion limit

Diffusion of the drifting charge is due to multiple scattering. This diffusion process sets a fundamental limit to the accuracy of position measurements in ionization chambers. Consider charge drifting with a velocity $\langle v_d \rangle$ in field E for time t or distance x. The distribution, dN, of transverse coordinate, x_T, is defined by the diffusion coefficient D, with units of cm²/s, $[D] = [L^2/T]$.

The diffusion equation is well known in classical physics in the study of heat. For a density $\rho(x, t)$, the equation without external forces is Eq. 8.12. A one dimensional solution in the case that particles appear instantaneously and tightly localized spatially at $x = 0$ at $t = 0$ is:

$$\frac{\partial \rho}{\partial t} = -D\partial^2 \rho/\partial x^2 \tag{8.12}$$

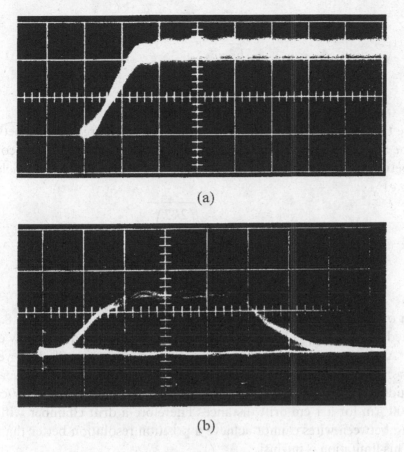

(a)

(b)

Fig. 8.4.   Collected charge in a liquid argon detector as a function of time, (a) point ionization, 100 ns/division, (b) line ionization, 200 ns/division. (From Ref 8.9, with permission.)

$$\rho(x, t) \sim \frac{1}{\sqrt{Dt}} e^{-x^2/4Dt}$$

The solution is a Gaussian distribution of particles with a width; $\sigma \sim \sqrt{2Dt}$ which grows with time as $\sqrt{t}$. In the case where there is a uniform applied electric field the mean moves with a drift velocity brought about by the replacement, $x \rightarrow x - \langle v_d \rangle t$ in Eq. 8.12. This is a stochastic process, as indicated by the fact that the rms of the distribution goes as $\sqrt{t}$. Recall the similar multiple scattering behavior, $\langle \theta_{MS} \rangle \sim \sqrt{t}$.

We assert that dimensional arguments lead to the conclusion that $D \sim v_T \langle L \rangle$ showing the relation between diffusion, thermal velocity, and collision mean free path. The rms of the diffusion distribution is

$$\sigma_{x_T} \sim \sqrt{2Dt} \sim \sqrt{2v_T\langle L\rangle(x/\langle v_d\rangle)}$$

$$\sigma_{x_T} \sim \sqrt{\frac{v_T^2 x}{a}} \tag{8.13}$$

$$\langle v_d\rangle \sim a\langle L\rangle/v_T$$

We use the relationship between v_T and the thermal energy kT and between a and the applied field E. This 'thermal limit' for diffusion shows the competition between random thermal energy, kT, and the imposed drift field with energy eV_o.

$$\sigma_{x_T} \cong \sqrt{\left(\frac{2kT}{eE}\right)x}$$

$$\sigma_{x_T} \cong \left[\sqrt{\frac{2kT}{eV_o}}\right] \tag{8.14}$$

A plot of σ_{x_T} as a function of reduced electric field is shown in Fig. 8.5. The expected behavior, $\sigma_{x_T} \sim 1/\sqrt{E}$, is roughly followed by the data at low electric fields. At higher electric field our assumption about small drift velocity, $v_T \gg \langle v_d\rangle$ breaks down. We ignore here the difference between transverse and longitudinal diffusion. A typical value of σ_{x_T} at a drift field of 1 kV/cm is $\sim$ 100–300 μm for a 1 cm drift distance. Therefore a drift chamber with 2 cm spacing between wires cannot achieve a position resolution better than ~ 100 μm. This limitation is intrinsic.

An example of the geometry of a drift tube, with 'field shaping' electrodes which are used to provide a drift field of >0.5 kV/cm, is shown in Fig. 8.6. Also shown are the equipotentials. By design there is a reasonably uniform electric field across most of the drift space. The relationship of drift distance to drift time for this structure is shown in Fig. 8.7 for cosmic ray traversals at two angles of incidence, 0° and 45°. In this example a mixture of argon and carbon dioxide is used as the drift gas. The phenomenon called 'saturation' also roughly obtains in this gas mixture for drift fields greater than about 0.5 kV/cm. This implies a linear relationship between the drift distance and the measured drift time (the time between particle passage and the collection of the ionization on the wire). We see from Fig. 8.7 that this is roughly the case for drift distances up to 5 cm with a maximum drift time of about 800 ns. The average drift velocity is about 6.2 cm/μs.

As a numerical example, electrons in gaseous argon in a drift field of 1 kV/cm have a drift velocity of $\langle v_d\rangle \sim 5$ cm/μs. In a 1 cm drift distance the transverse diffusion distance is ~ 1000 μm, while in pure carbon dioxide it is only 100 μm.

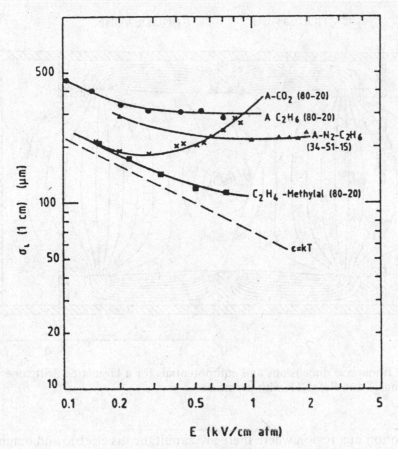

Fig. 8.5. Measured values of the longitudinal diffusion coefficient σ_L as a function of applied reduced electric field for different commonly used gases compared to the thermal limit, $\sigma_{x_T}/x \sim \sqrt{2kT/eV_0}$, indicated by the dashed line. (From Ref. 8.5, with permission.)

In the mixture Ar:CO$_2$ = 80:20 it is 300 μm (see Fig. 8.5). The width of the lines in Fig. 8.7 gives some indication of the spatial resolution. Diffusion limits the resolution for the longest drift distances. The observed resolution is $\sim$100 μm near the wire and $\sim$400 μm at the maximum 5 cm drift distance.

8.5 Motion in **E** and **B** fields, with and without collisions

So far we have only considered random thermal motion and motion in an applied electric field. In Chapter 7 we have also discussed free motion in a purely magnetic field. We concluded that the path was circular with a 'cyclotron frequency', $\omega_c = eB/m$ (Table 1.1). Now we consider the more complex case

CATHODE PADS   ANODE WIRE

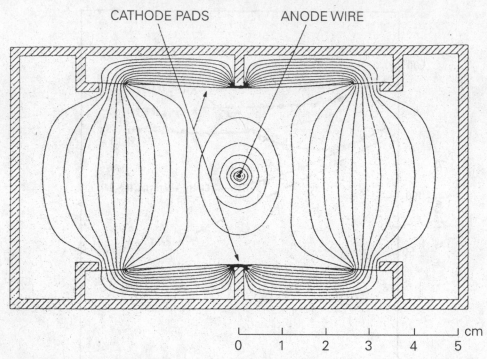

Fig. 8.6.   Transverse dimensions and equipotentials for a large area drift tube with field shaping. (From Ref. 8.11, with permission.)

of free motion in a region where there are simultaneous electric and magnetic fields. The dynamics in the absence of collisions is controlled by the Lorentz force equation. (CGS units here.)

$$\mathbf{F} = q(\mathbf{E} + \boldsymbol{\beta} \times \mathbf{B})$$

$$\boldsymbol{\beta}_d = (\mathbf{E} \times \mathbf{B})/|\mathbf{B}|^2$$

(8.15)

The general case of non-relativistic motion in a vacuum with an arbitrary angle θ between **E** and **B** is treated in Appendix G. In the special case that the electric and magnetic fields are transverse to one another we can see that in a direction transverse to both the electric and magnetic fields, the force is 0 if $\beta = \beta_d$. A particle can thus move at a constant drift velocity, β_d, perpendicular to crossed electric and magnetic fields in vacuum. This is the principle of electrostatic separation where we prepare a beam of particles of fixed momentum by bending in a magnet, focusing parallel, and then collimating (see Chapter 7). We then put the beam through a 'separator' which has crossed electric and magnetic fields. Only particles of the right mass, and therefore the right velocity, β_d, will be undeflected whereas the unwanted particles will be physically

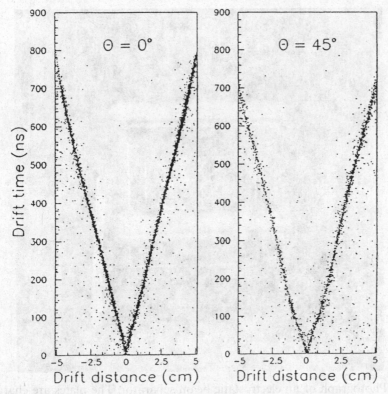

Fig. 8.7. Relationship of drift distance to drift time for the structure shown in Fig. 8.6 for particles incident at 0° and 45°. (From Ref. 8.11, with permission.)

deflected and subsequently can be removed, again by collimation. This technique is a form of 'destructive' particle identification.

A photograph of a working '**E** × **B** separator' is shown in Fig. 8.8. It is hard to maintain a voltage greater than a megavolt in air due to corona and other discharge effects. Typical operating parameters are $E \sim 1$ MV/cm and $\beta_d = 0.5$. To achieve that drift velocity a magnetic field of 7 kG is required.

In the case where there are collisions, in a medium such as a gas, they can be approximated as an effective frictional force that is characterized by a mean time between collisions τ. The non-relativistic relationship between the frictional force $\mathbf{F}_F$ and momentum $\mathbf{p}$ follows because force is the time rate change of momentum.

$$\mathbf{F}_F = \mathbf{p}/\tau, \quad \mathbf{F} = q(\mathbf{E} + \boldsymbol{\beta} \times \mathbf{B}) + mc\boldsymbol{\beta}/\tau = m\mathbf{a}$$

$$\mathbf{a} = 0 \text{ when } \mathbf{v} = -\frac{q\tau}{m}(\mathbf{E} + \boldsymbol{\beta} \times \mathbf{B})$$

(8.16)

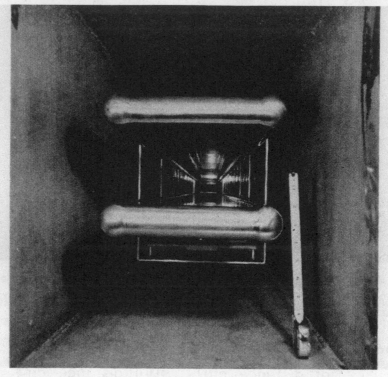

Fig. 8.8.   Photograph of an electrostatic beam separator. The plates are charged to establish the electric field. The vacuum tank surrounding the plates is made of soft iron and acts as the pole tips of the magnet supplying the magnetic field. The line of sight is the beam axis, $\otimes$. (From Ref. 8.10, with permission.)

We can solve Eq. 8.16 for the situation where the force is zero and we have a uniform drift velocity. The drift velocity $\langle \mathbf{v}'_d \rangle$ is now a function of the charge, mass, fields and τ.

$$\langle \mathbf{v}'_d \rangle = \frac{-e\tau/m}{[1+(\omega_c\tau)^2]}[\mathbf{E}+(\omega_c\tau)\mathbf{B}\times\mathbf{E}+(\omega_c t)^2(\hat{B}\cdot\mathbf{E})\hat{B}] \tag{8.17}$$

$$\omega_c = eB/m$$

The drift velocity has, in general, components along the $\mathbf{E}$, $\mathbf{B}$, and $\mathbf{B}\times\mathbf{E}$ directions.

   The vectors appropriate to drifting in perpendicular electric and magnetic fields are shown in Fig. 8.9. The 'Lorentz angle', ϕ_L, is defined to be the angle which the drift velocity $\langle \mathbf{v}'_d \rangle$ makes with the applied electric field. In the vacuum case ϕ_L is 90° as it is perpendicular to both $\mathbf{E}$ and $\mathbf{B}$. In general the drift velocity for perpendicular $\mathbf{E}$ and $\mathbf{B}$ fields is in the $\mathbf{E}$ and $(\mathbf{B}\times\mathbf{E})$ plane.

$$\mathbf{E} \perp \mathbf{B}, \quad \tan\phi_L = \omega_c\tau \tag{8.18}$$

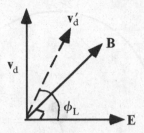

Fig. 8.9. Vectors for drift in perpendicular electric and magnetic fields. The Lorentz angle ϕ_L is in the $(\mathbf{E}, \mathbf{B} \times \mathbf{E})$ plane.

The quantity $\omega_c \tau$ is a measure of the rotation angle of the momentum in the $\mathbf{B}$ field between collisions. Therefore if $\omega_c \tau \gg 1$ we expect small collisional effects. If $\tau \to \infty$, the angle ϕ_L becomes 90° and we recover the case of the vacuum drift velocity. For a mean free path between collisions of 0.1 μm (gas) with a 10 cm/μs drift velocity (argon–methane), τ is ~1 ps and for a 10 kG field $\omega_c \tau$ is 0.17 or ϕ_L is ~11°. Note that there are serious practical consequences. Wire chambers operated in magnetic fields often have their electrode structure modified to follow the Lorentz angle.

In general, the drift velocity $\langle v_d' \rangle$ is less than the velocity $\langle v_d \rangle$ appropriate to the $|\mathbf{B}| = 0$ case because of the additional helical path length ($\langle \mathbf{v}_d \rangle = e\mathbf{E}\tau/m$, Eq. 8.2).

$$\langle v_d' \rangle = \frac{[\mathbf{v}_d + \omega_c \tau (\hat{B} \times \mathbf{v}_d)]}{[1 + (\omega_c \tau)^2]} \qquad (8.19)$$

$$\langle v_d' \rangle = \langle v_d \rangle / \sqrt{1 + (\omega_c \tau)^2}$$

The effective diffusion coefficient in the direction perpendicular to B is lowered by a factor $1/[1 + (\omega_c \tau)^2]$ due to the tight helical orbits. This effect is sometimes used to improve the position resolution for long drift distances where diffusion errors dominate.

8.6 Wire chamber electrostatics

We now consider a simplified description of a 'wire chamber'. We consider only the simplest case of a cylindrical tube with outer radius b coaxial with a small diameter wire of radius a, Fig 8.10. A particle is localized to a position resolution ~b if ionization is detected on a given wire assuming a linear array of such tubes. This is a simple problem in two dimensional electrostatics. We use Gauss's law (CGS units). In terms of λ, the charge per unit length on the wire, we find the electric field and the potential V.

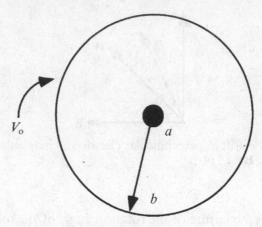

Fig. 8.10.   Cylindrical proportional tube of outer radius b at voltage V_o and inner (wire) of radius a at voltage zero.

$$E = 2\lambda/r$$

$$V_o = 2\lambda\ln(b/a), \quad V(r) = V_o\ln(r/a)/\ln(b/a) \tag{8.20}$$

$$E(r) = \frac{(V_o/r)}{[\ln(b/a)]}$$

The passage of a charged particle causes ionization. The electrons then drift towards the anode. Near the wire the high electric fields cause a multiplication of the electrons by collisions since at a small radius the energy gain can exceed the ionization potential, I, of the gas (see Chapter 1). The newly created electrons in turn will be accelerated and cause ionization. This is a runaway process similar to what we already discussed in the operation of a photomultiplier tube. Note that the 'gas gain' follows from the change in the number of electrons $dN(r)$ at a given position r in the multiplication process. This change depends on the number of electrons present $N(r)$ and the multiplication factor, α per unit length. The parameter α is the inverse of the distance over which a multiplication occurs. It is called the first Townsend coefficient and is identifiable as the inverse mean free path for ionization, $\alpha \sim 1/\langle L \rangle_{\text{ion}}$.

$$dN(r) = N(r)\alpha dr$$

$$N(r) = N_o e^{\alpha r} \tag{8.21}$$

$$\langle I \rangle_A \sim 26 \text{ eV}$$

The amplification or 'gas gain' is $e^{\alpha r}$. A typical gas gain might be 10^5. Above that level we approach the 'Geiger region' which is a true runaway process where the whole tube is discharged. We concentrate in what follows on the 'proportional region' where the output signal is roughly proportional to the input signal. The transition region between these operational modes occurs experimentally for $\alpha r \sim 20$ pairs or for gas gain $\sim 10^8$.

8.7 Pulse formation in a wire chamber

The time development of the signal is different from that of the 'unity gain' ionization chamber we discussed previously. The electrons from the primary ionization are multiplied by the high fields near the wire, $E(r) \sim 1/r$, and are collected on the anode. The ions remain, 'sheathing' the wire and moving slowly toward the cathode. The motion of all the charges in the applied field causes a change of system energy and a resultant capacitively induced signal, dV.

Suppose the multiplication takes place at N wire radii, i.e. $r = Na$. The voltage signal on the anode is due to the motion of the electrons, V^-, and the ions, V^+. If q_s is the source charge and C is the capacity, then $dU = CV_0 dV = q_s E\, dr$ (see Eq. 8.6). The induced voltage due to electron and ion motion is

$$dV = \frac{q_s}{CV_0} E(r) dr$$

$$V^- = \int_a^{Na} dV, \quad V^+ = \int_{Na}^{b} (-dV) \tag{8.22}$$

$$V^- = \frac{q_s}{C} \frac{\ln(N)}{\ln(b/a)}, \quad V^+ = \frac{q_s}{C} \frac{\ln(b/Na)}{\ln(b/a)} \gg V^-$$

Clearly, in the case of the wire chamber the signal due to the ions, V^+, dominates, basically because they move all the way to the cathode. Therefore, the ion motion is responsible for the majority of the anode signal. The electron contribution will be ignored from now on.

The signal can be found using the technique already applied to unity gain devices. Assuming a constant ion mobility, μ, $dr = \mu E(r) dt$. Thus, $dr = \mu[2\lambda] dt/r(t)$. Assume that the ions all appear at $r = a$ at $t = 0$ due to the final multiplication.

$$\int_a^r r\, dr = \mu[2\lambda] \int_0^t dt$$

$$r = a\sqrt{1 + 4\mu\lambda t/a^2} \tag{8.23}$$

$$\equiv a\sqrt{1 + t/\tau_0}, \quad \tau_0 = a^2/4\mu\lambda$$

The fact that the field goes as $1/r$ implies that the ion position r goes as $\sim \sqrt{t}$. Using the general equation, Eq. 8.7, we can then find the current induced on the wire.

$$I(t) = q_s \mu E^2 / V_o$$

$$= \frac{q_s \mu (2\lambda/r^2)}{\ln(b/a)} \quad \text{(Eq. 8.20)} \tag{8.24}$$

Substituting in the expression for $r(t)$ we find $I(t)$.

$$I(t) = \frac{q_s(2\lambda\mu/a^2)}{\ln(b/a)(1 + t/\tau_o)}$$

$$I(t) = \left[\frac{q_s/2\tau_o}{\ln(b/a)}\right] / (1 + t/\tau_o) = I(0)/(1 + t/\tau_o) \tag{8.25}$$

The key parameter τ_o is the characteristic time for the pulse to form. It can be thought of as roughly the time it takes for the ions to move away one wire radius, a, towards the cathode under the influence of the electric field, $E(a) = 2\lambda/a$, which exists near the surface of the anode wire.

$$\mu = \langle v_d \rangle / E \sim a(\tau_o E(a))$$

$$\tau_o \sim a/\mu E(a) = a^2/2\mu\lambda \tag{8.26}$$

We define $\underline{C}$ to be the capacity per unit length of the anode while the capacity is C. The relationship of voltage, charge and capacity is $V = Q/C = \lambda/\underline{C}$. From Eq. 8.20 we find $\underline{C} = 1/[2 \ln(b/a)]$ so that C is dimensionless in CGS units. It can be thought of as ε in ε_o units (Table 1.1). Typically for wire chambers, $\underline{C} \sim (0.01 - 0.1)$ pF/cm, $\underline{C}V_o \sim \lambda \sim 5 \times 10^{-10}$ coulomb/cm $= 500$ pC/cm. Recall that (Table 1.1), $\varepsilon_o = 8.85$ pF/m in MKS units.

The charge Q is derived by integrating Eq. 8.25.

$$Q(t) = q_s[\ln(1 + t/\tau_o)/[2 \ln(b/a)] \tag{8.27}$$

As a numerical example consider a 1 cm radius tube strung with a 20 μm wire, $b = 1$ cm, $a = 20$ μm. The field is about 160 kV/cm at the surface of the wire, $\lambda = 180$ pC/cm, and the pulse formation time is a few ns for argon ions. This is the kind of speed that we observe in the rise time of pulses from proportional chambers. (See Fig. 9.15)

$$E(a) = 160 \text{ kV/cm}$$

$$\mu_A \sim 1.5 \text{ cm}^2/(\text{V} \cdot \text{s}) \tag{8.28}$$

$$\tau_o = 8.3 \text{ ns}$$

Note that a gold clad tungsten wire of length ~ 5 m has a resistive time constant of $\tau_w \sim (300 \,\Omega)(100 \text{ pF}) = RC \sim 30$ ns. Thus long wires have slower pulses than what is implied by the intrinsic rise time τ_o.

Fig. 8.11. Photograph of a PWC with 1 mm wire spacing and without wire supports. The connectors for the readout electronics are evident. (Photo – Fermilab.)

8.8 Mechanical considerations

A photograph of a real proportional wire chamber, PWC, not operated as a simple linear array of coaxial tubes but as a planar series of detecting wires is shown in Fig. 8.11. This particular chamber has a 1 mm spacing between the wires and the wires are about 20 cm long. The cathode is a plane of aluminum foil. This chamber has been used to measure the position of particles to an accuracy $\sim \pm 0.5$ mm in prepared beams at Fermilab. A photograph of a large area drift chamber is shown in Fig. 8.12. In this case there is a large spacing, about 2 cm, between the anode detection wires. If we want precise position localization with this 'drift chamber' we need to accurately measure the drift time (Fig. 8.7).

Constructing proportional wire chambers with parallel wires, we need to keep track of the fact that the wires are charged to a typical value of 2×10^{-10} C/cm. The wires all being charged the same, will repel. There will be a deflection unless the wires are maintained under a certain tension, T. The tension has to be greater than the repulsive force between the wires if the PWC is to be

Fig. 8.12.   Photograph of a large area drift chamber. Note the large spacing between anode detection wires. (Photo – Fermilab.)

'stable'. Crudely, for a wire spacing, d, wire length L and a charge per unit length λ the stability condition is

$$T \geq (L\lambda/d)^2 \qquad (8.29)$$

For example, with a 20 μm diameter wire, a PWC built with a 1 mm spacing, d, having a typical charge per unit length, λ, of 2×10^{-10} C/cm has a maximum unsupported wire length, L, of about 13 cm for a wire tension, T, set to the wire yield strength of about 50 g. The natural limit to T for a given wire diameter comes at or below elongation or breaking tension, T_{max}. Clearly there is a competition between high spatial resolution, d small, rapid pulse formation time τ_o, (large λ – small a), and the ability to make a stable large area chamber, (small λ – large d – large a). Typically, the existence of these incompatible requirements is evaded by supporting the wires mechanically in some fashion at a separation between supports which is less than the maximum length allowed for stability. In this way we can build a large area, finely spaced, high speed, proportional wire chamber.

Fig. 8.13.   Schematic of a wire chamber wire array for both an unperturbed and an unstable condition.

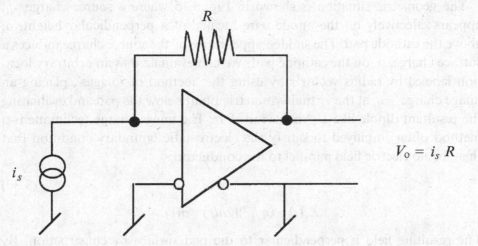

Fig. 8.14.   Schematic of a 'transimpedence amplifier'. The voltage output is proportional to the source current.

A schematic of the spatial pattern for typical wire chamber instability is shown in Fig. 8.13. For tension too small, the wires repel one another, and the uniform plane of wires which is desired is not a stable configuration. The distorted plane is a configuration with reduced energy (charges farther apart) so that unless sufficiently restrained, the system will adopt this configuration. These wires in motion are unstable, and such a chamber cannot be operated reliably.

A typical fast amplifier configuration used to process the signal pulse which forms on the anode is shown in Fig. 8.14. The detector current source signal i_s appears as a voltage output V_o proportional to the source current i_s, $V_o = i_s R$. Some algebra for this and other operational amplifier circuits is given in Appendix I. Front end electronics is relegated to this appendix, and references are given in Chapter 13 while front end noise is discussed in Chapter 9.

8.9 The induced cathode signal

So far we have not discussed the cathode signal. The almost instantaneous appearance of charge near the anode wire capacitively induces charge on the cathode. Therefore, we have a second detectable signal. Note that we are free to cut the cathode into whatever shape we wish, as long as all parts of it are maintained at the same voltage. This freedom of design has made cathode readout popular.

The geometric situation is shown in Fig. 8.15 where a source charge, q_s, appears effectively on the anode wire located at a perpendicular height, a, above the cathode pad. The sudden appearance of the source charge induces a surface charge, σ, on the cathode pad. We can evaluate σ at an arbitrary location labeled by radius vector r by using the 'method of images', placing an image charge $-q_s$ at the virtual symmetric point below the pad and evaluating the resultant dipole-like $1/r^3$ behavior of σ. The image charge technique is a method often employed to satisfy the electrostatic boundary condition that there be no electric field parallel to the conductor.

$$E_{\parallel} = 0 \tag{8.30}$$

$$E_T(r) \sim (q_s/r^2)(2a/r) \sim \sigma(r)$$

The resulting field is perpendicular to the pad surface by construction. By Gauss's law the perpendicular field (CGS) is also the induced surface charge density, $\sigma(r)$. The surface charge density, if sampled on several different cathode electrodes, allows us to infer the centroid location of the instantaneous charge which has appeared on the anode. This technique offers a rather precise second coordinate measurement if pulse amplitudes are well measured. For example, anode wires may have associated cathode strips at right angles. Wire pulses then localize in one coordinate, while strip centroids yield the second orthogonal coordinate.

As an example, we can use the geometry shown in Fig. 8.15 and integrate over a strip which is infinite in the z direction and is of width d in the y direction. We find that the charge induced on the pad, q_p, is simply related to the source charge, q_s, and the angles of the wire subtended by the pad boundaries. In the limit that we come very close to the pad, $a \ll d$, the ratio of q_p to q_s approaches 1/2, as expected by geometry, since half the charge is induced below the wire, half above. In general, the induced charge is approximately proportional to the angle subtended by the pad $\sim (\theta_2 - \theta_1)$.

$$q_p = \int_{-d/2}^{d/2} \int_{-\infty}^{\infty} (2q_s a/r^3) dy \, dz$$

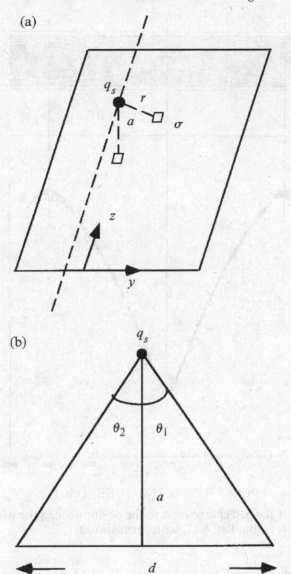

Fig. 8.15. (a) Geometry for deriving the induced surface charge on a pad, σ, for source charge q_s at height a above the pad. (b) Geometry for infinitely long strip electrode of width d.

$$= q_s(\theta_1 - \theta_2)/\pi,$$

$$\pm\, d/2a = \tan(\theta_{1,2})$$

$$(8.31)$$

$$q_p/q_s = \frac{\tan^{-1}(d/2a)}{\pi} \rightarrow \frac{1}{2}$$

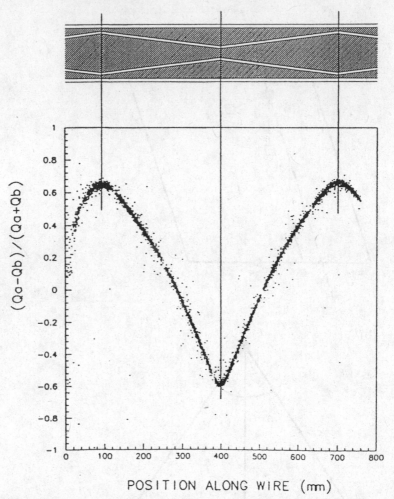

POSITION ALONG WIRE (mm)

Fig. 8.16.   Relationship of the pad charge ratio to the position along the wire for the structure shown in Fig. 8.6. (From Ref. 8.11, with permission.)

An example of an application of 'pads' is to use the induced charge on two different pad electrodes to infer the location of the charge along the wire. We use the drift time to infer the location transverse to the wire. We can therefore achieve full three dimensional readout in a single plane. A concrete example of such a detector is shown in Fig. 8.16 where the charges on the inner pad, Q_A, and outer pad, Q_B, are both measured. The charge ratio is a measure of the position along the wire. Data taken with an external high resolution device yields a precise value for the position along the wire. When this is compared to the charge ratio, as in Fig. 8.16, we see that we can get quite good spatial resolution by the use of induced pad charges. As a rule of thumb the technique is noise limited to a resolution ~ 1% of the geometric feature size of the pads. That limit is evident in the line width visible in Fig. 8.16.

Other uses of the induced charge exploit the complete freedom to configure the cathodes and their geometry. For example, in gas calorimetry we use the cathode signals so as to easily make projective 'towers' which match the outgoing paths of particles from a collision (see Fig. I.1). Because layers of calorimetric readout (see Chapter 11–13) occur at different depths, each tower layer has a different size, which is easily achieved by cutting a different pad on the readout circuit board.

Exercises

1. Use the density of air and assume $A = 14$. Find the mean free path for collisions in air if the cross section is 10 Å^2.
2. At room temperature, what is the electron thermal velocity?
3. Ions are created in the gaps of wire chambers and must ultimately be removed. Using the argon ion mobility of Eq. 8.5 and a field of 1 kV/cm estimate the ion drift velocity and drift time for a 1 cm gap ionization chamber.
4. Assuming an electron thermal velocity of 10 cm/μs and a mean free path of 0.2 μm, find the transverse size of the point ionization after 1 μs drift time using Eq. 8.13.
5. Assume we have a 10^6 V/cm electric field to use in a 5 GeV kaon beam. Find the magnetic field required to drift the beam undeflected in this field.
6. For a mean free path of 0.2 μm, and a thermal velocity of 10 cm/μs, find the mean time between collisions. Evaluate the cyclotron frequency for electrons in a 20 kG field. What is the Lorentz angle?
7. For a 2 mm tube strung with 20 μm wire, operated at 2000 V, what is the field at the surface of the wire?
8. For a 20 μm radius wire, with a field of 200 kV/cm at the wire surface, what is the pulse formation time in argon gas?
9. Explicitly make the image charge construction to show that $E_x = 2qa/[a^2 + z^2 + y^2]^{3/2}$.
10. Integrate E_x over a strip very long in y and from 0 to d in z – similar to the results of Eq. 8.31.

References

[1] *Drift and Diffusion of Electrons in Gases: A Compilation*, A. Peisert and F. Sauli, CERN 84-08 (1984).

[2] *Principles of Operation of Multiwire Proportional and Drift Chambers*, F. Sauli, CERN 77-09 (1977).

[3] V. Radeka, 'Signal, noise and resolution in position-sensitive detectors', *IEEE Trans. Nucl. Sci.* **21** 51 (1974).

[4] *Spark, Streamer, Proportional and Drift Chambers*, P. Rice-Evans, RicheLieu (1974).

[5] G. Charpak and F. Sauli, 'High resolution electronic particle detectors', *Annu. Rev. Nucl. Part. Sci.* **34** 285 (1984).

[6] *Basic Data on Plasma Physics*, S.C. Brown, Wiley (1959).

[7] *Atomic and Molecular Radiation Physics*, L.G. Christophorou, Wiley (1971).

[8] *Electrons in Gases*, J. Townsend, Hutchinson (1947).

[9] *Experimental Techniques in High Energy Physics*, T. Ferbel, Addison-Wesley Publishing Co., Inc. (1987).

[10] *Methods of Experimental Physics*, L. Yuan and C.W. Wu, Academic Press (1963).

[11] S. Abachi, *et al.*, 'The D0 detector', *Nucl. Instrum. and Methods* **A338** 185 (1994).

9

Silicon detectors

All composite things decay, strive diligently
Buddha Gauthama
To see a World in a grain of sand
William Blake

The gas detectors which we discussed in Chapter 8 are intrinsically limited due to diffusion, where the limit on the localization of the drifting ionization is of order 100 μm for drift distances ∼1 cm. For the study of heavy quark decays (see Chapter 13) the characteristic decay length ranges from 100 μm to 400 μm. The length scale is set by τ, the lifetime of the particle in its rest frame. Therefore, $c\tau$ is the decay length. Hence, if we want to resolve the production vertex (Fig. I.1) as separated from the subsequent quark decay vertices, we need a detection element with better spatial resolution than the proportional wire chamber. (See the discussion of 'vertex detectors' in Chapter 13.)

9.1 Impact parameter and secondary vertex

A bubble chamber photograph of the production of a pair of charmed mesons by an incident photon is shown in Fig. 9.1. Note that the laboratory decay lengths are time dilated by a factor of γ and are of order 1000 μm or 1 mm. If we draw the daughter momentum vectors back to the production vertex we can see that the impact parameter, or the transverse distance between the path of the decay particles and the production point are of the same order as the lifetime, $c\tau$. Note that the bubble chamber technique, although it contains an enormous amount of information per event, is much too slow a technology to use in the study of the relatively rare production of heavy quarks. It has, in fact, not been discussed in this text. It can, in principle however, resolve the production and decay vertices as illustrated in Fig. 9.1.

Silicon detectors are now the high rate devices of choice in the study of heavy quarks at both fixed target and colliding beam machines. Let us consider the resolution which is required of such a detector. A particle which is produced and subsequently decays into a secondary track which has an impact parameter b is shown in Fig. 9.2. The distance along the track is increased by a factor

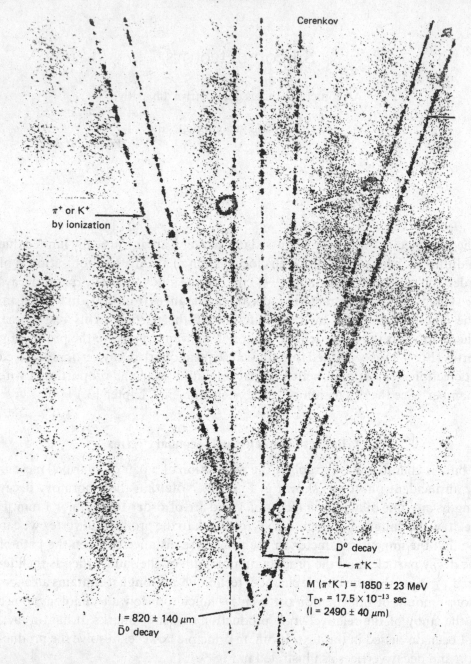

Cerenkov

π^+ or K$^+$
by ionization

D^0 decay
$\quad\longrightarrow \pi^+K^-$

M (π^+K$^-$) = 1850 ± 23 MeV
T$_{D^0}$ = 17.5 × 10^{-13} sec
(l = 2490 ± 40 μm)

l = 820 ± 140 μm
$\bar{D}^0$ decay

Fig. 9.1. Bubble chamber event showing pair production and subsequent decay of charmed D mesons. Note that the decay lengths are ~ 1000 μm = 1mm. (From Ref. 9.9, with permission.)

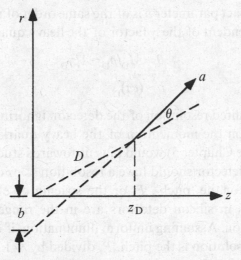

Fig. 9.2. Geometry of a short-lived particle D decaying into a secondary track a with impact parameter b.

of γ over the intrinsic 'lifetime' $c\tau$. The impact parameter is proportional to that distance, D, and the angle, θ, which the decaying track makes with respect to the heavy quark parent.

$$z_D \sim \gamma_D (c\tau)_D \sim D \qquad (9.1)$$

$$b \sim \theta z_D$$

The decay angle is approximately the transverse momentum divided by the longitudinal momentum. (See Appendix A.)

$$\theta \sim p_T^a / p^a \qquad (9.2)$$

$$b \sim \frac{p_T^a}{p^a} (c\tau) \gamma_D$$

For two body decays the transverse momentum impulse is again set by the physics (see Appendix A). The intrinsic physics in this case is fixed by the mass of the parent quark, M_D, decaying into daughters, which are here assumed to be essentially massless. The longitudinal component is set by the sharing of the momentum, p_D, of the heavy quark parent by the two daughters.

$$D \rightarrow a + b$$

$$p_T^a \sim M_D / 2 \qquad (9.3)$$

$$p^a \sim p_D / 2$$

Therefore, the impact parameter b is of the same order of magnitude as the lifetime, $(c\tau_D)$, independent of the γ factor of the heavy quark.

$$\theta \sim M_D/p_D \sim 1/\gamma_D \tag{9.4}$$

$$b \sim (c\tau)_D$$

Therefore, the required resolution of the detector, ignoring multiple scattering, does not depend on the momentum of the heavy quark. Multiple scattering considerations (see Chapter 5) would bias us towards studying higher momentum quarks. The detector should have a resolution $\ll c\tau$ or ~ 10–50 μm.

Typical values for the 'pitch', P, or the spacing of electrode strips which collect the signals in silicon detectors are in the range 20–100 μm, which satisfies this criterion. Assuming uniform illumination, if only which strip is hit is recorded, the resolution is the pitch, P, divided by $\sqrt{12}$. (See Appendix J.)

$$\sigma^2 \equiv \langle (y - \langle y \rangle)^2 \rangle, \quad \langle y \rangle = 0$$

$$= \int y^2 dy / \int dy$$

$$= \int_{-P/2}^{P/2} y^2 dy / \int dy \tag{9.5}$$

$$= P^2/12$$

If the charge on the strips is recorded, we may do better (see Chapter 8 on cathode strip charge measurements). For pitch $P = 50\,\mu$m, the resolution, $P/\sqrt{12}$, is 15 μm. A basic limitation is again set by diffusion. With a drift voltage of 50 V and a drift distance of 300 μm (these values will be justified later in this chapter), we find $\sigma_x \sim 10\,\mu$m for the diffusion error (see Chapter 8).

As an aside, not all measurements are resolution limited. Consider the measurement of heavy quark decay with lifetime τ, $N(t) \sim e^{-t/\tau}$. If the decay time t is measured with an error σ, the probability to observe time (t') is $N(t') \sim \int dt N(t) P(t,t')$ where $P(t,t') \sim \exp(-(t-t')^2/2\sigma^2)$, (see Appendix J). The shape of $N(t')$ has the same time structure with the same lifetime, $N(t') \sim e^{-t'/\tau}$, but the overall rate at the same time value t' as t is increased by a factor $\exp\left(\frac{1}{2}(\sigma/\tau)^2\right)$. Therefore, Gaussian measurement errors of the decay time do not change the shape of the decay distribution, at least for $t \gg \sigma$. The proof of this is left to the reader.

$$N(t) = e^{-t/\tau}, \quad N(t') \sim e^{-t'/\tau} e^{(\sigma/\tau)^2/2} \tag{9.6}$$

9.2 Band gap, intrinsic semiconductors and ionization

The band gap, E_g, is defined to be the minimum energy needed to excite an electron into the conduction band. In silicon it has a value of 1.12 eV. In a solid the sharp atomic energy levels are smeared into 'bands' by the mutual interactions among the electrons (see Chapter 2). In a semiconductor such as silicon the 'valence band' is full, and it is separated from the allowed states in the empty 'conduction band' by a 'band gap' E_g. This gap can be thought of as a residue of the sharp energy difference between atomic states.

The intrinsic conductivity, σ_i, of pure silicon is due to the thermal excitation of charge carriers across the gap, E_g, into the conduction band. The intrinsic carrier density, $n_i = p_i$, at room temperature is about $10^{11}/cm^3$ (see Ref. 9.1). The n_i refers to negative electrons in the conduction band while p_i refers to the mobile 'holes' remaining in the valence band. Note that this carrier density is rather small in that the full electron density is $n_{Si} = \dfrac{N_o \rho}{A} \sim 5.01 \times 10^{22}/cm^3$, $n_i/n_{Si} \sim 10^{-12}$. The small value of the ratio is due to the fact that $E_g/kT \sim 40$, so the Boltzmann factor, $e^{-E_g/kT}$ is very small, $\sim 4 \times 10^{-18}$. Therefore, silicon has relatively few mobile charge carriers and is a poor conductor.

The intrinsic resistivity, ρ_i, is the reciprocal of the intrinsic conductivity, σ_i. A high conductivity metal such as copper has a resistivity of $\sim 10^{-6} \ \Omega \cdot cm$ which is about 10^{11} times less than pure silicon, reflecting n_i/n_{Si}. In a metal there are allowed unfilled states in the valence band so that the valence electrons can participate in conduction without paying a big penalty due to a small Boltzmann factor.

The conductivity of semiconductor material is controlled by 'doping' the crystals with small levels of impurities. Typical doping levels are chosen, e.g. parts per million, such that the doped conductivity of the majority carriers (set by impurity levels) is >100 times the intrinsic conductivity while that for the minority carriers is consequently <100 times less since the product $n_i p_i = np$ remains constant under doping. For example, $n \sim 10^{13}/cm^3$ and $p \sim 10^9/cm^3$ for a lightly doped n-type semiconductor. Thus, doping with parts per million impurities will dominate the intrinsic charge carriers while still keeping the resistivity high with respect to good conductors. The unit of charge, the electron charge, is taken to be q in this chapter to avoid confusion with the exponential function. The carrier mobility is μ (see Chapter 8). Electrons have a mobility of $\sim 1400 \ cm^2/(V \cdot s)$.

$$n_i = p_i = 10^{11}/cm^3 \big|_{T=300K}$$

$$\rho_i = 1/q\mu n_i = 200 \ k\Omega \ cm \tag{9.7}$$

$$= 1/\sigma_i$$

When we consider the detection of ionization energy, the energy needed to create an electron–hole pair in silicon is 3.6 eV. That is greater than the band gap energy, E_g, because of the existence of competing dissipative processes such as phonon, or lattice vibration, excitation. The silicon detector acts as a 'unity gain' ionization detector operating similarly to those discussed in Chapter 8. In a detector of active depth 300 μm, the energy loss in silicon due to ionization is about 116 keV which liberates a source charge, q_s, of 5.1 fC.

$$E_{pair} \sim 3.6 \text{ eV}$$

$$d \sim 300 \text{ μm}$$

$$q_s \sim q(\Delta E_{ion}/E_{pair})$$

$$q_s \sim 5 \text{ fC} \sim 32000q$$

(9.8)

Silicon detectors are 'unity gain' devices. Compared to gas filled detectors the yield per unit length of ionization is increased by the density ratio and the ionization energy ratio, I_o/E_g for a total factor of ~ 10000. This factor approximately offsets the 'gas gain', G, available in PWC devices. The drift time and diffusion error both depend on the drift distance, d, which means that we expect silicon devices to be fast and capable of precise position measurements.

9.3 The silicon diode fields

The charge quoted in Eq. 9.8 is a rather small signal charge, and consequently it is fairly difficult to detect. The idea is to set up a situation where the other currents are small, so that the ionization current pulse can be detected. Consider a p-n junction diode consisting of silicon doped with p-type impurities on one side and n-type impurities on the other. The dielectric constant for silicon is $\varepsilon \sim 1$ pF/cm (recall $\varepsilon_o = 0.089$ pF/cm and $\varepsilon \sim 11.9\varepsilon_o$, Table 1.1). The current, I, voltage, V, relationship for a p-n junction is given below.

$$I(V) = I_o[e^{qV/kT} - 1]$$

$$R_F \equiv (dI/dV)^{-1} = \frac{kT}{qI} = 25 \text{ Ω}\Big|_{300K, 1 \text{ mA}}$$

(9.9)

Here I_o represents the thermal 'reverse' current due to the thermal excitation of the minority carriers in reverse biased operation, $V < 0$. The relevant scales are the applied energy qV with respect to the thermal energy kT. If $V < 0$ and $|qV| \gg kT$ then $I(V) \rightarrow -I_o$ and we have a small reverse current. If $V > 0$, then $I(V) \sim I_o e^{qV/kT}$ and we have large forward currents. The diode acts as a 'switch' with 'position-on/off' controlled by V.

The forward biased resistance, R_F, is estimated using a differential form of

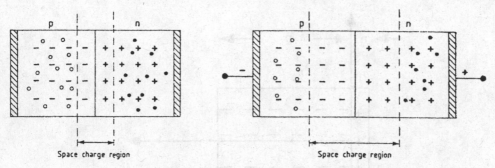

Fig. 9.3. Schematic of a reverse biased p–n semiconductor diode showing the depletion region. (From Ref. 8.5, with permission.)

Ohm's law which gives us a 25 Ω resistance for a forward biased diode at room
temperature with a 1 mA standing current. Fluctuations in the forward current
(shot noise is discussed later in this chapter) would swamp the ionization signal.
Therefore, we need to operate the diode as a reverse biased p-n junction. As we
will see, the reverse current at room temperature is ~10 nA.

A schematic of the location of the charge carriers in reverse mode is shown
in Fig. 9.3 where the n-type and p-type are shown with an applied voltage $V<0$
leading to a 'depletion' region centered on the junction. The applied voltage
sweeps the mobile charge carriers out of the junction region leaving a uniform
ion number density n. As the applied voltage is increased, the mobile charge
carriers are swept further out of the junction region leaving the fixed ion charge
which is positive for the n-type region and negative for the p-type region.

For an applied reverse potential, V_D, the geometry at 'full depletion', where
all the mobile charge carriers are just swept out, is shown in Fig. 9.4a. We have
assumed a very thin n-side of the junction diode. The number density of
doping impurities is n_A on the p-side and n_D on the n-side. Charge conservation requires that the n-type is highly doped as it is thin.

$$n_A A = n_D D \tag{9.10}$$

This static situation can be treated using Maxwell's equations. The electric field
vanishes at $x = A$ and $x = -D$.

$$\nabla \cdot \mathbf{D} = \rho \tag{9.11}$$

$$\nabla \cdot \mathbf{E} = [qn]/\underline{C}, \quad \underline{C} \text{ or } \varepsilon \text{ (CGS)}$$

In Eq. 9.11 ρ is the charge density and not the resistivity and n is the ion number
density. The charge in the depletion region is assumed to be static and uniformly distributed. Thus, the electric field is linear in x.

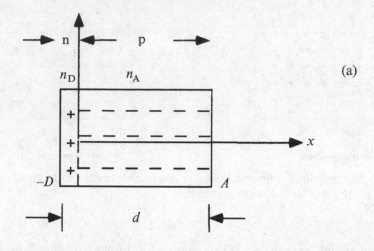

(a)

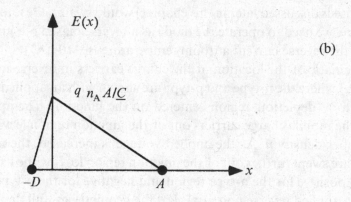

(b)

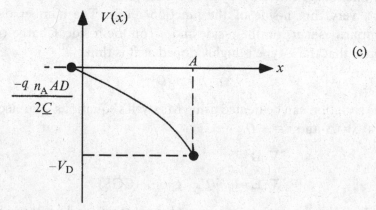

(c)

Fig. 9.4. (a) Geometry of a p-n junction. The static charge number density is n_D and n_A. The full depletion region is $d = A + D \sim A$. (b) Electric field for a p–n junction. (c) Electric potential for a p–n junction.

$$E(x>0)=-qn_A(x-A)/\underline{C}$$

$$E(x<0)=qn_D(x+D)/\underline{C} \tag{9.12}$$

$$=q\left[\frac{n_A A}{D}\right](x+D)/\underline{C}$$

The slope discontinuity in E is due to the discontinuity of charge at the junction seen in Fig. 9.4a. We integrate a second time to get the potential, applying the boundary conditions $V(x=A)=-V_D$, $V(x=-D)=0$.

$$\mathbf{E}=-\boldsymbol{\nabla}V$$

$$V(x<0)=-\frac{qn_A A(x+D)^2}{2D\underline{C}} \tag{9.13}$$

$$(V(x>0)+V_D)=\frac{qn_A(x-A)^2}{2\underline{C}}$$

Note that we have ignored the 'built in potential' due to thermal excitation of the charge carriers. It is $V_{bi}\sim kT/q=25$ mV at $300\,°C$ (Eq. 9.9) times a factor of order one, $\ln(n/n_i)$. For heavy doping, 10^6 times the intrinsic level, we have a ~ 0.36 V drop due to thermal excitation. Compared to a depletion voltage, ~ 50 V, this 'diode drop' is negligible.

A plot of $V(x)$ is provided in Fig. 9.4c. Note that V and its slope E are continuous across the $x=0$ junction. Let us make a simplifying assumption. Continuity of $V(x<0)$ and $V(x>0)$ at $x=0$ requires that $V_D=qn_A A(A+D)/2\underline{C}$. If the doping levels are much different, $n_D\gg n_A$ then, $A\gg D$. In that case, $d\equiv A+D\rightarrow A$ and

$$V_D\sim\frac{qn_A d^2}{2\underline{C}} \tag{9.14}$$

$$d\sim A$$

The n layer, in this approximation, is very thin. Defining $n_A=p$ for simplicity, we find that $E(x)\sim-qp(x-d)/\underline{C}$, $V(x)\sim-V_D+qp(x-d)^2/2\underline{C}$. The relationship of the applied voltage and the depletion thickness, is:

$$V_D=[qp]d^2/2\underline{C}, \quad \left(\frac{2\pi}{\varepsilon}[qp]d^2, \text{MKS}\right) \tag{9.15}$$

For example, a 300 μm depletion region, d, with a doped resistivity of $\rho=6$ kΩcm has a depletion voltage, V_D, of 51 V and a capacity per unit area, $\underline{C}/d$, of 35 pF/cm^2. Note that in comparison with the intrinsic values there are ~ 30 times more majority charge carriers in this example, so that the doping is fairly light. A detector of area 25 mm^2 has a source capacity, C_s, of ~ 8 pF.

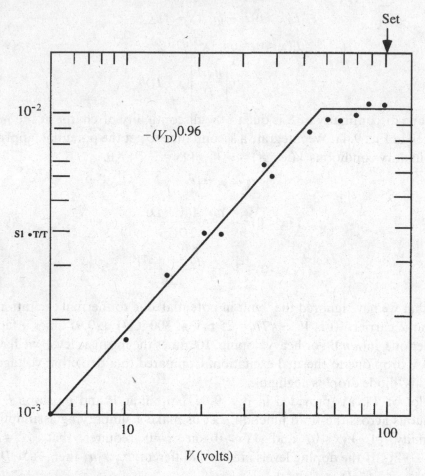

Fig. 9.5. Data taken with a 50 μm pitch silicon microstrip plane on the coincidence ratio of strip S1 with external telescope T as a function of applied voltage V.

9.4 The silicon diode: signal formation at depletion

A plot of data taken with a silicon strip detector, Fig. 9.5, shows the coincidence rate of a silicon strip, S1, with an exterior scintillator array of multiple counters or 'telescope', T, triggered on minimum ionizing particles as a function of the applied voltage. It is clear that as the voltage increases the coincidence rate rises because the depleted region is expanding rapidly and more charge is being collected, $V_D \sim d^2$. At a fixed detection threshold, V_T, more charge means higher efficiency. The amount of charge rises until a voltage is reached where the detector is 'fully depleted'. In that case the efficiency comes to a 'plateau' without further increase. This occurs at ~ 50 V applied voltage as estimated from Eq. 9.15.

In Fig. 9.6 is shown a photograph of a real detector used in an experiment.

Fig. 9.6.   Photograph of a 50 μm strip spacing silicon detector mounted as a planar detector. Note the 'fanout' of silicon strips to high density connectors which accommodate the electronics. (Photo – Fermilab.)

Fig. 9.7.   Photograph of a silicon detector mounted as an azimuthal coordinate detector in a cylindrical geometry. (Photo – Fermilab.)

A 50 μm pitch silicon detector is shown mounted as a planar detector for a 'fixed target' geometry. Note that transverse diffusion and delta ray emission argue against a smaller 'pitch' size. In Fig. 9.7 the silicon detectors are mounted as azimuthal coordinate detectors in a cylindrical geometry more appropriate to the solenoidal field used by colliding beam detectors. (See Chapter 13.)

What about the time for pulse formation? First we find the electric field as a function of depth just at depletion, $E(x) = -\left(\dfrac{2V_D}{d^2}\right)(x-d)$. For $V_D = 50\,\text{V}$ and $d = 300$ μm the maximum field is $E(0) = 3\,\text{kV/cm}$.

We use the methods developed in Chapter 8 to find the charge position as a function of time. For the special case of point ionization created at $x=0$ at $t=0$ (see Fig. 9.4b) the charge will reach $x=d$ at $t \to \infty$.

$$dx = \mu E dt = -\mu\left(\frac{2V_D}{d^2}\right)(x-d)dt$$

$$x(t) = d[1 - e^{-t/\tau_D}], \quad \tau_D = \underline{C}/\mu[qp] \tag{9.16}$$

$$x(0) = 0, \quad x(\infty) = d$$

See Appendix H for more general results and more details. Note that $\tau_D = C\rho$ so that high speed operation favors low resistivity, highly doped silicon while a desire for low leakage currents favors high resistivity.

The current is found using the formalism developed in Chapter 8 and Eq. 9.16.

$$I(t) = dQ/dt = \mu q_s E^2/V_D$$

$$= \frac{4\mu q_s V_D}{d^2} e^{-2t/\tau_D} = 2q_s/\tau_D [e^{-2t/\tau_D}]$$

(9.17)

The characteristic charge collection time τ_D can be thought of as the time for the charge to go a distance d in a field $E(0)$ with charge carriers having a mobility μ.

$$I(0) = 2q_s/\tau_D$$

$$I(\infty) = 0$$

$$\tau_D = [d^2/2V_D\mu]$$

$$= d/[\mu E(0)]$$

(9.18)

With an electron mobility, μ_e, of 1400 cm^2/(V/s), the drift velocity is of order 4.2×10^6 cm/s or $= 42$ μm/ns. This velocity should be compared to a gas where the electron drift velocity (see Fig. 8.2) is comparable, i.e. a few cm/μs. However, the drift distances are much smaller so that the charge collection time is shorter for silicon devices. For a drift distance of 300 μm we get $\tau_D \sim 7$ ns. The holes have a mobility about three times less and therefore the hole signal is about three times slower.

Note that τ_D was derived just at depletion. Clearly if we can raise the applied voltage it will drive the electric field into the diode and reduce the collection time. (See Appendix H.) A limitation to this method of speeding up the detector comes because there exists a reverse breakdown voltage. Normally, we do not operate far above depletion $|V| \leq 100$ V in order to avoid breakdown. However, new detectors operating in high radiation environments have been developed with a maximum ~ 500 V reverse voltage operating point.

A plot of the coincidence rate between a silicon strip S1 and a scintillation counter 'telescope', T, as a function of the relative delay between them is shown in Fig. 9.8. The basic falloff time is ~ 12 ns. From Chapter 2, a coincidence rate will persist on a time scale equal to the intrinsic 'time jitter' of the device. One component of that time jitter is the resolving time of the silicon detector. Clearly, the silicon strip is experimentally confirmed to be a high speed detector.

Recalling that the source charge (q_s) is 5 fC, we find a peak electron (hole) current of 710 nA (240 nA).

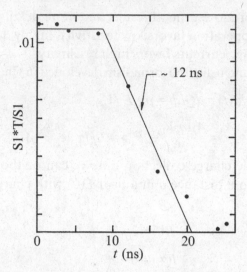

Fig. 9.8.   Plot of coincidence rate between a silicon strip S1 and an external scintillation counter T array as a function of the time delay inserted in the coincidence logic.

$$q_s = 5 \text{ fC}$$

$$\tau_D \sim 7 \text{ ns (20 ns)} \qquad (9.19)$$

$$I(0) \sim q_s/\tau_D = 710 \text{ nA (240 nA)}$$

Having established the size of the current signal, let us turn to issues of noise.

9.5 Noise sources – thermal and shot noise

Typical reverse currents for junctions as a function of temperature are shown
in Fig. 9.9. Note the exponential relationship between the operating temperature and the reverse leakage current, I_o. Since this current is due to the thermal
excitation of minority carriers, such behavior is not unexpected. The scale for
reverse currents at room temperature for the lowest leakage junctions is 2–20
nA. Therefore the signal to noise ratio in such detectors is adequate if high
quality high resistivity detectors are used (see Eq. 9.7, high resistivity means
small intrinsic current). Note that 'bulk damage' of the crystal lattice, say from
irradiation due to detector operation at a high luminosity accelerator, induces
defects which act as carriers, thus increasing the reverse current. For example,
a neutron 'fluence' of $\sim 10^{14}$ n/cm^2 may cause discernible degradation in performance. Looking at Fig. 9.9, cooling the silicon devices is an option to restore
low noise operation, and indeed, solid state detectors are routinely run at low
temperature in many applications.

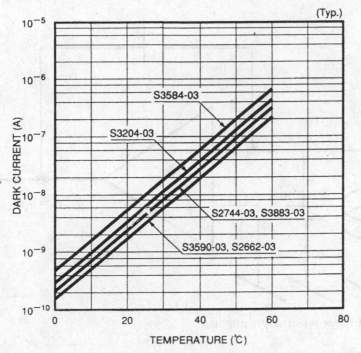

Fig. 9.9.   Reverse leakage current as a function of temperature. (From Ref. 9.11, with permission.)

Now let us consider briefly methods of detection for these small pulses. It is the smallness of the signal compared to a PWC which dictates a less cavalier treatment of front end electronics in this case. A schematic of an operational amplifier realization of a charge integrator is shown in Fig. 9.10 (see Appendix I). The ionization detectors we have discussed are all a capacitive source of charge; the signal can be thought of as a current source. Using $Q = CV$, $I = CdV/dt$, we find that the output voltage V_o is proportional to the source charge q_s. If the amplifier gain G is $\gg C_s/C$ then the device acts as an ideal operational amplifier $(G \to \infty)$, $V_o = q_s/C$. A typical output pulse of such a device is seen in Fig. 8.4. A more detailed schematic showing the noise sources, e_n, for a signal with source charge q_s, driven by source capacity C_s and source resistance R_s into a generic amplifier with gain G and output filter with 'transfer function' $f(\omega)$ is shown in Fig. 9.11.

Let us now look in more detail at the behavior of the noise sources. There are dissipative elements such as resistors which are a source of intrinsic 'thermal noise'. The size of the thermal currents is set by the thermal energy, kT, in the resistor. We assume that the frequency spectrum of this noise is 'white'; it is uniform over all frequencies. That means, physically, that the noise

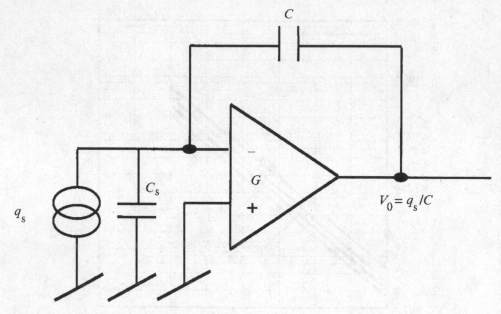

Fig. 9.10.   Charge sensitive preamplifier.

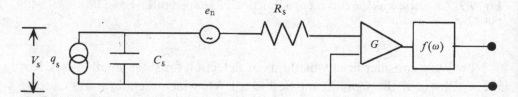

Fig. 9.11.   Amplifier and bandwidth limiting filter, $f(\omega)$ with source capacitance, C_s, source resistance, R_s, source charge, q_s, and input noise voltage, e_n.

occurs impulsively, or as a delta function, in time. Recall that the Fourier transform of a plane wave with fixed momentum, $\delta(k)$, is e^{ikx} whose intensity is spatially uniform. The relationship to time and frequency is similar.

The noise power in a resistor is the thermal energy, kT, spread uniformly over all frequencies, ω. This thermal noise power leads to a thermal resistor current I_T, since power $= \underline{P} = VI = I^2 R$.

$$dI_T^2 = \frac{2kT}{R}\left(\frac{d\omega}{\pi}\right)$$

(9.20)

Additionally there is 'shot noise' which occurs because standing currents, I, are in fact due to the motion of discrete charge carriers since charge is quantized in multiples of q. Fluctuations in the numbers of these charge carriers lead to

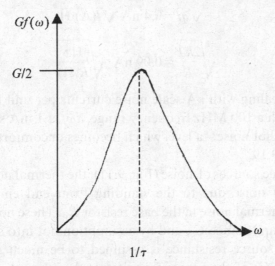

Fig. 9.12.   Plot of the transfer function $Gf(\omega)$ which $\rightarrow 0$ as $\omega \rightarrow 0$ and as $\omega \rightarrow \infty$ and peaks at $\omega = 1/\tau$.

the 'shot noise' of the standing current, $dI_s \sim q\, d\omega$. If charge were not quantized, $q \rightarrow 0$ and $dI_s \rightarrow 0$.

$$dI_s^2 = qI\left(\frac{d\omega}{\pi}\right) \tag{9.21}$$

We need a filter to limit the bandwidth. Otherwise we will have an infinite amount of noise since all frequencies are hypothesized to be equally probable. We assume a simple shaping network which cuts off linearly at low frequencies and falls as $1/\omega$ at high frequencies with a single 'shaping time', τ. A schematic plot of $Gf(\omega)$ as a function of ω is given in Fig. 9.12.

$$f(\omega) = G\left[\frac{\omega\tau}{1 + (\omega\tau)^2}\right] \tag{9.22}$$

We define the transconductance, g_m, to be the inverse of the base resistance of a hypothetical front end transistor in the amplifier. We assume a front end transistor with a base resistance characteristic of a forward biased diode. The relationship between standing emitter current I_E and base resistance R_B is one we have already seen in Eq. 9.9.

$$R_B = kT/qI_E = 1/g_m \tag{9.23}$$

To set the scale for these quantities, we evaluate the coefficients numerically (at room temperature)

$$\sqrt{qI} = 0.4 \text{ nA } \sqrt{I(\text{A}) \cdot \text{Hz}},$$

$$\sqrt{\frac{2kT}{R}} = 0.09 \text{ nA } \sqrt{\frac{\text{Hz}}{R(\Omega)}} \qquad (9.24)$$

Clearly we are dealing with nA scale noise currents per unit frequency range. For example, with a 100 MHz frequency range, $d\omega$, at 1 mA standing current, we have 126 nA shot noise – a level which becomes uncomfortably close to the signal level, Eq. 9.19.

We assume three sources of noise (Fig. 9.11): the thermal noise in the source resistor R_S, shot noise due to the standing front end emitter current I_E, (amplifier) and thermal noise in the base resistor R_B. These noise currents flow into the source capacitance C_S, and by assumption not into source resistance R_S, because the source resistance is assumed to be much greater than the impedance, $\sim 1/\omega\, C_S$ of the source capacitor at the relevant peaking frequencies, $\omega \sim 1/\tau$. Note that $I = CdV/dt = Ci\omega V$ means that the capacitor impedance Z_C is equal to $= V/I = 1/i\omega C$.

These assumptions lead to an expression for the total square summed input noise voltage $\bar{V}^2$. Since the noise sources are statistically independent we add them in quadrature (see Appendix J). The first two terms go as $\bar{V}^2 \sim 1/\omega^2$ while the last goes as $\bar{V}^2 \sim$ const. The source capacity is important at low frequencies.

$$d\bar{V}^2 = d\bar{I}_T^2 Z_{Cs}^2 + d\bar{I}_S^2 Z_{Cs}^2 + d\bar{I}_T^2 R_B^2$$

$$= \left[\frac{2kT}{R_s(\omega C_s)^2} + \frac{qI_B}{(\omega C_s)^2} + \frac{2kT}{g_m} \right] \frac{d\omega}{\pi} \qquad (9.25)$$

9.6 Filtering and the 'equivalent noise charge'

The noise voltages are referred to the amplifier input. They go through the amplifier and through the output low and high pass filter network, $f(\omega)$, to produce a mean square voltage at the output terminals, $\langle V^2 \rangle$.

$$\langle V^2 \rangle = \int_0^\infty |f(\omega)|^2 \, (d\bar{V}^2/d\omega) d\omega$$

$$= G^2 \left[\left(\frac{kT}{2R_s} + \frac{qI_B}{4} \right) \tau/C_s^2 + \frac{kT}{2g_m\tau} \right] \qquad (9.26)$$

Note that without the shaping network this integral would diverge. The noise going into C_S diverges at low frequencies because the C_S impedance diverges. In comparison, the base thermal noise needs to be cut off at high frequencies. Thus $\langle V^2 \rangle$ has terms that go as τ and as $1/\tau$.

For the signal charge itself, q_s, there is an optimized output from the filter if the frequency distribution of the source is well matched to the shaping time, $\omega \sim 1/\tau$, of the filter.

$$\text{input } q_s \to (q_s G/C_s)/e = V, \text{ at } \omega \sim 1/t \text{ at the output} \tag{9.27}$$

In Eq. 9.27 and below e is the base of the natural log not the electronic charge which we have defined to be q in this chapter in order to avoid confusion.

The behavior of the signal allows us to define an equivalent noise charge, ENC, referred to the input, since q_s appears on C_s.

$$(\text{ENC} \cdot G/C_s e)^2 \equiv \langle V^2 \rangle \tag{9.28}$$

The ENC is the noise charge that occurs due to the noise sources referred to the input with respect to the output signal that arises for optimal shaping, $V \sim (q_s G/C_s e) \sim G V_s/e$. It is convenient to further define 'parallel' and 'series' equivalent noise charges. Those noise charges are proportional to the square root of the shaping time τ (parallel) and $1/\tau$ (series).

$$\text{ENCP} = (C_s e/G)\sqrt{\langle V^2 \rangle_P} = e\sqrt{\tau\left(\frac{kT}{2R_s} + qI_B/4\right)} \tag{9.29}$$

$$\text{ENCS} = (C_s e/G)\sqrt{\langle V^2 \rangle_S} \equiv eC_s\sqrt{kT/2g_m\tau}$$

Clearly, for low noise operation we want low temperature, large source resistance (ideal current source) and small standing base current in order to minimize the parallel noise. Similarly, we try to obtain small source capacity and small base resistance to reduce series noise. In addition, high speed operation argues for short shaping times, τ, but a compromise with enhanced series noise may be required, $\text{ENCS} \sim 1/\sqrt{\tau}$. Short shaping time implies a large bandwidth, $\Delta\omega$ which inevitably puts more noise into the system. We should then, if possible, operate at reduced temperature because thermal noise in the resistors is reduced.

A plot of the noise as a function of source capacitance for typical preamplifier parameters is shown in Fig. 9.13. You can see the existence of a 'parallel' part of the equivalent noise which is independent of source capacitance and a 'series' part which becomes important at large source capacitance and which increases linearly as C_s. Clearly, we try to minimize the capacitance of the source because it causes noise. For example, long silicon strips are economically desirable, but we may incur too high a penalty due to increased source capacitance if high speed (small τ) operation is planned.

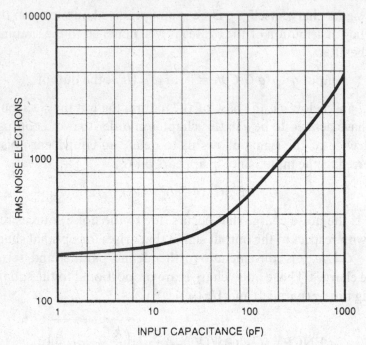

INPUT CAPACITANCE (pF)

Fig. 9.13.   Noise in RMS electrons referred to the input as a function of source capacitance for typical preamplifier parameters. The linear dependence for series noise dominance is evident for > 100 pF capacity. (From Ref. 9.11, with permission.)

9.7 Front end transistor noise

So far we have neglected the transistor as an active element and as a noise source. For front end transistors, the equivalent noise voltage referred to the input, e_n (see Fig. 9.11), is often quoted in the manufacturer's specification. For a typical transistor, $e_n \sim 1 \text{ nV}/\sqrt{\text{Hz}}$, and a series noise charge can be defined.

$$\text{ENCS} \sim e_n C_s / \sqrt{\tau} \tag{9.30}$$

As an example, a source capacitance of 25 pF using a 25 ns shaping time yields a front end transistor noise of 1000 electrons, to be compared to a silicon signal of 31 000 electrons.

The plot of the ENC in rms electrons for various types of front end transistors is shown as a function of shaping time in Fig. 9.14 for a 10 pF source capacitance C_s. For short shaping times the expected $1/\sqrt{\tau}$ behavior is observed. At the point, $\tau \sim 10$ ns, we find ENC ~ 800 electrons or $e_n \sim 1.3$ nV/$\sqrt{\text{Hz}}$ for bipolar transistors. For high speed applications bipolar transistors are favored (GaAs if money is no object), while for slower operation junction FETs are the best choice. Clearly, the devil is in the details.

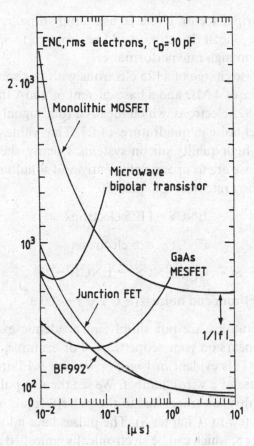

Fig. 9.14. Equivalent noise charge in rms electrons as a function of shaping time τ for a 10 pF source capacitance and a variety of front end transistors. (From Ref. 9.12, with permission.)

Can transistors improve dramatically? Well, it is often remarked upon that the speed of computer chips doubles every year and this is quoted as a 'rule'. However, all rules only work in limited regions of the parameters. The speed of a transistor is determined by the 'feature size'. That size is currently of the order of 1 μm or 10000 atoms. If the trend toward shrinking feature size were to continue for 7 years, the size would go to 0.01 μm or 100 atoms. At that size quantum effects are crucial and the extrapolation collapses.

9.8 Total noise charge

As a numerical example, assume a source charge of 5.1 fC which is appropriate to a silicon detector. Use a transconductance, g_m, of 25 Ω^{-1}. Pick a source capacitance of 30 pF which is not unreasonable for the leads and connections of a typical silicon strip detector. Use a 10 ns shaping time, τ, which, given the intrinsic speed of the silicon detector, would be a match to τ_D and would not compromise the silicon high rate performance.

We find a series noise charge of 1125 electrons with this source capacitance. For a source resistance of 1 MΩ and a base current of 1 mA the parallel equivalent noise charge is 825 electrons, which means a total signal to noise, folding the series and parallel noise in quadrature, of 23. That value of S/N is typical of those achieved in high quality silicon systems. Clearly, the S/N value must be made to be $\gg 1$ if we are to operate efficiently and simultaneously run with low accidental counting rates (see Chapter 2).

$$\text{ENCS} \sim 1125 \text{ electrons}$$

$$\text{ENCP} \sim 825 \text{ electrons}$$

$$\text{S/N} = q_s / \sqrt{\text{ENCS}^2 + \text{ENCP}^2} \sim 23$$

$$\text{Front end noise} = (e_n C_s) / \sqrt{\tau} \sim 800 \text{ e}$$

(9.31)

These noise considerations are not simply an academic exercise. They are symbols for what appears on your scope trace. For example, the noise due to source capacitance, C_s, is evident in Fig. 8.4. In Fig. 9.15 are shown oscilloscope traces in the case of a wire chamber. We see the raw pulses which have a short rise time (5 ns/division trace) due to the pulse formation time, τ_o, which we evaluated to be a few ns (Chapter 8). The pulses have a long $1/t$ tail, as we mentioned in Chapter 8, which can be electronically cancelled as shown in Figs. 9.15b and 9.15c. We see that pulse shaping does indeed achieve a shorter pulse. However, the 'base line', which is the signal to noise is becoming worse. The equivalent noise charge basically tells us what the level of 'hash' is on the base line of our scope trace. It is that level of 'hash' which defines the possible lower levels for pulse discrimination, V_T (Chapter 2).

In Fig. 9.16 there is a photograph of some Applications Specific Integrated Circuit chips, or 'ASICs', which are mounted directly on a detector. We can see that the level of technology which is used in such applications is rather high. The electronics is mounted very close to the detector in order to minimize electromagnetic noise pickup as an additional noise source and also in order to minimize the source (leads) capacity. We are constantly attempting to hold down the sources of noise. These tight requirements are an example of the fact that detector design is something of an artform.

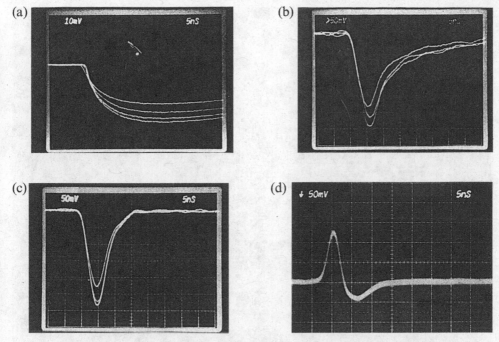

Fig. 9.15.   Oscilloscope traces for a wire chamber: (a) raw pulses, (b) tail cancelled, (c) pole/zero cancelled, (d) impulse response. Note that the width of the baseline is a direct measure of the ENC for the system. (From Ref. 9.5, with permission.)

9.9 Hybrid silicon devices

Recently 'hybrid photodetectors' have become commercially available. These devices combine aspects of photomultipliers and silicon devices which open up new applications. An example is the hybrid photodiode which has a photo-cathode in a vacuum vessel along with a silicon diode. The photocathode, held at ~ 10 kV, emits photoelectrons which are accelerated across a small gap, ~ 1 mm, in a short, < 1 ns, transit time and strike the diode. The released charge, $G \sim 10\,000/3.6 = 2778$ electrons per photoelectron, is collected rapidly by the diode. Thus the device acts like a compact, fast photomultiplier tube with gain ~ 2800. In addition, the device can work in magnetic fields up to 5 T if the **E** and **B** axes are aligned, since then the path is a helix with axis along the **E** direction and with small radius of curvature (see Appendix G). Finally, the silicon diode can be cut into many independent devices, allowing for operation with multiple channels per device making for low cost per channel. These devices are to be used in the next generation of high energy physics experiments.

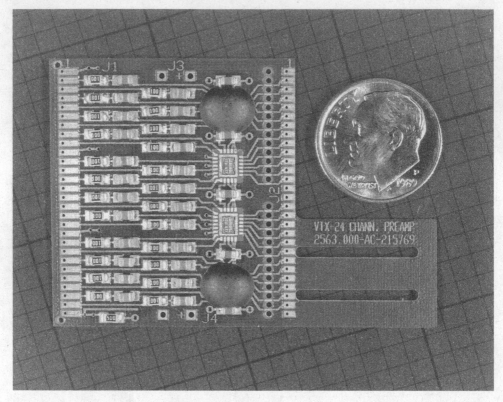

Fig. 9.16.   Photograph of integrated electronics for detectors. (Photo – Fermilab.)

Exercises

1. Explicitly work out the RMS for a uniformly illuminated strip of pitch P. What is the rms error using strips 25 μm wide?

2. Show by performing the integration that an exponential shape smeared by a Gaussian retains the shape but with a factor $\exp[(\sigma/\tau)^2/2]$. Draw a picture of what the symbols mean in order to justify this factor.

3. Use the V, I relationship for a diode, Eq. 9.9, to show that $dI/dV \sim I(q/kT)$ for large forward bias voltages. Evaluate the effective forward biased resistance at a standing current of 10 mA.

4. Show that the voltage just at depletion (MKS) can be expressed as $V_D = 2\pi(1/\mu\rho\varepsilon)d^2$. For a diode with 300 μm depth, and a resistivity $\rho = 10$ kΩ cm, find the voltage at depletion.

5. For a diode driven well over depletion, a field 10 kV/cm can be attained. Find the drift velocity and signal collection time for the holes under these conditions (assuming a uniform field) if the drift distance is 300 μm.

6. Consider the charge sensitive amplifier shown in Fig. 9.10. Suppose the

source charge is 5 fC. For a source capacitance of 10 pF and a feedback capacitor of 1 pF what gain, G, is needed at all frequencies $<1/\tau \sim 100$ MHz so that source capacitance does not affect the feedback. What is the output voltage caused by the signal charge?

7. For a bandwidth of 100 MHz $= d\omega/\pi$ evaluate the thermal noise current caused by a 1 kΩ resistor. Do the same for the shot noise current due to a 10 μA standing current.

8. For the case of shot noise, use Eq. 9.22 and Eq. 9.25 to explicitly perform the integration indicated by the final result, Eq. 9.26. Look at the limit of large shaping time and confirm that the noise contribution diverges at low frequencies without the shaping filter.

9. Find the ENC in electron units for standing current of 10 μA if a 20 ns shaping time is used in the filter network.

10. Find the ENC in electron units for a source resistor of 1 MΩ if a 20 ns shaping time is used.

References

[1] *Introduction to Solid State Physics*, C. Kittel, John Wiley and Sons (1953).

[2] *Physics of Semiconductor Devices*, J.M. Sze, John Wiley and Sons (1981).

[3] *Semiconductor Detectors*, G. Bertolini, A. Coche, North-Holland Publishing Co. (1968).

[4] *Semiconductor Detectors for Nuclear Radiation*, G. Dearnaley and D.C. Northrop, John Wiley & Sons (1966).

[5] V. Radeka, 'Low noise techniques in detectors', *Annu. Rev. Nucl. Part. Sci.* **38** 217 (1988).

[6] P.G. Rancoita and A. Seidman, *Rev. Nuovo Cimento* **7** 5 (1982).

[7] H.M. Heijne, CERN Report, 83–06 (1983).

[8] G. Lutz, A.S. Schwarz, 'Silicon Devices for Charged-Particle Track and Vertex Detection' *Annu. Rev. Nucl. Part. Sci.* **45** 295 (1995).

[9] *Elementary Particles*, I.S. Hughes, Cambridge University Press (1972).

[10] *Experimental Techniques in High Energy Physics*, T. Ferbel, Addison-Wesley Publishing Co., Inc. (1987).

[11] *Silicon Photodiodes and Charge Sensitive Amplifiers for Scintillation Counting and High Energy Physics*, Hamamatsu Photonics (1993).

[12] *Instrumentation in Elementary Particle Physics*, C.W. Fabjan and J.E. Pilche, Trieste 1987, World Scientific Publishing Co. (1988).

Part III

Destructive measurements

Having examined, in Chapters 2 through 9, the categories of particle detection where the energy lost by incident charged tracks is minuscule, we now turn to detection where the lost energy is a significant fraction of the kinetic energy that is carried by the incident particle. The exploration begins in Chapter 10 with considerations of the radiation of photons. Radiative processes occur at all energies, but tend to dominate at the highest energies since radiation is relativistically enhanced.

The applications of destructive readout are looked at in Chapters 11 and 12. First, in Chapter 11, we look at the total absorption of the energy of particles whose interactions are dominated by electromagnetic processes (electrons, photons). The 'shower' in the detecting medium is driven by electron Bremmstrahlung and photon pair production. Second, in Chapter 12, we look at the total absorption of the energy of strongly interacting or hadronic particles. Note that, for neutral hadronic particles, e.g. K_L or n, there is no ionization deposit nor is there a force exerted by E and B fields. Thus, total energy absorption is the measurement of choice. Therefore, this detection technique is of major importance for designers of general purpose detectors where the goal is to detect all emitted particles, (see Fig. I.1) both charged and neutral.

Part IIIA

Radiation

The emission of radiation by a charged particle is a fundamental process. An isolated free particle cannot radiate and conserve energy and momentum (see Chapter 3). Therefore, since another electromagnetic vertex, which reduces the rate by a factor of α, is needed to soak up energy (see Chapter 6, Fig. 6.10), the emission of radiation dominates all other processes only at high energy, $E > E_c$ (critical energy).

Radiation is often used to provide a signal in particle detectors. Low energy photons are detected by observing the electron recoil energies in γ–e elastic scattering (Compton scattering). High energy electrons which are accelerated emit radiation (Bremsstrahlung). The probability to emit radiation is typically enhanced by a power of γ with respect to non-relativistic considerations. Hence high energy processes are often dominated by radiation. The processes that are needed to understand particle detectors are explored in Chapter 10.

10

Radiation and photon scattering

My candle burns at both ends; it will not last the night; but ah, . . .
it gives a lovely light

Edna St. Vincent Millay

Many aspects of radiation and photon scattering are used in particle detectors. The attempt is made in this chapter to 'derive' the relevant formulae. In order to describe the processes, dipole radiation is first 'derived' using the static solutions and dimensional scaling. Thomson scattering is the first application and it is then extended to higher energies. The kinematics at high energies (Compton scattering) is derived. Relativistic four dimensional velocity, acceleration, and momentum are invoked to extend dipole radiation into the relativistic regime. The kinematics of photon emission and the virtual photon frequency spectrum lead to the concept of radiation length and Bremsstrahlung/pair production. These latter two processes are the engines that drive EM 'showers' and thus are at the heart of EM calorimetry (see Chapter 11).

10.1 Non-relativistic radiation

As the first task in this chapter we will try to heuristically 'derive' the formula for non-relativistic radiation by making dimensional substitutions to the well known static solutions for a dipole in order to go from statics to dynamics. We deal only with the far zone, where an observation point, O, is at a long distance from the dipole. The scale for 'long' is set by the distance of separation of the charges of the dipole, the source size. We ignore any higher order multipoles, since we assume $kr \ll 1$. As we show in Fig. 10.1, the static field is transverse which leads to a $\sin \theta$ factor in E where θ is the angle between the axis of the dipole and the vector from the source point to the observation point.

The static dipole field falls as the cube of the distance between the source and the observation point. This leads to static solutions where the energy density falls off as the sixth power of the radius. Clearly, this can not be a radiative solution.

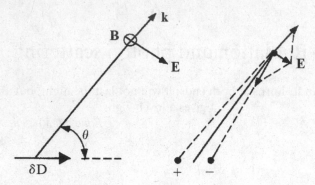

Fig. 10.1.   Vector diagram in the far zone for plane waves, (**E**, **B**, **k**), caused by dipole acceleration. Static situation shown for comparison.

We assume that for dipole radiation the electric and magnetic fields are transverse. If the charge distribution is accelerated, with acceleration a, we make the dimensionless substitution, (ar/c^2), to the static dipole field. This substitution leads to an energy density, u, going as the inverse square of the radius, and a power, P, which is independent of radius, as required for a freely propagating radiative solution. The radiated power is proportional to the charge q squared (or the fine structure constant) and to the acceleration of the charge squared.

$$E \sim \frac{(q \sin \theta)}{r^2} \left[\frac{ar}{c^2} \right]$$

$$u \sim E^2 \sim (qa\sin\theta/rc^2)^2 = \text{energy density}$$

$$\underline{P} \sim cr^2 E^2 \tag{10.1}$$

$$d\underline{P}/d\Omega = \frac{1}{4\pi} \frac{(qa \sin \theta)^2}{c^3}$$

$$\underline{P} = \frac{2}{3}(qa)^2/c^3$$

This expression connects the radiated power and the acceleration a of the charge q. We can recast the expression for power in terms of the frequency ω of the driving acceleration. We replace the static dipole b, $E \sim qb/r^3$, by the dynamic displacement, d.

$$b \to d, \quad \text{dipole moment} = D = 2q[b + d \sin \omega t] \tag{10.2}$$

There is a length scale related to the frequency ω. We find that the electromagnetic dipole power goes as the fourth power of the frequency, ω, and the square of the displacement (or alternatively the acceleration of the dipole).

The factor of 1/3 comes from the integration of the dipole angular distribution, $\int \sin^2\theta \, d\Omega \sim 1/3(8\pi)$. Note that the $\langle \rangle$ means time averaged in Eq. 10.3, and the dot means time derivative. We use Eq. 10.1, making the replacement $a \to d\omega^2$.

$$\langle P \rangle \to [q^2 d^2 \omega^4 / c^3]$$

$$\langle P \rangle = \omega^4 (\delta D)^2 / 3c^3, \tag{10.3}$$

$$= \langle \ddot{D} \rangle^2 / 3c^3 \quad (\ddot{D} \equiv \partial D / \partial t^2)$$

Consider an example on the atomic scale. We can estimate the radiated power for an electron and proton oscillating at a frequency corresponding to a visible wavelength of $\lambda = 3000$ Å. The e–p separation, d, is taken to be an atomic size scale, ~ 1 Å, which leads to $\langle P \rangle = 0.1$ nW. It only takes ~ 10 ns to radiate the ~ 10 eV binding energy E_0, leaving the atom collapsed to zero size. Indeed, the stability of atoms was a mystery for classical physics.

10.2 Thomson scattering

Having 'derived' the radiated power, we can now consider the problem of charges accelerated by incident radiation. Thomson scattering is the non-relativistic scattering of incident photons by electrons of mass m, charge e. The incident photon energy flux is the energy crossing unit area in unit time and is given by the Poynting vector, $\mathbf{S}$. Since the plane wave electric and magnetic fields are perpendicular to each other and to the wave vector, $\mathbf{k}$, the time averaged flux is simply proportional to the amplitude E_0 of the electric field squared. (Note that the time average of $(\sin \omega t)^2$ which is $\sim \langle |\mathbf{E}| \rangle^2$ is 1/2.)

$$\mathbf{S} = \frac{c}{4\pi} (\mathbf{E} \times \mathbf{B}), \quad \text{(CGS)}$$

$$\langle |\mathbf{S}| \rangle = \frac{c}{8\pi} |E_0|^2 \tag{10.4}$$

The angular distribution of the radiated power follows from Eq. 10.1.

$$\frac{dP}{d\Omega} = \frac{e^2}{4\pi c^3} |a|^2 \sin^2\theta \tag{10.5}$$

$$a = eE_0/m$$

Note that θ is the angle between the dipole vector or acceleration vector and the vector to the observation point as in Fig. 10.1. The Lorentz force, for non-relativistic motion of the target, is just the incident electric force divided by the target mass. Note that we ignore the proton charges as sources of power because they are too heavy to attain significant acceleration.

The cross section (Chapter 1) is defined to be the probability of scattering per unit incident flux. We first derive the scattering cross section in its differential form. Integrating, we then find the total cross section for the scattering of incident photons by electrons in subsequent non-relativistic motion. We average over initial photon polarizations assuming the incoming photons are unpolarized. This is historically called the Thomson cross section, σ_T (not to be confused with the total cross section).

$$\frac{d\sigma_T}{d\Omega} = (e^2/mc^2)^2 \sin^2\theta \equiv \langle d\underline{P}/d\Omega \rangle / \langle |\mathbf{S}| \rangle$$

$$\sigma_T = \frac{8\pi}{3}(\alpha\lambdabar)^2$$

(10.6)

Note that the cross section is proportional to the square of the product of α and the Compton wavelength of the target. The Compton wavelength of the electron is 0.004 Å (Table 1.1). The Thomson cross section is 0.665 barns. Note that, as seen in Fig. 1.10, at low energies the photoelectric effect (see Chapter 2) dominates while at higher energies Thomson/Compton scattering with $\sigma \sim Z\sigma_T \sim 10$ b is important.

It is instructive to manipulate the formula for the cross section and to put it in terms of the geometric cross section, which we discussed in Chapter 1. To do that we use the Bohr radius, a_0, which we know is roughly the atomic size.

$$\sigma_T = \frac{8}{3}[\pi a_0^2]\alpha^4$$

$$a_0 = \lambdabar/\alpha \quad \text{(Chapter 1)}$$

(10.7)

$$\sigma_T/\pi a_0^2 \sim \alpha^4 \sim 10^{-8}$$

What we observe is that the Thomson cross section is reduced by the fourth power of α with respect to the geometric cross section. In Chapter 1 we saw that the atomic geometric cross section was roughly 10^8 barns, so that estimating α as 1/100 we expect about a 1 barn cross section for σ_T. Numerically, low energy photons incident on a liquid hydrogen experimental target would have a Thomson mean free path of $\langle L \rangle \sim 0.35$ cm.

It is traditional to cast the Thomson cross section in terms of the geometric cross section of a fictitious 'electron size' defined by the classical self-energy or the classical electron radius (see Table 1.1). Since at our present level of understanding electrons behave as point particles down to a length scale of

10^{-16} cm, we will simply omit this standard treatment as somewhat misleading. Interested readers are referred to the literature given in the references.

10.3 Thomson scattering off objects with structure

At this point we would like to extend the discussion of Thomson scattering to the case of a target with some internal structure. We briefly formulate non-relativistic scattering theory and look at the radial solution for a spherically symmetric potential (see Appendix B for slightly more detail). The radial Schroedinger equation in this situation is given below. The Y_ℓ^m are the spherical harmonic solutions that arise in central force problems as the solution of the angular equations.

$$u \equiv Rr, \quad R = \text{radial wave function}, \quad \Psi \sim RY_\ell^m$$

$$\frac{d^2u}{dr^2} + \left[k^2 - U(r) - \frac{\ell(\ell+1)}{r^2} \right] u = 0$$

(10.8)

We identify the terms of the differential equation as the radial kinetic energy, the potential energy, the centrifugal potential and the total energy, which is proportional to k^2. Near the origin (see Appendix B) $|\Psi|^2 \sim r^{2\ell}$ so that only $\ell = 0$ (S wave) states are non-zero at $r = 0$.

$$u|_{r \to 0} \to r^{\ell+1}$$

(10.9)

Let us now consider the scattering of an incoming wave by a square well of radius a and of depth U_0. The outgoing scattered solutions are found by applying boundary conditions requiring the continuity of interior $(r<a)$ and exterior $(r>a)$ solutions and first derivatives. (See Appendix B.) In the simplified case where $ka \ll 1$, we have S wave dominance of the phase shifts (Chapter 1).

$$\delta_\ell \sim (ka)^{2\ell+1} = \text{partial wave shift}$$

$$\delta_0 \sim -ka \gg \delta_\ell$$

(10.10)

Using the results of Chapter 1, we find that the total cross section is simply proportional to πa^2.

$$\sigma \sim \frac{4\pi}{k^2} \sin^2 \delta_0 \quad \text{(see Chapter 1)}$$

(10.11)

$$\sigma \sim 4\pi a^2$$

Let us now try to generalize the situation. In the limit where the depth of the well, U_o, becomes small, the solution is the Mie–Debye–Rayleigh law (Appendix B).

$$k^2 + U_o = K^2 \quad \text{(interior solution } K)$$

$$\delta_o = -ka\left[1 - \frac{\tan(Ka)}{Ka}\right] \sim -ka\left[\frac{(Ka)^2}{3}\right]$$

$$\sigma \rightarrow 4\pi a^2[(ka)^2/3]^2, \quad K \sim k$$

(10.12)

$$\sigma_{RA} \sim a^6/\lambda^4 \sim a^2(a/\lambda)^4, \quad \sigma_{RA}/\pi a^2 \sim (a/\lambda)^4 \sim (ka)^4$$

The previous result is modified by a factor $(ka)^4$. This factor can be considered to be the 'form factor' which describes the non-pointlike behavior of the target scattering center.

For example, if visible light, $\lambda \sim 5000$ Å, is incident on a typical atom, $a \sim 1$ Å, the suppression factor, ka, in the Rayleigh cross section is enormous. Rayleigh scattering is something which is evident to us every day as we look up at the sky and ask, 'If the sun is yellow then why is the sky blue?'. The answer is that the Rayleigh scattering of the sun's light in the atmosphere is very dependent on the wavelength of the incident light. The atmosphere preferentially scatters short wavelength light (blue) into our eyes. This phenomenon also tells us why the setting sun reddens. As more atmosphere is traversed by the sunlight, the shorter wavelengths have an increased relative chance to be scattered.

10.4 Relativistic photon scattering

At higher energies, where relativistic effects come into play, the Thomson formula, is no longer valid. One of the contributing Feynman diagrams for high energy elastic scattering is given in Fig. 10.2. We can count powers of e in the amplitude and conclude that it goes like α. The cross section will then go as α^2. There are now two scales of energy, the center of mass energy, $\sqrt{s}$, and the target mass, m. Thus a cross section going like α^2/s is possible.

The high energy cross section, called the Klein–Nishina cross section, σ_{KN}, is proportional to the Thomson cross section, σ_T, but with a multiplicative factor which at high energies is the dimensionless ratio of the target electron mass, m, to $\sqrt{s}$ squared. Thus at high energies, $\sqrt{s} > m_e$, the elastic cross section in Fig. 1.10 falls with energy. There are also logarithmic factors which are simply indicated in parenthesis.

$$\sigma_{KN} \sim \frac{3}{8}\sigma_T\left(\frac{m}{\sqrt{s}}\right)^2[1 + \ln(\ )] \sim \alpha^2/s$$

(10.13)

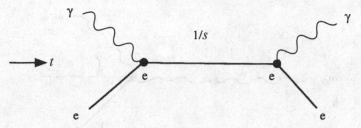

Fig. 10.2.   Feynman diagram for photon scattering off electrons.

As another example, the $e^+e^- \rightarrow \mu^+\mu^-$ cross section, $\sigma_{\ell\ell}$, also goes as α^2/s.

$$\sigma_{\ell\ell} = \frac{4\pi\alpha^2}{3s} = \frac{87 \text{ nb}}{[s(\text{GeV}^2)]}, \quad 1 \text{ nb} = 10^{-33} \text{ cm}^2 \quad (10.14)$$

For example, at a 1 GeV center of mass energy, the cross section is 87 nb. An electron incident on a liquid hydrogen target would then have a 2900 km mean free path to make muon pairs by this process. This value for $\langle L \rangle$ is much larger than that for photon Thomson scattering. The fall of the elastic photon–electron cross section with increasing energy is visible as the behavior of $\sigma_{\text{coherent}} + \sigma_{\text{incoherent}}$ in Chapter 1.

10.5 Compton scattering

Having looked at non-relativistic elastic scattering, we now consider the kinematics of the fully relativistic case. To this point in the discussion we have assumed that the frequency of the incoming photons is unchanged in the scattering, e.g. Thomson scattering. This is the principle which allows for clear radio transmission, since reflection in this case will not cause frequency shifts. However, the constancy of frequency is no longer valid at high energies. The quantitative scale for what we mean by high energy is set by the mass of the targets, or by the 'Compton wavelength' of the electrons which scatter the incident radiation.

The kinematic definitions for the scattering of photons of frequency ω_0 off a particle of mass m are given in Fig. 10.3. As hypothesized by Einstein, we consider the photon to be a particle. We find the quoted relation by squaring the two equations, for energy and momentum conservation, Eq. 10.15, and subtracting them (see Appendix A).

$$\omega_0 + m = \varepsilon + \omega$$
$$\mathbf{k}_0 = \mathbf{p} + \mathbf{k}$$
(10.15)

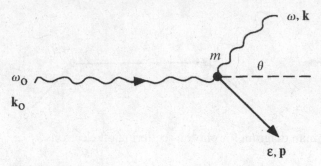

Fig. 10.3.   Kinematic definitions for photon scattering off a particle of mass m.

The incident photon, frequency ω_0, strikes the target which is at rest in the laboratory frame. This causes the target to recoil with energy ε after scattering the photon into final energy ω. We know that energy conservation requires a decrease in the energy of the photon, which means an increase of the wavelength. The scale for the increase is set by the Compton wavelength of the electron, $\lambdabar$.

$$\left(\frac{1}{\omega}-\frac{1}{\omega_0}\right)=\frac{1}{m}(1-\cos\theta)$$

(10.16)

$$\lambda-\lambda_0=2\pi\lambdabar(1-\cos\theta)$$

Since the wavelength of radio waves is meters and the Compton wavelength of the electron is $\sim 10^{-13}$ m, the frequency shift can be ignored in radio transmission. However, at incoming photon energies which are comparable to the rest energy of the electron, 0.511 MeV, (Table 1.1) the frequency shift becomes a large fraction of the incident energy and can no longer be ignored. The dynamics also contributes to the observability of the frequency shift. As seen in Fig. 10.4, one of the Feynman diagrams for Compton scattering has an exchanged electron. Since the 'virtual' electron would like to be as 'real' as possible, i.e. have small momentum transfer, the outgoing electron will take off most of the energy of the incident photon. Therefore, the dynamics leads to fast forward electrons, or large energy loss of the photons.

In dealing with photon radioactive sources (see Table 6.1) the appearance of the almost full incident photon energy 'Compton peak' recoil electron is an everyday experience. The kinematic limit for back-scattering, $(1/\omega-1/\omega_0)=2/m$, is that the maximum e energy, or 'Compton edge', is $T_{max}=\varepsilon_{max}-m=\omega_0[1-m/(m+2\omega_0)]\rightarrow\omega_0$ if $\omega_0\gg m$.

The exact angular distribution is quoted in Appendix A. Part of the forward peaked nature of the distribution is purely kinematic due to the transformation of CM solid angle, $d\Omega^*$, to lab solid angle $d\Omega$.

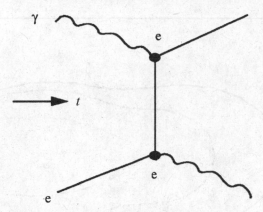

Fig. 10.4. Feynman diagram for Compton scattering. Electron exchange leads to fast forward electrons.

$$p^* \sim \sqrt{s}/2 \sim \sqrt{\omega_o m/2}, \quad d\Omega = d\Omega^*(p^*/p)^2 \sim d\Omega^*\left(\frac{\omega_o m}{2\omega^2}\right) \qquad (10.17)$$

The angular distribution at several values of ω_o is shown in Fig. 10.5. Note the forward peaking at high energies.

Some of the actual data taken by Compton showing the appearance of the primary photon energy and the scattered photon energy is shown in Fig. 10.6. Of course, this was also historically one of the first experiments which showed the particle nature of light. Referring to Eq. 10.16, we have $\Delta\lambda = 2\pi\lambdabar$ at $\theta = 90°$. Since $\lambda_e = 0.0038$ Å we expect a shift in wavelength of ~ 0.024 Å which, Fig. 10.6, is exactly what was observed.

10.6 Relativistic acceleration

The fundamental formula for non-relativistic radiation is given in Eq. 10.1. Since acceleration is not a Lorentz invariant, this is not a formulation which is easily generalized to the relativistic case. Recall that in special relativity the normal three dimensional vector equations of classical physics are generalized to four dimensions. For example, the position of an event in space and time is labeled by a position vector, x^μ, which includes both the spatial position, $\mathbf{x}$, and time, ct, in units of distance.

$$x^\mu = (\mathbf{x}, ct) \qquad (10.18)$$

The relativistic velocity, U^μ, is the proper time derivative of the position.

$$U^\mu = dx^\mu/d(s/c)$$
$$= \gamma(\mathbf{v}, c) \qquad (10.19)$$

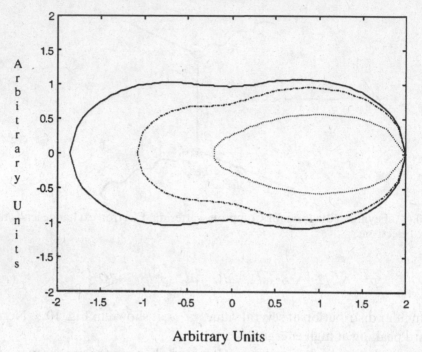

Arbitrary Units

Fig. 10.5.   Angular distribution shown as (x, y) contour for Compton scattering at several different incident photon energies, $\omega_o = 0.01$ MeV (——), 0.1 MeV (–·–) and 1.0 MeV ($\cdots$).

The proper time, ds, is the relativistically invariant 'distance' between two events. The factor of γ (see Appendix A) that appears is near to 1 at low velocities, $\beta \to 0$, and increases without limit as the velocity approaches that of light, $\beta \to 1$.

$$ds^2 = dx_\mu dx^\mu = dx^2 - (cdt)^2 = (cdt/\gamma)^2$$
$$\gamma = 1/\sqrt{1 - \beta^2}, \quad \beta = v/c \tag{10.20}$$

The relativistic momentum p^μ is the mass measured with the particle at rest, the rest mass, m, times the relativistic velocity. The particle energy is ε and the momentum is $\mathbf{p}$.

$$p^\mu = mU^\mu = (\mathbf{p}, \varepsilon/c)$$
$$= \gamma m(\mathbf{v}, c)$$
$$\mathbf{v} = d\mathbf{x}/dt \tag{10.21}$$
$$\varepsilon = \gamma mc^2$$
$$\mathbf{p} = \gamma\boldsymbol{\beta}mc = \boldsymbol{\beta}\varepsilon/c$$

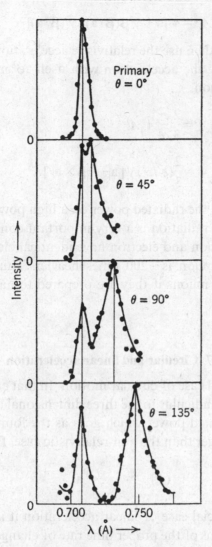

Fig. 10.6.   Data taken by Compton showing the appearance of the scattered photon at higher wavelength (lower energy) for different scattering angles. The shift in λ increases with scattering angle, as expected. (From Ref. 10.9 with permission.)

The four dimensional generalization of acceleration, A^μ, is the proper time derivative of the relativistic velocity.

$$A^\mu = dU^\mu/d(s/c)$$

$$= \gamma \frac{d}{dt}[\gamma(\mathbf{v}, c)] \tag{10.22}$$

$$\frac{d\gamma}{dt} = \gamma^3(\boldsymbol{\beta} \cdot \mathbf{a}), \quad \mathbf{a} = d\mathbf{v}/dt$$

$$A^\mu = \gamma^3[\mathbf{a} + \gamma^2\boldsymbol{\beta}(\boldsymbol{\beta}\cdot\mathbf{a}),\ \gamma^2(\boldsymbol{\beta}\cdot\mathbf{a})]$$

Using Eq. 10.1, we can then use the relativistic acceleration to form a Lorentz invariant by contracting the acceleration with itself to get the square of the 'length' of the acceleration.

$$P = \frac{2}{3}\frac{e^2}{c^3}A_\mu A^\mu$$

$$= \frac{2}{3}(e^2/c^3)\gamma^6[|\mathbf{a}|^2 - |\boldsymbol{\beta}\times\mathbf{a}|^2] \tag{10.23}$$

Note the dependence of the radiated power on a high power of the relativistic γ factor. This enhanced radiation is a very important consequence of relativity. For example, the muon and electron have identical electromagnetic interactions. However, the muon is ~ 200 times heavier. Thus, electrons radiate much more power than muons if they are prepared to have the same energy, since $\gamma = E/mc^2$.

10.7 Circular and linear acceleration

Let us look at the special case of circular motion. In that case the three dimensional velocity $\mathbf{v}$ is perpendicular to the three dimensional acceleration $\mathbf{a}$ which leads us to a total radiated power which goes as the fourth power of γ. The power radiated is γ^4 larger than the non-relativistic case, Eq.10.1.

$$(P)_{\text{circ}} = \frac{2}{3}\frac{e^2}{c^3}\gamma^4|\mathbf{a}|^2 \tag{10.24}$$

In looking at the special case of linear acceleration it is convenient to formulate the power in terms of the proper time rate of change of the four dimensional momentum. We can then use the relationship between proper time, ds, and coordinate time, dt, and the relationships between momentum and energy given above to formulate the radiated power in terms of the energy actually supplied by external forces. In making that substitution we note that in going a linear distance dx in time interval dt the particle instantaneously has velocity β since all these quantities are defined by clocks and rulers set up in the laboratory reference frame.

$$A^\mu = \left(\frac{\gamma\beta}{m}\right)\frac{d}{dx}[\varepsilon(\boldsymbol{\beta}, 1)], \quad dx = \beta c\,dt$$

$$(P)_{\text{LIN}} = \frac{2}{3}\frac{e^2}{m^2c^3}\left(\frac{d\varepsilon}{dx}\right)^2 \tag{10.25}$$

Thus, in the case of linear acceleration the radiated power is simply related to the energy supplied by the external forces per unit length. This fact has immense practical implications in that circular colliders, such as e^+e^- storage rings, pay an enormous power penalty as γ increases. Therefore, current electron accelerator research is in the area of linear colliders where we need not pay that added radiative power bill. This factor first became important in the design of the Stanford Linear Accelerator Center, SLAC, which is a linear machine and is the pioneer in the study of linear acceleration and linear colliders. Note, however, that in terms of three dimensional acceleration, Eq. 10.23, $(P)_{LIN} = 2/3(e^2/c^3)|\mathbf{a}|^2\gamma^6$, with acceleration measured by lab clocks and rulers.

10.8 Angular distribution

The non-relativistic dipole radiation pattern is strongly altered in the highly relativistic case. However, the details are quite technical and will not be discussed here. Suffice it to say that much of the basic physics can be extracted by looking at a photon emitted by a charge with velocity β. In the rest frame of the charge the photon has energy ε^* and polar angle θ^*. In the laboratory frame the factor relating ε and ε^* is, by Lorentz transformation, $\varepsilon = \gamma(\varepsilon^* - \beta p^* \cos\theta^*) = \gamma\varepsilon^*(1 - \beta\cos\theta^*)$ since photons have $\varepsilon = p$ or $v = c$, in all reference frames. This factor, $(1 - \beta\cos\theta)$, appears in the angular distributions as, for example, Eq.10.17. In the limit $\beta \to 1$ we use the approximations

$$\frac{1}{\gamma^2} = 1 - \beta^2$$

$$\beta \equiv 1 - \delta\beta$$

$$\beta^2 \cong 1 - 2\delta\beta \tag{10.26}$$

$$\frac{1}{\gamma^2} \cong 2\delta\beta$$

$$\cong 2(1 - \beta)$$

We have previously used these approximations in Chapter 3 for the discussion of Cerenkov radiation.

If this factor alone determined the angular distribution, the characteristic angle for emission of radiation is $1/\gamma$ (see the discussion on TRD, Chapter 4). This same factor can be found by looking at an isotropic emission distribution, in the CM frame, and boosting it to a rapidly moving frame, $\theta \sim P_T^*/P_{\parallel} \sim \varepsilon^*/\gamma\varepsilon^*(1 - \beta^*\cos\theta^*)$, $\langle\theta\rangle \sim 1/\gamma$. This is called the 'searchlight effect' in special relativity.

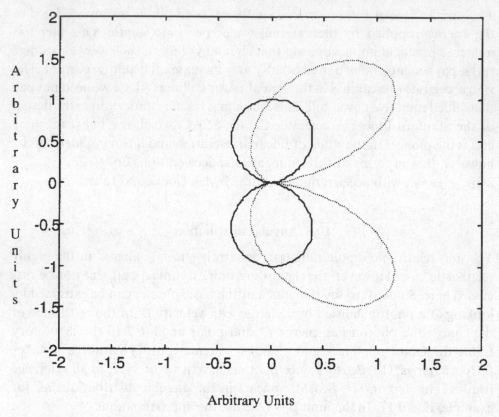

Fig. 10.7.   Dipole angular distribution as $(x_{\parallel}, y_T)$ contours for the case of linear acceleration, showing the tipping forward due to the 'searchlight' effect, for $\beta = 0.01$ (——) and $= 0.3$ (·····). The growth in the radiated power, $\underline{P} \sim \gamma^6$, is evident.

$$1 - \beta\cos\theta \sim 1 - \beta(1 - \theta^2/2)$$

$$\sim \frac{1}{2}\left(\frac{1}{\gamma^2} + \theta^2\right) \tag{10.27}$$

$$\langle\theta\rangle \sim 1/\gamma$$

This factor appeared already in our discussion of transition radiation detectors in Chapter 4 as a factor in the vacuum phase angle.

In general the angular distribution is ugly and complex. In the special case of linear acceleration we find the dipole radiation pattern thrown forward by the searchlight effect. The angular distribution is shown in Fig. 10.7 for various values of β. The searchlight effect is very obvious. The general linear motion result approaches, for non-relativistic motion, $\beta \rightarrow 0$, the dipole result:

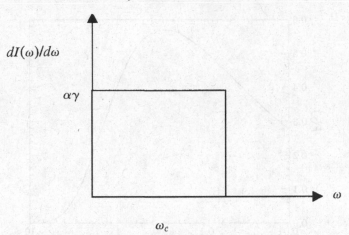

Fig. 10.8. Schematic representation of the frequency distribution of the intensity for circular motion.

$$dP_{\text{LIN}}/d\Omega = \alpha a^2/4\pi c^3[\sin^2\theta/(1-\beta\cos\theta)^5]$$

$$dP_{\text{LIN}}/d\Omega \rightarrow \alpha a^2/4\pi c^3(\sin^2\theta) \qquad (10.28)$$

10.9 Synchrotron radiation

Let us now turn to the special case of circular motion. We have already learned that the power radiated in circular motion with respect to the non-relativistic case is increased by a factor of γ^4. We have also learned that in general there is a searchlight effect which throws the radiation forward into a cone of typical angular size $1/\gamma$. Now we would like to look at the frequency spectrum of the radiated photons since we know already from the Compton effect that the radiated frequency, ω, need not be the same as the frequency of the driving acceleration, ω_0.

We simply assert that the relativistic effect is to stretch the emitted frequency ω by a factor $\sim \gamma^3$ beyond ω_0. The circular radius is a and β refers to the circular velocity.

$$\omega_0 \sim c\beta/a$$

$$\omega_c \sim \gamma^3\omega_0 \qquad (10.29)$$

$$\equiv 3\gamma^3 c/2a$$

A rough schematic representation of the frequency distribution of the intensity of radiation is shown in Fig. 10.8. The quantity $dI(\omega)/d\omega$ is the intensity as

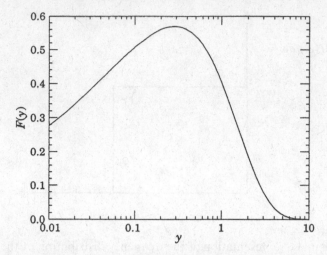

Fig. 10.9.　Frequency distribution for synchrotron radiation where y is ω/ω_c. (From Ref. 1.1.)

a function of frequency defined to be the radiated energy crossing unit area per unit frequency. The radiative energy loss per revolution, ΔE, is simply the radiated power, $\underline{P}$, times the period of revolution. The power is proportional to the characteristic frequency, ω_c, which has a γ^3 factor, times another power of γ. Very roughly we can think of synchrotron radiation, or radiation emitted in relativistic circular motion, as the radiation per turn of $\alpha\gamma$ photons with energies up to $\hbar\omega_c$. Hence, in every turn we lose energy ΔE which goes as γ^4.

$$\Delta E = \left(\frac{2\pi a}{c\beta}\right)\underline{P}$$

$$\sim \left(\frac{4\pi}{3}\,\alpha\gamma\right)(\gamma^3\omega_o) \sim \gamma^4$$

$$\sim \alpha\gamma\omega_c$$

$$\sim [dI(\omega)/d\omega]\Delta\omega, \text{ where } \Delta\omega \sim \omega_c$$

(10.30)

An exact calculation of the intensity of radiation is shown in Fig. 10.9. The sharp fall off in emitted frequencies above the characteristic frequency ω_c survives as a feature of the exact calculation, as does the basically flat spectrum below ω_c.

Synchrotron emission is very important in certain applications. A representation of the emission of synchrotron radiation by an electron spiraling in a magnetic field is shown in Fig. 10.10. The photons are emitted largely in the plane of the orbit. A contour plot of the radiation pattern in circular motion

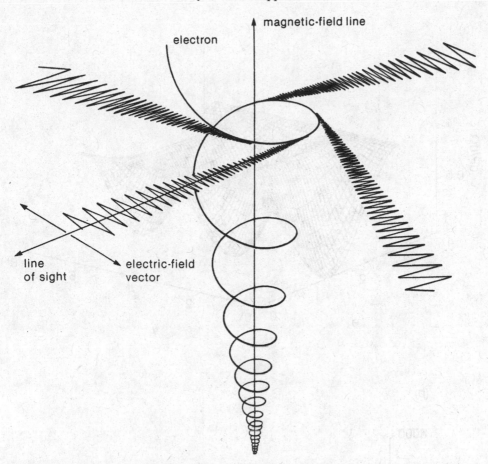

Fig. 10.10. Schematic representation of the emission of synchrotron radiation by an electron spiraling in a magnetic field.

is given in Fig. 10.11. The two plots are for classical and extremely relativistic synchrotron radiation. The 'searchlight' effect, $\theta \sim 1/\gamma$, is very evident in the case that $\gamma \gg 1$.

10.10 Synchrotron applications

There are now several facilities in existence which create synchrotron light for scientists who use it as a probe of the properties of materials. Examples in the United States are at Brookhaven National Laboratory, Argonne National Laboratory and Cornell University. A dedicated accelerator which magnetically 'wiggles' electrons is used to provide the radiation.

Synchrotron radiation can also be used for 'particle identification'. Consider a prepared beam of pions and electrons of the same momentum. If the beam

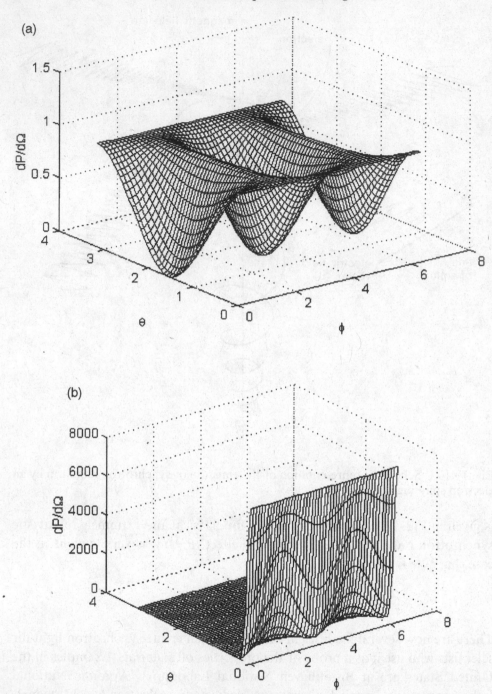

Fig. 10.11.   The 'searchlight' effect in synchrotron radiation. The x-axis is along **a**. The z-axis is along **β** and the motion is circular. Shown are $d\sigma/d\Omega$ surfaces with $z(\theta, \phi)$ proportional to $d\sigma/d\Omega$. (a) $\beta = 0.01$. Note the dipole like pattern. (b) $\beta = 0.95$. Note the growth in power which goes as γ^4. Note also the strong forward peaking with $\theta \sim 1/\gamma$.

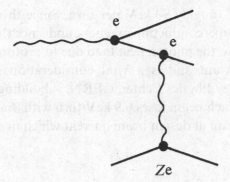

Fig. 10.12. Feynman diagram for photon pair production in the field of a nucleus with atomic number Z.

is bent in a magnet and forward produced x-ray energy photons are detected, their presence indicates electrons (large γ) and not pions. Assume a 6 m beam line magnet with a field of 20 kG and a 200 GeV incident beam. The radius of curvature (see Chapter 7) is, $a = 300$ m. Therefore, the 6 m magnet represents only $\Delta\phi/2\pi = 0.003$ of a revolution. The electrons in the beam will radiate ~ 24 photons with energy $\hbar\omega_c \sim 53$ MeV for a total energy loss of ~ 1.26 GeV. Such photons are radiated almost straight ahead (see Fig. 10.10), while the charged beam is bent away by the magnetic field. Photons of this energy will make e^+e^- pairs which are then detectable (see Chapter 1 and Fig. 10.12).

As a practical matter the emission of radiation places a limit on the ease of operation of circular accelerators. For example, 1 GeV electrons have a loss per turn in a circular orbit of radius 1 m of 90 keV. The driving frequency, ω_o, is of order 10^8/s. The γ factor is 2000, which means that the characteristic frequency, ω_c, is of order of 10^{17}/s. Since $\hbar$ is 6.6×10^{-16} (eV·sec), Table 1.1, the energy of the photon is ~ 2 keV, i.e. in the x-ray range. We have roughly 45 photons emitted per turn with energy 2.2 keV for a total energy loss per turn of 90 keV. Thus there is a substantial radiated power to be supplied by external energy sources.

$$\Delta E \text{ (keV)} = 90[E(\text{GeV})]^4/a(\text{m})$$
$$\hbar\omega_c(\text{keV}) = 2.2[E(\text{GeV})]^3/a(\text{m})$$
(10.31)

So far we have only talked about electrons, which is perfectly reasonable since they are the lightest stable charged particle, which implies that at a given energy they will have the largest γ factor. However, recently, in designing high luminosity multi-TeV proton–proton colliders, the effects of proton radiation are also beginning to become important. For example, a 20 TeV proton has a γ factor of 20 000. If the radius of such a machine is 10 km, scaling as in Eq. 10.30, we find the same loss of roughly 90 keV per turn. Since this new generation of accelerators utilizes superconducting magnets, and since that the heat capacity of such magnets is low, the photon heat load due to proton synchrotron radiation is indeed important and is a vital consideration in the design. The European High Energy Physics Center, CERN, is building a 7.5 + 7.5 TeV pp collider (the LHC). Each beam loses 6.9 keV/turn with $(\hbar\omega_c) \sim 45$ eV. The heat load is 3.7 kW per beam at design beam current which must be removed from the cryogenic magnets.

10.11 Photon emission kinematics

Let us now turn to the photon emission probability and the kinematics of the radiation of a photon by a charged particle. We have already seen that radiation requires acceleration. Thus, it is not possible for an isolated free particle to radiate as we proved in Chapter 3, where we showed that a particle which does radiate must have $\omega = ck_\parallel\beta$. Since $\beta < 1$ and $\omega = c|\mathbf{k}|$ for a real photon, this relation can not be satisfied. It is most nearly satisfied when $k \sim k_\parallel$ or when the photons are emitted in the forward direction.

Since the amplitude, A, in perturbation theory is proportional to the inverse of the energy difference between the initial and final unperturbed states, the largest amplitudes will have the smallest energy difference. The favored scenario is, therefore, to have soft and colinear photon emission,

$$A \sim 1/(\varepsilon - \varepsilon') \sim 1/\omega \qquad (10.32)$$

10.12 Photon frequency spectrum

In order to quantify the expectation of soft emission, we estimate the intensity of photons for different frequencies. The energy density, u, is proportional to the square of the electric field. The impact parameter is b. The peak electric field of a non-relativistic charged particle seen at a transverse distance b is e/b^2. The moving charge makes a pulse of field in time which therefore implies a frequency spectrum. The characteristic frequency ω_c is the inverse of the collision time at low velocities. The fall off of the field with distance means that there is a finite collision time which is $\Delta t \sim 2b/v$ for non-relativistic collisions. (See Chapter 5.)

The Fourier components of the field, $[dE/d\omega]^2$, are $\sim E\Delta t \sim 2e/bv$. The field in frequency space is fairly uniform for $\omega < \omega_c = v/b = 1/\Delta t$. The total energy, U, is found by integrating the energy density, u, over all spatial volume.

$$dU = \int du = \int\int [dE/d\omega]^2 d\omega 2\pi b \, db$$
$$= \int [dU/d\omega] \, d\omega \qquad (10.33)$$

Ignoring the logarithmic details, we find a simple behavior for $dU/d\omega$ by explicitly substituting for $dE/d\omega$.

$$dU/d\omega = 4\alpha/\beta^2[\ln(\ )] \tag{10.34}$$

We identify the total field energy of the charged particle as the total photon number density N_γ and the energy of the photon, $\hbar\omega$.

$$dU = \hbar\omega \, dN_\gamma \tag{10.35}$$

Therefore, the number of photons emitted per unit frequency due to the passage of a charged particle with velocity v goes as $1/\omega$ which is a reflection of the fact that the emitted photons would 'like to be' soft, Eq. 10.32.

$$\frac{dN_\gamma(\omega)}{d\omega} \sim \frac{\alpha}{\beta^2}\left(\frac{1}{\omega}\right)[\ln(\ )]$$
$$= \frac{2\alpha}{\pi}\left(\frac{1}{\beta^2}\right)\left(\frac{1}{\omega}\right)[\ln(\ )] \tag{10.36}$$
,

The implication of Eq. 10.36 is that the electromagnetic field of a moving charge can be represented as the frequency spectrum of its associated 'virtual photons' which is called the Weissacker–Williams method of virtual quanta. We will ignore the modifications to the virtual photon spectrum due to relativistic motion.

10.13 Bremsstrahlung and pair production

Now let us consider Bremsstrahlung and pair production. First the definition: Bremsstrahlung is the emission of photons by charged particles accelerated in the Coulomb field of a nucleus. The Feynman diagram for photon pair production in the field of a nucleus of atomic number Z was given in Fig. 10.12 while the topologically similar diagram for Bremsstrahlung by charged particles in the field of a nucleus is given in Fig. 10.13. Note that Fig. 10.13 is just the radiation of a photon by an electron with an additional interaction with the nuclear Coulomb field to satisfy energy–momentum conservation and thus allow the reaction to proceed.

The energy distribution of the positron in photon pair production is shown in Fig. 10.14, for two photon energies. Clearly, the energy spectrum is roughly uniform from about zero to the maximum allowed energy. We can think of this shape as arising from the isotropic 'decay' of a 'virtual' photon of some non-zero mass ($> 2m_e$) acquired by interaction with the nuclear Coulomb field. Isotropic 'decay' can easily be shown to yield a uniform secondary energy spectrum. (See Appendix A.)

Clearly these diagrams are higher order in the coupling constants than the

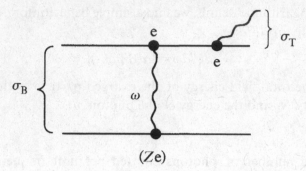

Fig. 10.13.   Feynman diagram for Bremsstrahlung in the field of a nucleus of atomic number Z.

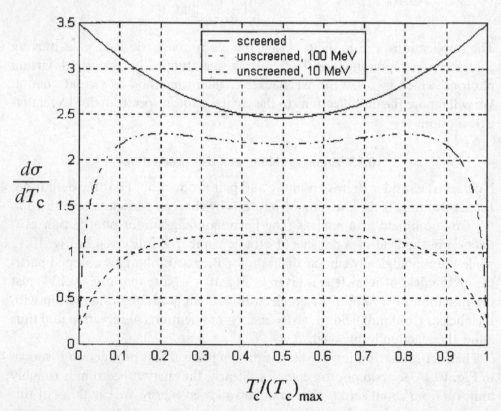

Fig. 10.14.   Kinetic energy spectrum of the positron scaled to the maximum allowed energy for incident photon energies of 10 and 100 MeV. The spectra for nuclear charge Z screened by the atomic electrons and unscreened are shown. (See references 10.10 and 10.1 for a definition of the screening approximations.)

photon particle scattering with which we began this chapter. The idea is to now use what we have already learned and factorize the problem of radiation by a charged particle into distinct pieces. Consider high speed incident particles scattering off a nucleus. In the rest frame of those particles, the nucleus is incident at high velocity. It has Z protons of charge e. We view the electric field of the nucleus as a distribution of soft virtual photons described by the Weissacker–Williams formula given in Eq. 10.36. Since the photons are soft, their wavelengths are larger than the size of the nucleus. Thus, the fields are coherent over the nucleus. By 'coherent' we mean that the phase of the field does not vary over the size of the nucleus. Therefore the quantum mechanical amplitude squared has a factor of $(Ze)^2$ in adding up the Z individual amplitudes which are in phase.

The incoming nucleus is equivalent to a distribution of soft photons. They will Thomson scatter off the projectile. We have already derived the Thomson cross section, Eq. 10.6, and we know the Weissacker–Williams distribution, Eq. 10.36. We put them together to describe the Bremsstrahlung cross section, σ_B, as the Thomson scattering, σ_T, of the soft virtual photons, N_γ, describing the coherent field of the nucleus off the projectile charge.

$$\frac{d\sigma_B}{d\omega} \sim Z^2 \frac{dN_\gamma}{d\omega} \sigma_T$$

$$\frac{d\sigma_B}{d\omega} \sim \frac{(Z^2\alpha)}{\omega} (\alpha\lambda)^2 [\ln(\ )]$$

(10.37)

Therefore the Bremsstrahlung cross section as a function of frequency displays the coherent factor $Z^2\alpha$, and goes as $1/\omega$. There are also logarithmic factors which again we ignore. The cross section is $\sigma_B \sim [Z^2\alpha^3/m^2]\ln(\ )$. Numerically $\sigma_B \sim Z^2[0.58 \text{ mb}]$ which is larger than the typical hadronic cross section (see Chapter 1).

10.14 The radiation length

We integrate the Bremsstrahlung distribution over all frequencies and weight by the energy of the emitted photon to find the total emitted energy or total radiative energy loss, dE.

$$dE \sim \int_0^E (\hbar\omega) \left(\frac{N_o \rho \, dx}{A}\right) \left(\frac{d\sigma_B}{d\omega}\right) d\omega$$

(10.38)

Cast in terms of a fractional energy loss per path length in g/cm^2, $\rho \, dx$, the radiative mean free path is just X_o, the radiation length.

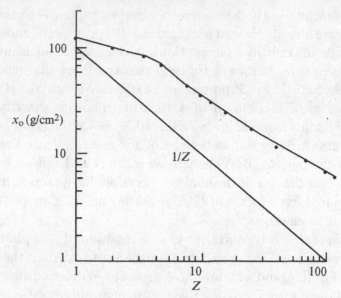

Fig. 10.15.   Plot of the radiation length X_o in g/cm² as a function of the atomic number Z. The straight line has the functionality $1/Z$ for comparison.

$$\frac{1}{E}\left(\frac{dE}{\rho\,dx}\right)\equiv\frac{1}{X_o}, \quad E(x)=E(0)e^{-\rho x/X_o} \tag{10.39}$$

Using Eq. 10.38, $dE\sim(Z^2\alpha)(\alpha\lambda)^2(N_o\rho\,dx/A)E$, and the radiation length, X_o, is

$$X_o^{-1}=\frac{16}{3}\left(\frac{N_o}{A}\right)(Z^2\alpha)(\alpha\lambda_p)^2[\ln(\ )]$$
$$\sim(Z/A)Z\sim Z \tag{10.40}$$

The radiation length, X_o is defined to be the mean free path for emitting Bremsstrahlung radiation. Note that the Compton wavelength refers to the wavelength of the projectile, λ_p.

We can re-express the Bremsstrahlung cross section in terms of the mean free path (Chapter 1).

$$\sigma_B\equiv[A/(N_o\rho X_o)]$$
$$X_o=\langle L\rangle \tag{10.41}$$

A plot of the radiation length in g/cm² as a function of the atomic number (data given in Table 1.2) is shown in Fig. 10.15. Roughly speaking the functional dependence of X_o, expressed in g/cm², is to decrease as $1/Z$. A reasonable representation of the numerical value is, X_o (g/cm²) $=[180(A/Z)]/Z$. This dependence has an immediate implication. We should use heavier materials in order to shield against photons. For example, lead is a standard material used in such shielding applications. Photon sources are routinely housed in lead 'pigs'. The radiation length of lead is 0.56 cm, so that the 'pigs' are fairly compact.

The radiation length goes like the square of the mass of the projectile. Thus electrons will easily radiate whereas muons, which have the same coupling and are nearly 200 times heavier, will not tend to radiate, because it is acceleration which is important in radiation. Since the forces are the same on the electrons and muons (because they have the same charge), the difference is in the acceleration^2 (since this is a radiative process) which explains the appearance of the square of the masses in Eq. 10.40. We return to the particle identification of muons in Chapter 13.

For lead, Eq. 10.40 yields a rough estimate of 17 g/cm^2 for the radiation length. Table 1.2 gives us the exact answer of 6.37 g/cm^2 or 0.56 cm for lead and incident electrons. Scaling to the muon by the square of the mass, we estimate that the radiation length for muons on lead would be 236 m, which is certainly not a practical way to shield against muons. The only possibility is to let them ionize and 'range out'.

10.15 Pair production by photons

Comparing Figs. 10.12 and 10.13 we can see that photon pair conversion and Bremsstrahlung are very similar in terms of Feynman diagrams. In fact, the cross sections are almost equal.

$$\sigma_{\text{pair}} = \frac{7}{9}\sigma_{\text{B}} \tag{10.42}$$

The roughly energy independent cross section, $\sigma_{\text{B}} \sim (Z^2\alpha)(\alpha\lambda_{\text{p}})^2[\ln(\ )]$, is responsible for the constant high energy photon cross section already displayed in Chapter 1.

We can think of the nucleus as supplying a small 'virtual mass' to the photon, 'm_γ' $> 2m_{\text{e}}$. The opening angle for the electron-positron pair in this process is roughly the electron mass divided by the electron energy. This is typical; masses or other basic physics quantities provide scales for the transverse momentum, whereas the only longitudinal energy scale in the problem is the incident energy, $\theta_{\text{e}} \sim m_{\text{e}}/\varepsilon_\gamma \sim p_{\text{T}}/p$.

The angles for photon conversion into pairs can thus be very small. Often the photon conversion is not even resolved into individual electrons and positrons (see Chapter 6). A photograph of a photon conversion in a bubble chamber is shown in Fig. 10.16. We can see, indeed, that the opening angle is

Fig. 10.16.   Bubble chamber photograph showing photon conversion into a particle-antiparticle pair. (From Ref. 10.11, with permission.)

initially rather small and is then increased due to the opposite sense of the rotation of the momenta of the electron and positron in the magnetic field.

In the next chapter on electromagnetic calorimetry we will describe an electromagnetic cascade. Basically an accelerated electron emits photons. Given sufficiently energetic photons, they will pair produce electrons and positrons, which will in turn make more Bremsstrahlung photons. This is clearly a runaway process leading to a large number of electrons and photons of reduced energy. We expect, then, that an electromagnetic calorimeter which fully absorbs the particles might be $\sim 20\ X_0$ 'deep' or about 11 cm of lead. Glancing back at Chapter 1 we can see that when it becomes energetically possible for photons to make an electron–positron pair, which means photons with energies above twice the electron rest mass or 1 MeV, the cross section for pair production rises rapidly. Pair production is more important for high Z elements than low Z elements as shown by the comparison of carbon and lead shown in Fig. 1.10 and by Eq. 10.40.

10.16 Pair production by charged particles

A closely related process is pair production by a charged particle. It may be thought of as Bremsstrahlung of a 'virtual' photon which 'decays' into an e^+e^- pair. From a diagrammatic viewpoint, there is another vertex with respect to Bremsstrahlung where $\gamma \rightarrow e^+e^-$. We expect that the pair production cross section is reduced by a factor α. However, integrating over the distribution of the energies of the pairs, and comparing the two processes at the same value of y, the ratio of the photon energy to the incident energy, we find (in the high energy 'completely shielded' approximation):

$$\frac{d\sigma_{pair}/dy}{d\sigma_B/dy} \sim \frac{\alpha/\pi(1/y^3)}{1/y} \tag{10.43}$$

$$y \cong {}'E_{\gamma}'/E$$

Therefore, the pair production cross section is much larger at low y then the Bremsstrahlung cross section. This effect is sufficient to compensate for the power of α that is the cost of the additional $\gamma \rightarrow e^+e^-$ vertex. For example, the integral of Eq. 10.43 for $y > 5\%$ is $(\sigma_{pair}/\sigma_B) = 1.24$. Thus in Fig. 6.12 the energy loss in iron for pair production is comparable to that lost due to Bremsstrahlung.

10.17 Strong and electromagnetic interaction probabilities

The cross section for photons incident on nucleons at high energies approaches a constant as seen in Chapter 1. At high energies the main photon process is pair production.

$$\sigma_N \sim A^{2/3} \lambda_p^2$$

$$\sigma_B \sim (Z\alpha)^2 \alpha \lambda_e^2 \tag{10.44}$$

$$\lambda_I / X_o \sim (Z/A)/5.1 A^{2/3}$$

It is interesting to note that the nuclear cross section, as we saw in Chapter 1, goes as $A^{2/3}$ times the Compton wavelength of the proton squared. By comparison, the Bremsstrahlung cross section goes as Z^2 because of coherence. Since this is an electromagnetic process with three vertices, (Fig. 10.13), we pay the penalty of a factor α^3. Even so, the electromagnetic Bremsstrahlung cross section is comparable to the nuclear cross section for $A \sim 3$ (lithium). A glance at Table 1.2 shows that for Be and heavier elements the characteristic length for radiation, $(X_o)_{Be} = 65.2$ g/cm^2 is less than the characteristic length for nuclear interaction, $(\lambda_I)_{Be} = 75.2$ g/cm^2. It is amusing that this particular electromagnetic process is stronger than the 'strong' interaction because of coherence effects. We will explain how the fact that $X_o \gg \lambda_I$ for heavy elements is exploited to provide calorimetric 'particle identification' in Chapter 13.

Exercises

1. Suppose an electron falls freely with acceleration of g near the surface of the earth. What is the power (dipole) radiated? Is it significant?

2. Work out the problem of scattering off the well, with phase shift given in Eq. 10.12. Use Appendix B to get started. Match the wave function and the first derivative at $r = a$.

3. Work out the details of the kinematics going from Eq. 10.15 to Eq. 10.16. Use Appendix A to get started.

4. Show that the maximum Compton electron recoil energy is $T_{max} = \omega_0[1 - m/(m + 2\omega_0)]$. For a 1 MeV photon incident, what is the ratio of T_{max}/ω_0?

5. Use Eq. 10.17 and the expression for the minimum deflected photon energy to find the maximum ratio of CM and laboratory solid angles, $d\Omega/d\Omega^* \sim (\omega_0 m/2\omega^2)$. For a 1 MeV photon what is the ratio?

6. Calculate the wavelength shift for angles of 45, 90, and 135 degrees. Check Fig. 10.6 to show that the expected $(1 - \cos\theta)$ behavior is observed in the data.

7. Show that $\dfrac{d\gamma}{dt} = \gamma^3(\boldsymbol{\beta} \cdot \mathbf{a})$, as quoted in Eq. 10.22.

8. Explicitly work out the algebra connecting Eq. 10.22 and Eq. 10.23.

9. Consider a 1 GeV electron in circular motion with a radius of 1 m. What is ω? What is $\omega_c = 3/2(\gamma^3\omega_0)$? Using Table 1.1 for the Planck constant, confirm the numerical result quoted in Eq. 10.31.

10. Consider isotropic emission of a photon in the CM frame. In the lab frame show that this implies a uniform distribution of energy in the lab frame for monochromatic CM photons (see Appendix A). What is the maximum laboratory energy of the photon? The minimum?

11. Using the approximate expression in Eq. 10.36, integrate over frequency and ignore logarithmic factors. Show that the Bremsstrahlung cross section in lithium is ~ 5.4 mb. Since the pp cross section is ~ 40 mb, the Z^2 vs $A^{2/3}$ behavior of the cross sections implies dominance of EM over strong sections in heavy elements.

12. Use Eq. 10.40 to estimate the radiation length for lead, ignoring logarithmic factors. Use Table 1.2 to check that the approximate result is in reasonable agreement with the exact result.

13. What is the physical size of a 20 radiation length absorber used to almost totally absorb a high energy electron if made of carbon, aluminum, iron, lead and air?

References

[1] *Classical Electrodynamics*, J.D. Jackson, John Wiley & Sons, Inc. (1962).
[2] *Modern Physics*, R.L. Sprooll, John Wiley & Sons, Inc. (1956).
[3] *Relativistic Kinematics*, R. Hagedorn, W.A. Benjamin, Inc. (1964).

[4] *Classical Theory of Fields*, L.D. Landau and E.M. Lifshitz, Addison-Wesley (1951).

[5] *Conduction of Electricity Through Gases*, J.J. Thomson and G.P. Thomson, Cambridge University Press (1928).

[6] *X-rays in Theory and Experiment*, A.H. Compton and S.K. Allison, van Nostrand Company, Inc. (1935).

[7] A.H. Compton, *Phys. Rev.* **21** 175 (1923).

[8] *Quantum Theory of Radiation*, W. Heitler, Oxford University Press (1954).

[9] *Quantum Physics of Atoms, Molecules, Solids, Nuclei and Particles*, Robert Eisberg and Robert Resnick, John Wiley & Sons, Inc. (1974).

[10] *Principles of Modern Physics*, Robert B. Leighton, McGraw-Hill Book Company, Inc. (1959).

[11] *Special Relativity*, A.P. French, W.W. Norton & Company, Inc. (1968).

Part IIIB

Energy measurements

The energy of a particle can be destructively detected by initiating a series of reactions whereby the total particle kinetic energy is deposited in the detector. This technique is called calorimetry. The energy is locally deposited so that the particle position is measured to a degree of accuracy specified by the transverse fluctuations in the energy deposition. The particle energy is measured to a level of accuracy specified by the uniformity of the detector medium (the 'constant term') and the level of active sampling of the detector with respect to total volume (the 'stochastic term'). Thus, the calorimetric technique provides a measurement of **p**, as defined in Fig. I.1. However, particle identity is lost so that subsequent redundant measurements are not possible.

11
Electromagnetic calorimetry

A hen is only an egg's way of making another egg.
Samuel Butler
It grows – it must grow; nothing can prevent it.
Mark Twain

In this chapter we will begin the discussion of calorimetry by looking at electromagnetic showers. The reason to begin with the electromagnetic processes is that they are somewhat better understood and better modeled than the strong interaction processes which we will study in the subsequent chapter. Calorimetry itself refers to the destructive detection of the energy of an incident particle. Its root comes from the Greek word for heat. The idea is to absorb all the energy in a detecting medium and by that means to record the energy of the incident particle. We can make redundant measurements of charged particle momenta by first tracking them in a magnetic field and then absorbing them in a calorimeter. This redundancy is often very useful in cleaning up the backgrounds which are always present in the study of a rare process. Detection in a low mass medium allows for a measurement of the velocity (Chapters 2, 3, 4), momentum (Chapters 5, 6, 7, 8) and secondary decay vertex (Chapter 9) of a charged particle. Subsequently the energy can be absorbed in the calorimeter.

11.1 Radiation length and critical energy

The underlying physics of electromagnetic calorimetry has been touched on previously. The radiative process for electrons is defined by the Bremsstrahlung cross section and the characteristic length for that is the radiation length, X_o. (See Chapter 10.)

$$(dE_B/E) \sim (\rho dx)/X_o$$

$$X_o \sim [180 \text{ g/cm}^2][A/Z^2]$$ (11.1)

$$t = x/X_o$$

The radiation length in g/cm^2 goes as A/Z^2. We characterize the depth in material by expressing it in radiation length units, t. Note that we implicitly do

237

'particle identification' by observing radiative processes, since at energies <100 GeV only electrons and photons are sufficiently relativistic to emit significant radiation (see Chapter 13).

The other quantity, which is vitally important for describing the electron energy loss mechanism, is the critical energy, E_c.

$$(dE/dx) \sim -E_c/X_o$$

$$E_c \sim [550 \text{ MeV}]/Z \qquad (11.2)$$

$$y = E/E_c$$

The critical energy is that energy above which radiative processes dominate. We discussed E_c in Chapter 6, showing loss mechanisms in lead as a function of energy and the $1/Z$ dependence of the critical energy. We define the electron energy in units of the critical energy, y. As an example, in lead the radiation length is 0.56 cm, the critical energy is about 8 MeV, and the minimum ionization loss is 1.13 MeV g/cm^2 (Table 1.2). The ionization energy losses are expected to be effectively universal (see Chapter 6).

$$dE_I/dx \sim [3 \text{ g/cm}^2][Z/A] \qquad (11.3)$$

Minimum ionizing particles should lose about 1.5 MeV g/cm^2 in all elements.

11.2 The electromagnetic cascade

Now consider the cascade process which occurs when a high energy electron hits a block of material. The incoming electron will first Bremsstrahlung. If the energy of the Bremsstrahlung photon is sufficiently high it will, in turn, produce an electron–positron pair. The pair partners will then each Bremsstrahlung. Clearly, the processes of Bremsstrahlung and pair production imply that a runaway 'shower' process will occur leading to a rapid, geometric, increase in the number of particles with depth.

This 'cascade' process will continue until the secondary particles are no longer energetically capable of multiplying. At that point the maximum number of shower particles, N_{max}, exists. Beyond the depth of 'shower maximum', t_{max}, the number of particles dies away due to ionization range out (Chapter 6) in the case of the electrons and Compton scattering (Chapter 10) and photoelectric absorption (Chapter 2) in the case of the photons.

A 'cloud chamber' photograph of an electromagnetic cascade developing in lead plates spaced by a gaseous detection medium is shown in Fig. 11.1. We can see the shower begin in the upper plates, build up to shower maximum, and

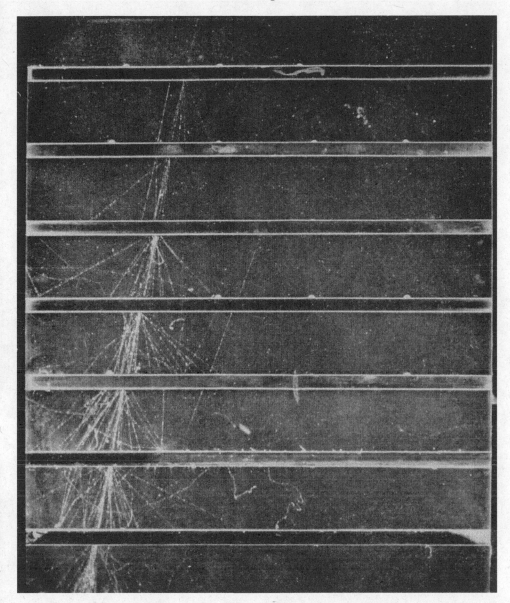

Fig. 11.1.   Cloud chamber photograph of an electromagnetic cascade developing in spaced lead plates. (From Ref. 11.11, with permission.)

then begin to die away. This is a visual realization of the words just used to describe the cascade.

A very schematic view of an electromagnetic cascade is shown in Fig. 11.2. The horizontal axis is the depth of the shower in radiation length units, t, which is, as we said, the characteristic length scale. In reality there are fluctuations

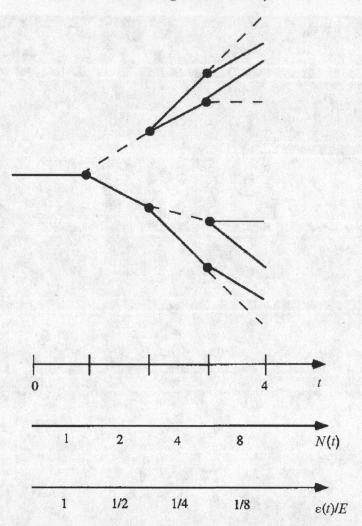

Fig. 11.2.   Schematic of an electromagnetic cascade for the first four generations indicating depth t, number in the cascade N, and fraction of the incident energy carried by each particle. Electrons and positrons are indicated as ——, while photons are represented as – – –.

in the interaction points. We use the mean which occurs at one X_o into the material for the first interaction and, on average, at depths spaced by one X_o for each subsequent interaction. We also show the number of shower particles, N, as the cascade develops as a horizontal scale and the fractional energy per particle, ε/E, in the cascade where E refers to the incident energy.

As the number of particles increases, their energy decreases. For example, in Chapter 10 we showed the energy spectrum for electrons in pair production. Since it is basically uniform from zero to the photon energy, we again ignore the fluctuations and assign the mean energy of $^1/_2$ the photon energy. A similar approximation is made for Bremsstrahlung. The fluctuations both in the interaction points and in the secondary energies are ignored in the interest of clarity.

Let us explore the 'decay' kinematics very briefly. Consider a particle of mass M, energy E, momentum p, which 'decays' into massless particles. The relativistic transformation from the CM (starred) frame to the lab frame is $\varepsilon = \gamma_M \varepsilon^*(1 + \beta_M \cos\theta^*)$, where $\varepsilon^* = M/2$ is the daughter CM energy, and $\gamma_M = E/M$. For isotropic decay, $d\varepsilon = \gamma_M \beta_M \varepsilon^* d(\cos\theta^*) = $ constant. The energy distribution of the daughter is uniform. Clearly, $\langle \varepsilon \rangle = \gamma_M \varepsilon^* = E/2$, since $\langle \cos^* \rangle = 0$, and ε is distributed uniformly from $\varepsilon \sim 0$ to $\varepsilon \sim E$. This discussion should serve to motivate the model which appears in Fig. 11.2. Other kinematic details are given in Appendix A.

We observe that the number of particles $N(t)$ in the cascade grows geometrically with the depth in the cascade. In what follows $\varepsilon(t)$ and $N(t)$ are the symbols used for the particle energy and number of particles as a function of depth.

$$\varepsilon(t) = E/2^t = E/N(t)$$

$$N(t) = 2^t \tag{11.4}$$

The multiplication processes go on until the mean energy of the particles in the cascade is equal to the critical energy. As the energy falls below the critical energy multiplicative processes are no longer possible and the maximum number of particles is reached. Beyond that depth the cascade dies off.

$$E_c = E/2^{t_{max}} = \varepsilon(t_{max}) = yE$$

$$t_{max} \sim \ln y \tag{11.5}$$

$$N_{max} \sim E/E_c = N(t_{max}) = y$$

For example, a 3.2 GeV shower in Pb has shower maximum at a depth of $t_{max} = 6$. At shower maximum there are about 400 particles in the cascade.

11.3 Energy – linearity and resolution

The depth at which 'shower maximum' occurs, t_{max}, increases logarithmically with the scaled incident energy y. The number of particles at shower maximum, N_{max}, increases linearly with energy, with a scale factor equal to the critical energy. Therefore, if the response of a 'calorimeter' is proportional to the number of shower particles, we expect that the calorimeter is a linear device.

This is not a completely trivial statement. Human beings do not have linear sensory transducers. Our ears and eyes respond logarithmically to the sound and light levels of our environment respectively. This method of response allows us to perceive a large dynamic range at the expense of nonlinear sensitivity. For example, the volume control on a radio appears to be linear, but that is simply the dial label; the actual power level is logarithmic.

Note that the length needed for full shower containment, t_{max}, goes as $\sim \ln E$ while for tracking the length required for fixed measurement error rises linearly with energy ($dp/p = ap$). This fact explains the preferred use of calorimetry in applications requiring the measurement of very high energy particles.

We now find a 'total path length', L, for all generations in the shower by converting the sum of all particles in the shower to an integral. We assume that each particle in the shower goes one X_o in depth before interacting. We find that the total path length in the shower is proportional to the incident energy.

$$L \sim X_o \sum_{i=1}^{t_{max}} N(t)$$

$$\frac{L}{X_o} \sim \int_0^{N_{max}} N(t)dt \sim \int_0^{t_{max}} 2^t dt \tag{11.6}$$

$$\sim (E/E_c)/\ln 2 = N_{max}[\ln 2]$$

The number of particles in the shower $N(t)$ is strongly peaked at shower maximum. Hence it is not too surprising that L is $\sim X_o N_{max}$. This result can be thought of as the N_{max} particles of the last 'generation' going one final X_o in depth before they range out by losing energy E_c to ionization. We assume that the detected energy signal is proportional to L.

Fluctuations in the shower development lead to fluctuations in the number of particles in the shower. Thus, the path length, L, exhibits stochastic fluctuations since the cascade is a random process. Therefore, the fluctuation in the number of particles should go as $dN \sim \sqrt{N}$ (see Appendix J). We expect that a measurement of the energy contains a 'stochastic term' due to those fluctuations and consequently that the fractional error in the calorimeter energy measurements should go as $1/\sqrt{E}$.

$$N_{max} = E/E_c \tag{11.7}$$

$$dE/E \sim 1/\sqrt{E} \sim dN/N$$

For a 3.2 GeV shower in lead, there is a 5% fluctuation in the 400 particles which exist at shower maximum.

This behavior of the error is in distinction to gaseous tracking devices, which have fractional error, $dp/p \sim p$. Therefore, the fractional momentum error degrades linearly with momentum in a tracking device of fixed length whereas the fractional energy error for a calorimeter, in principle, improves with energy. This is another basic reason why detectors of very high energy particles focus on calorimetry since it is one of the few detectors whose performance actually improves with increasing energy.

11.4 Profiles and single cascades

In Fig. 11.3 is shown the longitudinal energy distribution of six individual 170 GeV electron showers. This particular device consists first of a stack of 40 lead plates, each 1/8″, 0.57 X_o, thick interspersed with scintillator (see Chapter 2) as the active sampling element. Each depth segment is read out independently.

What is striking about these individual electron showers is that the shower development has few fluctuations. The shape in depth is essentially the same for each shower reflecting the large number of particles in the cascade, $N_{max} \sim E/E_C = 170$ GeV/8 MeV $\sim 21\,250$. We know there is a fluctuation in the first interaction point. Given the exponential nature of the distribution of free paths, Chapter 1, we know that the fluctuation in the path length to the first interaction, t_1, is X_o, while the mean first interaction point occurs at a depth X_o, i.e. $\langle t_1 \rangle = 1$, $\langle \sigma_{t1} \rangle = 1$. Looking at the events shown in Fig. 11.3 we see that the main fluctuation, on an event by event basis, is just the fluctuation in the interaction point which is roughly ± 2 plates. The shower shape at high energy has both energy sharing fluctuations and interaction point fluctuations of the first and subsequent generations. However, the shower contains so many particles that the latter are washed out.

These observations of real electromagnetic showers lead us to a one dimensional parameterization of the longitudinal behavior of the shower.

$$\frac{dE/E}{du} = [u^a e^{-u}]/\Gamma(a+1)$$

$$u = bt, \quad b \sim 1/2 \tag{11.8}$$

$$t_{max} \sim \ln y \sim (a-1)/b$$

The depth t is defined with respect to the initial interaction point. In the shower parameterization the variable u is, up to multiplicative constants, t. The variable a is proportional to $\ln y$. The variable b is weakly energy dependent and weakly dependent on material. The power law behavior, u^a, implies a fast geometric rise in the energy deposition due to the multiplication process in the

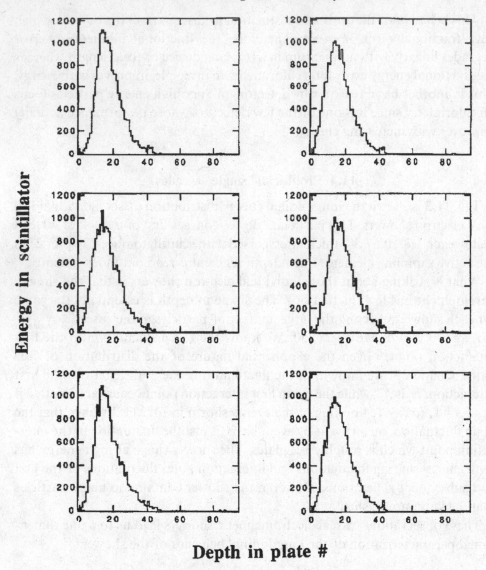

Depth in plate #

Fig. 11.3.   Longitudinal energy profiles of six individual 170 GeV electrons incident on a stack of 40 lead plates of 1/8″ interspersed with scintillator active elements.

cascade, whereas the exponential fall off, e^{-u}, is expected at larger depths beyond shower maximum where the mean particle energy is less than the critical energy. The gamma factor is there only to normalize the total shower energy to the incident energy, E.

The shower maximum location, t_{max}, moves to greater depths as the energy increases. For example, a 3.2 GeV shower has $t_{max} \sim 6$, while $t_{max} \sim 9$ for a 64 GeV shower. It is also weakly dependent on the critical energy which goes as

$1/Z$, Eq. 11.2. Thus, the depth in radiation length units of the location of shower maximum is somewhat dependent on the type of material of which the calorimeter is made, $t_{max} \sim \ln y = \ln (E/E_C) \sim \ln (ZE)$.

As a numerical example for shower development in lead, with the critical energy of 7.5 MeV, a photon with $\varepsilon = E_c$ has a mean free path of 10 g/cm^2, see Chapter 1. Since the radiation length in lead is 6.4 g/cm^2, see Table 1.2, the mean free path of a photon near the critical energy is about 1.6 radiation lengths. That explains why the exponential fall off given in Eq. 11.8 is basically in units of radiation lengths up to a numerical factor of order 1. The range of an electron with $\varepsilon \sim E_c$, is also roughly $\sim X_o$.

11.5 Sampling devices

Since the calorimeter is an energy measuring device, its most important physical characteristic is the energy resolution. Let us consider a 'sampling calorimeter' where the shower is developed in high Z plates where most of the energy is lost. The shower energy is sampled in thin plates of low Z material such as scintillator plastic or liquid argon. The small fraction of the energy which is actively sampled should be proportional to the total absorbed energy. The basic model for the sampling calorimeter is that the shower develops in the inert high Z material and is sampled in the active low Z material.

If we assume that the shower evolves rapidly, i.e. that there are high Z sampling plates with thickness $> X_o$, then there is no correlation between consecutive active sampling layers. Thus the total number of particles traversing the active layers, N_s, is the total number in the shower or the total path length, $L = E/E_c$, (Eq. 11.6) divided by the thickness of the inert layers, Δt, placed between the active samples. The geometry is shown schematically in Fig. 11.4.

$$N_s = L/\Delta t \qquad (11.9)$$

$$\sim (E/E_c)/\Delta t$$

The particles traversing the sampling layers are presumed to all register, either directly by charged particle ionization or indirectly with photon Compton or photoelectric absorption accompanied by recoil ionization. The error due to fluctuations in the fraction of the energy appearing in the active sampling layers, δE, should be the fluctuation in the total number of sampled particles, N_s. It is customary to define what is called the 'stochastic coefficient' in the energy resolution, a_{samp}. The stochastic term, due to sampling fluctuations, depends on the thickness of the inert plates and scales as $\sqrt{\Delta t}$. Obviously, if

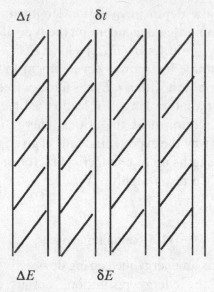

Fig. 11.4.   Definitions for sampling calorimetry with active samples, of thickness δt and sampled energy δE, and passive absorber, with Δt, ΔE.

the plates become very thick, the sampled energy is not a good representation of the full shower energy and the fluctuations are increased.

The energy deposited by ionization, ΔE, in going through a plate of thickness Δt at the critical energy E_c is just $E_c \Delta t$. Clearly low E_c means more shower particles, $N_{max} \sim E/E_c$, which implies a smaller stochastic term.

$$\left(\frac{dE}{E}\right)_{samp} = 1/\sqrt{N_s} \equiv a_{samp}/\sqrt{E}$$

$$a_{samp} \sim \sqrt{E_c \Delta t} \qquad\qquad (11.10)$$

$$\sim \sqrt{\Delta E}$$

As a simple example, a 1 GeV electron with a critical energy of 7.5 MeV, has a maximum number of shower particles, $N_{max} = 131$. With $^1/_2$ radiation length sampling for the thickness of the plates, $\Delta t = {}^1/_2$, we find the total number traversing the active layers to be 262 which yields a sampling stochastic term of 6.2% due to fluctuations in the sampling of the number in the shower. This will later be shown to be a reasonable scale for the errors observed in typical sampling calorimeters constructed to measure electromagnetic showers (see Fig. 11.5).

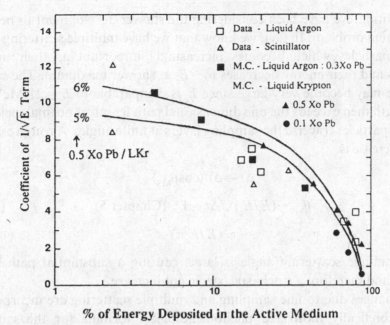

Fig. 11.5. Stochastic term coefficient as a function of sampling fraction W for a number of calorimeters. (From Ref. 11.9, with permission.)

11.6 Fully active devices

There are other effects that we need to consider in going to a situation with very fine sampling. The logical extension of fine sampling is the 'fully active' calorimeter where the shower developing medium is itself active. Examples of this type of calorimeter are crystals, such as BGO, lead tungstate (PbWO$_4$), or lead glass.

We have assumed previously that counts in the different active layers were uncorrelated. A photon of low energy which is in the cascade will Compton scatter (Chapter 10) or photoelectric effect (Chapter 2) as these are the two largest cross sections at low energies. Below about 1 MeV in lead the photo-electric effect dominates. To set the scale, the mean free path for a 1 MeV photon in lead is about 3 cm, or about 6 X_o. If we go to fine sampling, in order to reduce the stochastic term which goes as $\sim \sqrt{\Delta t}$, the photon can and does cross several sampling layers invalidating our assumptions.

Another assumption we made was that the shower development took place entirely in the absorber, while the detection took place in the thin active layers. Now, clearly, if we go to very fine sampling this assumption also begins to break down, since significant energy is deposited in the active medium, δE, rather than the inert absorber, ΔE.

In addition, we have been considering the shower development to be a one dimensional problem. In fact, we know that we have multiple scattering of the shower particles which becomes increasingly important as their number increases and their energy decreases to $\sim E_c$ at shower maximum. The scattering angle may become very large since E_c is comparable to $E_s = 21$ MeV. The path length then exceeds the one dimensional path length at normal incidence, Δt, since particles traverse the sampling layers at finite angles. A rough estimate of the increase is

$$\Delta t \rightarrow \Delta t / \langle \cos \theta_{MS} \rangle$$

$$\theta_{MS} \sim (E_s / E_c) \sqrt{\Delta t \sim 1} \quad \text{(Chapter 5)} \tag{11.11}$$

$$\sim (E_s / E_c \pi)$$

The multiple scattering angle is large, causing a substantial path length increase and therefore a stochastic coefficient increase.

The changes due to fine sampling and multiple scattering are incorporated in a theoretically motivated phenomenological formula for the sampling coefficient.

$$a_{samp} \sim \left[(1 - W) \left(\frac{\Delta E + \delta E}{\langle \cos \theta_{MS} \rangle} \right)^{\left(\frac{1 - W}{2} \right)} \right] \tag{11.12}$$

$$W = \delta E / (\delta E + \Delta E)$$

The ratio of the sampled energy to the total energy lost in a layer is a parameter, W. Clearly as the parameter $W \rightarrow 0$ we are ignoring the effects of the active sampling layers, and Eq. 11.12 collapses to Eq. 11.10 modified by Eq. 11.11. In the limit $W \rightarrow 1$ we have a fully active detector and $a_{samp} \rightarrow 0$. This unrealistic limit occurs because in this particular formulation the only errors that are being considered in the measurement of the energy are those of a stochastic nature. We will discuss other sources of error later.

Data on a_{samp} taken with different levels of sampling, W, in calorimeters are shown in Fig. 11.5 where the vertical axis is the stochastic coefficient in percent. We see that indeed a 2% stochastic term can be achieved for fine sampling detectors which are of order 80% active. If the sampling fraction becomes 10% there is a rapid increase to a $\sim 8\%$ stochastic coefficient. The lines drawn on Fig. 11.5 are just Eq. 11.12.

A photograph of a typical R&D sampling calorimeter, in this case installed in a test beam, is shown in Fig. 11.6. This object has lead plates interspersed with scintillator in its electromagnetic section and iron plates, rather thicker, interspersed with scintillator in the following section which is used to intercept

Fig. 11.6.   Photograph of a sampling calorimeter showing interspersed lead and scintillator (EM) and iron and scintillator (HAD) planes. (Photo – Fermilab.)

strongly interacting particles or hadrons (see Chapter 12 and Chapter 13). The data plotted in Fig. 11.3 come from the initial 40 lead plates of the most 'upstream' part of this detector. Each layer has independent photomultiplier (Chapter 2) readout in order that shower development can be studied. In contrast, a photograph of a fully active homogeneous calorimeter medium using lead glass is shown in Fig. 11.7. This is a transparent medium with high density with respect to ordinary glass. Other homogeneous media are other glasses and crystals such as BGO.

An energy level diagram for a 'generic' crystal detector is shown in Fig. 11.8. Energy deposition by ionization causes an electron to be excited from the valence band (VB) to the conduction band (CB). The electron is quickly trapped by an impurity level A^* and a radiative transfer causes the subsequent emission of a photon of energy, $E_\gamma = E_{A^*} - E_A$. Note that this photon energy is $< E_g$ so that it exits the crystal without absorption. The crystal is thus transparent to its own impurity state emission. An example is 'thallium activated'

Fig. 11.7.   Photograph of a fully active homogeneous calorimetric medium (lead glass). (Photo – Fermilab.)

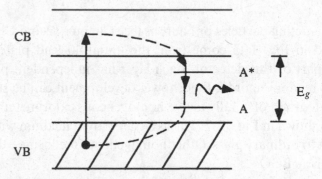

Fig. 11.8.   Energy level diagram for a crystal detector.

(i.e. A, A* states) NaI. In NaI the band gap is $E_g = 6$ eV so that visible light, since it has energy less than the gap energy, is not absorbed. The thallium impurity state has a decay time ~ 200 ns and the emission $A^* \rightarrow A + \gamma$ has a spectrum of photons peaked near $\lambda \sim 4100$ Å or $E_\gamma \sim 3.1$ eV.

11.7 Transverse energy flow

The shower has, so far, been considered a one dimensional object, which is roughly correct. The physics defines the transverse momentum scale and in this particular case the physics has to do with atomic electrons. The characteristic transverse momentum in these collisions is the mass of the electron, m_e. If we are early in the development of a high energy shower, typical angles are a few milliradians, $\theta \sim m_e/\varepsilon \sim 0.51$ mrad at $\varepsilon = 1$ GeV. However, as we get deeper in the shower the cascade particles become softer as they share the parents, energy among themselves. In this case the shower rapidly becomes more isotropic.

$$\langle p_T \rangle \sim m_e$$

$$\langle \theta \rangle \sim m_e/\varepsilon(t) \tag{11.13}$$

$$\langle \theta \rangle_{SM} \sim m_e/E_c$$

The typical angle of a particle at a depth t is defined by the energy of the particles in the shower at that depth, $\varepsilon(t)$. At shower maximum the typical energy is the critical energy, so that the typical angle of a particle at shower maximum is $\sim m_e/E_C$. For example, in lead, the pair production angle at shower maximum is about 4°.

In fact, multiple scattering dominates the transverse shower development rather than finite production angles since $E_s \gg m_e$. Hence, we expect multiple scattering to give the dominant contribution to the transverse size of a shower (see Eq. 11.11).

$$\langle p_T \rangle_{MS} \sim E_s \sqrt{t}, \quad \sqrt{t} = 1$$

$$\langle \theta \rangle_{SM}^{MS} \sim E_s/E_c \tag{11.14}$$

It is traditional to define a transverse size, the Moliere radius, which can be thought of as the transverse distance that a particle at the critical energy goes in traversing the last radiation length before it dies off.

$$r_M \sim E_s X_o/E_c = \langle \theta \rangle_{SM}^{MS} X_o$$

$$r_M \sim (7 \text{ g/cm}^2)(A/Z) \tag{11.15}$$

The Moliere radius, r_M, depends on E_s, the multiple scattering energy (Chapter 5), E_c the critical energy (Chapter 6) and X_o the radiation length (Chapter 10). Referring to previous chapters for numerical constants, we find that the Moliere radius is fairly constant expressed in g/cm^2. For most of the periodic table, where $A/Z \sim 2$, we have a Moliere radius of about 14 g/cm^2. Denser materials have physically smaller Moliere radii.

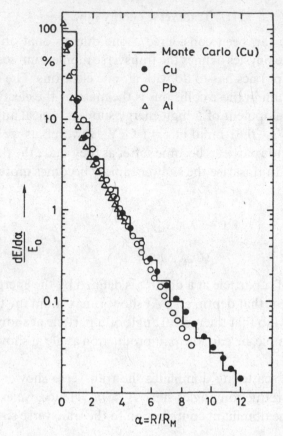

Fig. 11.9. Transverse electromagnetic cascade energy distribution as a function of the number of Moliere radii from the incident electron direction. (From Ref. B.1, with permission.)

A shower distribution of transverse radius in Moliere units is shown in Fig. 11.9. In this plot the entire shower has been integrated over in depth. Comparing Monte Carlo model (Appendix K) results with measurements on aluminum, lead and copper, we see that the shape in r_M units is approximately universal, as expected from Eq. 11.15, and that 90% of the energy is contained within about 1 Moliere radius while 95% is contained within about 2 Moliere radii, which, for example, is about 3.0 cm in lead.

The radius for energy containment clearly depends on where you are in depth in the shower. In Fig. 11.10 is plotted the transverse electromagnetic shower energy distribution in a Monte Carlo model as a function of the depth in the shower. We see that there is a sharp peak extending out to a depth of $\sim 0.1–0.3$ radiation lengths which broadens as we go deeper in the shower. This is a logarithmic plot so that even at a depth of 9 radiation lengths there is containment within $\sim 0.5 X_o r$.

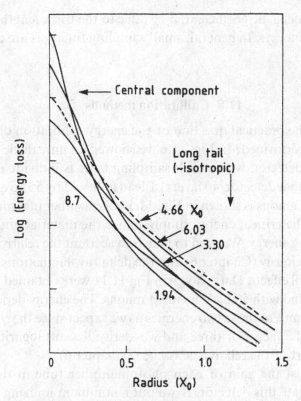

Fig. 11.10. Transverse electromagnetic cascade energy distribution in arbitrary units as a function of the number of radiation lengths from the incident electron direction. Curves are shown at different depths in the shower, displaying the broadening of the shower as the depth increases. (From Ref. 11.10, with permission.)

There are other subtle effects due to multiple scattering. As we follow particles down in energy, we find that $\sim 30\%$ of the energy in a shower is deposited by electrons with energies less than 1 MeV for high Z absorbers. These particles at the end of the shower are almost isotropic in angle due to multiple scattering; as many go backwards as forwards. Therefore, it is important in making computer models of showers (see Appendix K) to keep track of particles with quite low energies, as they are important to the total energy deposition mechanism in calorimeters.

In addition to the multiple scattering increase in the path length there are also delta rays, which we considered in Chapter 5. They lead to ionization fluctuations. This effect may dominate over the sampling fluctuations. It is clearly most important for thin active layers. For example, if we were sampling with a gaseous detector instead of a plastic or a liquid, and if the thickness of the sampling layer were only $10^{-3}\,g/cm^2$, (~ 1 cm of gas), we would have a 40%

increase in the stochastic coefficient, a_{samp}, due to the track length fluctuations caused by the delta rays. In general, small sampling fractions are dangerous (if inexpensive).

11.8 Calibration methods

Finally there is the practical question of the energy calibration of a calorimeter. How is that performed? In Fig. 11.6 we showed an unrealistic device, since this is an R&D detector, where each sampling layer is individually read out. Data taken with that detector, 40 layers of lead followed by 55 layers of 1″ thick iron, for incident muons is shown in Fig. 11.11. Remember that muons at energies well below the critical energy simply ionize the material and deposit the minimum ionizing energy. We need to take into account the relativistic rise that we discussed previously (Chapter 6) and the delta ray fluctuations (Chapter 5), but these are small effects. Data shown in Fig 11.11 were obtained with 15 GeV incident muons and with 50 GeV incident muons. The energy deposit per layer is effectively the same at those two energies as we expect since the γ of the muon has only changed a factor of three and we 'derived' a soft logarithmic dependence on γ as part of the relativistic rise (see Chapter 6).

We could adjust the gain of each photomultiplier tube in the individual samples to calibrate this detector. If we put a minimum ionizing particle into the detector, we would then get out the same signal in all layers. That is one way in which we could use external particles to calibrate the detector. Note that for financial reasons we normally construct a homogeneous detector, optically sum the sampling layers (light pipe, Chapter 2) and use only a few phototubes or other transducers. In that case, movable radioactive sources (Chapter 6) are sometimes used to deposit a fixed amount of energy in each sampling layer in order to monitor the individual responses.

We could also expose the detector to electrons of different known energies using a beam of particles prepared by momentum analysis in magnets (Chapter 7). Thus we can map out the energy response, resolution, and linearity of the device using 'test beams'. The use of muons would then monitor the initial beam calibration and insure that the system response is time independent. Recall that muons are available as the largest component of sea level cosmic rays. (See Table 6.1.)

Electromagnetic calorimeters are major subsystems of all collider detectors, because photons and electrons are thought to be pointlike fundamental particles (see Chapter 13). Therefore, the energy and position of these particles should be recorded (Fig. I.1). For both photons and electrons, the electromagnetic shower can be localized transversely to a size which is of order r_{M}.

(a)

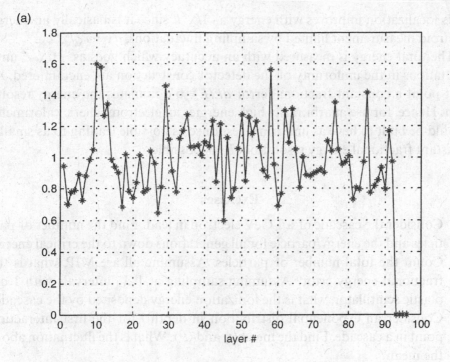

(b)

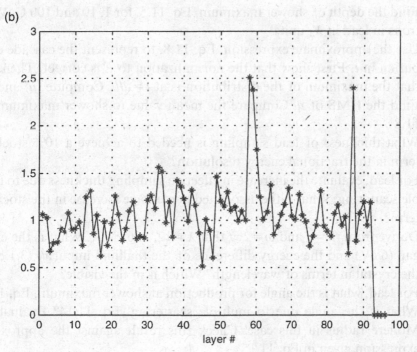

Fig. 11.11. Response of each sampling layer of the calorimeter shown in Fig. 11.6 to incident muons indicating the potential use of muons as a means of calibration. (a) 15 GeV muons, (b) 50 GeV muons.

This localization improves with energy as $1/\sqrt{E}$ since it is basically an energy centroid measurement limited by sampling fluctuations, $dx_r \sim r_M/\sqrt{E}$.

The total energy is measured with an accuracy which goes as $\sim 1/\sqrt{E}$ until limitations in the uniformity of the detector construction are encountered. At that point a constant fractional error $dE/E \equiv b$ dominates the energy resolution. Hence, for use at arbitrarily high energies an electromagnetic calorimeter should be built to have as uniform a response as possible leading to as small a constant fractional energy resolution, b, as possible.

Exercises

1. Consider the cascade of a 1 GeV electron in lead. Find the number of particles and the energy/particle for all generations down to the critical energy. Count the total number of particles. Assuming all are MIP, what is the fractional energy error? Assuming sampling at 1 X_o intervals with 1 cm plastic scintillator, what is the ionization energy deposited by the cascade?

2. Consider an exponential distribution in depth t for the first interaction point in a cascade. Find the mean $\langle t \rangle$ and $\langle t^2 \rangle$. What is the fluctuation about the mean?

3. Find the depth of shower maximum, Eq. 11.5, for 1, 10 and 100 GeV electrons in lead in X_o units.

4. Use the approximate expression, Eq. 11.8, to represent the cascade distribution in t. First show that the normalization to E is correct. Then show that the maximum of the distribution is at $t = a/b$. Compute $\langle u \rangle$ and $\langle u^2 \rangle$. Find the RMS of u. Compare the mean value to shower maximum, Eq. 11.8.

5. What thickness of lead sampling is needed to achieve a 10% stochastic term in the fractional energy resolution?

6. For lead, evaluate the increase in effective sampling thickness due to multiple scattering. What is the associated percentage increase in the stochastic term?

7. Derive from $f\lambda = c$ and $\omega = 2\pi f$ that $\lambda = 12$, 566 eV·Å. What is the energy gap (6 eV) and the energy difference of the thallium impurity (3.1 eV) in the crystal in terms of wavelength? Which is in the visible?

8. For lead, what is the angle for production at shower maximum, Eq. 11.13? What is the angle due to multiple scattering, Eq. 11.14? Evaluate the Moliere radius in this case. Check this result against the approximate expression given in Eq. 11.15.

9. In Exercise 1, the ionization energy deposited by the cascade was estimated. Suppose the calorimeter has 20 plates with 1 X_o samples of lead

sandwiched with 1 cm plastic scintillator. Find the ionization energy deposited by a high energy muon traversing the 20 layers of sampling. Assuming a calibration using electrons in a test beam, how much energy is estimated to correspond to a through going muon?

10. What is the expected transverse position resolution in lead for a 10 GeV electron? 100 GeV?

11. In Exercise 1, assume a gas detector with density appropriate to air. Estimate the ionization energy deposited by a 1 GeV cascade. Note that each MIP has a probability to emit a delta ray. Use the results of Chapters 5 and 6 to estimate the probability that some MIP in the gas emits a delta ray equal to the total shower ionization energy, indicating a major mis-measurement of the energy.

References

[1] *High Energy Particles*, B. Rossi, Prentice-Hall, Inc. (1952).
[2] *Proceedings of the Calorimeter Workshop*, M. Atac, Fermilab (1975).
[3] *Tristan eP Working Group Report*, S. Iwata, Nagota University, DPNU-3–79 (1979).
[4] G. Gratta, H. Newman and R.Y. Zhu, 'Crystal calorimeters in particle physics,' *Annu. Rev. Nucl. Part. Sci.* **44** 453 (1994).
[5] W. J. Willis and V. Radeka, 'Liquid-argon ionization chambers as total absorption detectors,' *Nucl. Instrum. Methods* **120** 221 (1974).
[6] *Calorimetry in High Energy Physics*, D.F. Anderson, M. Derrick, H.E. Fisk, A. Para and C.M. Sazama, World Scientific Publishing Co. (1991).
[7] U. Amaldi, 'Fluctuations in calorimetry measurements,' *Phys. Scr.* **23** 409 (1981).
[8] D.F. Crawford and H. Messel, *Phys. Rev.* **128** 2352 (1962).
[9] 'Performance and limitations of electromagnetic calorimeters', T.S. Virdee in *Calorimetry in High Energy Physics*, Capri, Italy (1991) World Scientific.
[10] *Experimental Techniques in High Energy Physics*, T. Ferbel, Addison-Wesley Publishing Co., Inc. (1987).
[11] *Principles of Modern Physics*, R.B. Leighton, McGraw-Hill Book Company Inc., New York (1959).

12

Hadronic calorimetry

All exact science is dominated by the ideal of approximation.
Bertrand Russell

We discuss in this chapter the calorimetric measurement of hadronic energy. A 'hadronic' particle is one which interacts strongly. Examples are protons, pions, and kaons. Note that calorimetry is useful in measuring all energy, even that which is invisible to tracking detectors such as neutral particles like γ, K_L^0, and n. There are several new physical concepts which we need to understand in order to make intelligent design choices for a calorimeter used to measure the energy of strongly interacting particles. In Chapter 1 the concept of the mean free path for a nuclear interaction was introduced and numerical values were given in Table 1.2. Since the geometric cross section goes as the square of the size of the nucleus, a_N^2, and since the nuclear radius scales as $a_N \sim A^{1/3}$, the nuclear mean free path in g/cm^2 units scales as $A^{1/3}$ (Chapter 1).

$$\lambda_I \sim [35 \text{ g/cm}^2]A^{1/3}$$

$$v = x/\lambda_I$$

(12.1)

Therefore, in analogy with the radiative mean free path (Chapter 11), we define a nuclear mean free path which is the depth, v, in units of nuclear interaction lengths. Numerically, in lead the inelastic nuclear mean free path is 194 g/cm^2 (Table 1.2), so that the radiation length divided by the interaction length is only about 3.3%. Therefore, if a photon is incident on a 20 X_o deep electromagnetic calorimeter it will be almost totally absorbed (Chapter 11). In comparison, an incident hadron would have a probability simply to interact once hadronically of only 48%. As we will see, those that do interact create a cascade which occupies several λ_I in depth. Thus we can do 'particle identification' by exploiting the differences in the longitudinal development of showers between electrons/photons and hadrons. The transverse size of electromagnetic and hadronic showers also differs. We will return to this theme in Chapter 13.

12.1 Properties of single hadronic interactions

A 'streamer chamber' photograph of a high energy single hadronic interaction, 300 GeV π^-–p, is shown in Fig. 12.1. Note that there are a large number of emitted secondary particles. Note also that the interaction must have a limited transverse momentum imparted to the secondary particles because they are typically moving forward, following the incident particle direction.

Data on the mean number of charged particles emitted in a hadronic collision as a function of the center of mass energy, $\sqrt{s}$, are shown in Fig. 12.2. Note that this is a semi-logarithmic plot which means that a first approximation to the data is that the mean multiplicity, $\langle N \rangle$, grows only logarithmically with the energy. Thus the CM energy, $\sqrt{s}$, is largely going into the kinetic energy of the secondary particles rather than into simple particle production, which would imply $N \sim \sqrt{s}$. Reading off Fig. 12.2, at $\sqrt{s} = 100$ GeV, the mean number of secondary charged particles, $\langle N \rangle$, is about 15.

The emitted particles are expected to be, on average, $2/3$ charged and $1/3$ neutral since they are mostly pions and we assert that all pions, π^+, π^-, π^0, are emitted with equal probability. Pions are the lightest hadrons and are therefore preferentially emitted in hadronic interactions. Neutral pions quickly decay into two photons which share the parent π^0 energy. These photons then participate in the hadronic cascade as an electromagnetic component which can be treated in ways already described in Chapter 11.

In describing electromagnetic cascades we said that the multiplication process continued until the electrons reached the critical energy. There is a similar situation in the hadronic interaction. Just as we need a minimum energy for a photon to create an electron–positron pair, a hadron needs a minimum laboratory energy in order to make a pion. The threshold lab energy, E_{TH}, for $\pi p \rightarrow \pi \pi p$ is roughly twice the pion mass, $E_{TH} \sim 2m_\pi$ (see Appendix A). Therefore, we have a characteristic energy below which the charged particles simply ionize since they are unable to multiply by making more pions.

$$E_{TH} \sim 2m_\pi = 0.28 \text{ GeV} \qquad (12.2)$$

As was seen in Chapter 11, there is a limited transverse momentum for electromagnetic reactions, $\langle p_T \rangle_{EM}$ which is set by the electron mass. For hadronic reactions it is set by some poorly understood collective phenomena and $\langle p_T \rangle_h$ is empirically about 0.4 GeV although it too depends logarithmically on $\sqrt{s}$.

$$\langle p_T \rangle_{EM} \sim m_e$$

$$\langle p_T \rangle_h \sim 0.4 \text{ GeV} \qquad (12.3)$$

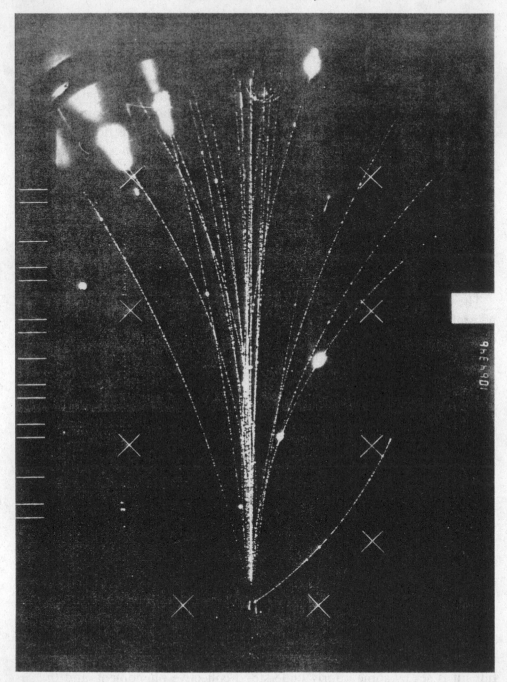

Fig. 12.1.   Streamer chamber photograph of a 300 GeV π^- hadronic interaction illustrating the large number of secondary particles and the limited value of the transverse momentum. (From Ref. B.1, with permission.)

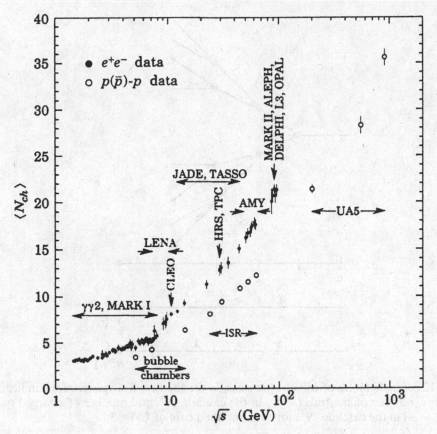

Fig. 12.2. Data on the mean number of charged particles as a function of the CM energy $\sqrt{s}$ showing the $\langle N\rangle \sim \ln s$ dependence. (From Ref. 1.1, with permission.)

For a single hadronic interaction we expect the multiplicity to depend logarithmically on the laboratory energy since $s \sim 2E_a m_b$ (Appendix A). For a dense high Z material we expect the radiation length to be much shorter than the interaction length. Thus, the electromagnetic part of the cascade develops quickly. We also expect the neutral to charged ratio, f_o, in a single interaction to be equal to 1/3, $(\pi^0/\pi^+ + \pi^0 + \pi^-)$.

$$N \sim \langle N\rangle \sim \ln E$$

$$X_o/\lambda_I \ll 1 \qquad (12.4)$$

$$f_o = 1/3$$

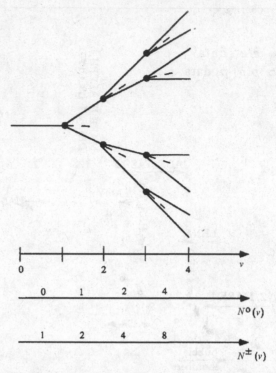

Fig. 12.3.   Schematic of a hadronic cascade for the first four generations indicating depth v, number of neutrals ($---$) in the cascade, N°, and number of charged particles (———) in the cascade, $N^\pm$, for the simplified case of $\langle N\rangle = 3$.

12.2 The hadronic cascade – neutrals

What do we expect for the full hadronic cascade? In each interaction we expect to produce, on average, $^2/_3$ charged pions and $^1/_3$ neutral pions. We expect the mean multiplicity to depend only logarithmically on the parent energy and we can, in lowest order, ignore this slow variation. It is very important to understand that π^0 production is an irreversible part of the hadronic cascade because, when π^0s are produced, they immediately decay to photons. Photons then develop as an electromagnetic cascade which has a characteristic length scale X_0 which is typically much shorter than λ_0. Therefore, the π^0s produced are quickly absorbed and 'drop out' of the shower. The energy transport to greater depths is thus carried out largely by the charged pions.

A schematic picture of a hadronic cascade for the first four generations is shown in Fig. 12.3. In the interest of clarity we assume that an incident charged hadron makes only three pions per interaction. The number of neutral particles, N°, and the number of charged particles, $N^\pm$, as a function of depth v in the cascade is plotted. The neutral pions are represented as dashed lines. The charged particles carry the shower deeper into the material and they deposit only ionization energy which is small.

Assuming equal partition of the energies (Chapter 11) and labeling the generations by an index ν, we can work out the number of charged and neutral particles in the cascade in this simple minded model. On average, the index ν is the depth in interaction length units, as defined above. Note that, as in the electromagnetic case, we ignore fluctuations and simply assign the mean values for the successive interaction points, ν, for the particle secondary multiplicity, N, and for energy fraction, $\varepsilon(\nu)$. We also ignore the logarithmic energy dependence of N.

$$N \equiv \langle N \rangle$$

$$N^0_{(\nu)} \sim N f_o [N(1-f_o)]^{\nu-1}$$

$$N^\pm_{(\nu)} \sim [N(1-f_o)]^{\nu 1} \tag{12.5}$$

$$N^0(\nu) = N f_o [N(1-f_o)]^{\nu-1} = N f_o [N^\pm(\nu-1)]$$

$$N(\nu) = N^0(\nu) + N^\pm(\nu) = N[N^\pm(\nu-1)] \sim N^\nu$$

Under these assumptions, the number of π^0s, $N^0_{(\nu)}$, scales as a power of generation number ν, as does the number of charged particles, $N^\pm_{(\nu)}$. The total energy going to neutral particles, E_o, is then straightforwardly summed up. The effective neutral fraction is 'f_o' $= E_o/E$.

$$E_o \sim \sum_\nu \varepsilon(\nu) N^0_{(\nu)}, \qquad \varepsilon(\nu) = E/N^\nu \tag{12.6}$$

$$E_o \sim E f_o \sum_\nu [1-f_o]^{\nu-1}, \text{ total neutral energy}$$

The number of generations, in exact analogy to the electromagnetic case, is defined by the condition at 'shower maximum' that the mean energy of charged particles in the cascade is equal to the threshold energy for pion production, E_{TH} or $\varepsilon(\nu_{max}) = E/N^{\nu_{max}} = E_{TH}$.

$$\text{'}f_o\text{'} \sim f_o \sum_1^{\nu_{max}} (1-f_o)^{\nu-1} \tag{12.7}$$

As a numerical example, a 250 GeV hadronic shower is shown in Table 12.1. The total number of secondary particles emitted per interaction at that center of mass energy, $\sqrt{s} \sim 22$ GeV, is taken to be $\langle N \rangle = 9$ which is compatible with the data given in Fig. 12.2. The total number of generations is rather small, $\nu_{max} = 3$. The particle energy per generation drops by a factor of 9 in each generation. The number of particles in the cascade increases at each interaction, with $1/3$ of the produced particles dropping out of the cascade at each generation as

Table 12.1.   *Simplified model for a hadronic cascade developed by a 250 GeV incident pion.*

Generation v	$\varepsilon(v)$ (GeV)	$N^{\pm}(v)$	$N^{\circ}(v)$	$E_{o}(v)$ (GeV)
0	250	1	0	0
1	28	6	3	84
2	3.1	36	18	56
3	0.35	216	108	38
				178 GeV

Note: $f_{o} = 1/3$, 'f_{o}' $= 0.71$

π^{0}s and $2/3$ continuing to initiate the next generation. Therefore, the energy deposited by the π^{0}s, which starts out at zero before the first interaction, falls off slowly and sums to 178 GeV. Although the fundamental interaction of a hadron gives a neutral energy fraction f_{o} of $1/3$, in this particular example the energy deposited by neutrals in the hadronic cascade is, 'f_{o}' $= 71\%$.

Clearly, in the limit where the incident energy becomes infinite, since the π^{0}s are the irreversible part of the cascade, all of the energy in the hadronic cascade will be neutral. The charged pions will merely serve to transport the shower from generation to generation, depositing energy only by ionization which is small (~ 0.3 GeV in going $\lambda_{I} \sim 200$ g/cm^2 in Fe – see Table 1.2). We estimate ionization losses in Table 12.1 as the total path length for charged particles, 259, times the MIP ionization loss in traversing λ_{I} or 78 GeV. In the low energy limit where $v_{max} \to 1$, there is only one generation and the neutral fraction is $f_{o} = 1/3$.

Some more serious and exact model (see Appendix K) results for the effective neutral fraction are shown in Fig. 12.4 where we see that the absolute number of produced π^{0}s increases as the log of the energy and the fraction of energy in the electromagnetic part of the cascade rises from $\sim 1/3$ at low energy, $E \sim 3$ GeV, to about 70% at 50 GeV. Therefore, the simple minded estimate given in Eq. 12.7 is in reasonable accord with a detailed Monte Carlo result.

12.3 Binding energy effects

There are still other new physics effects in hadronic interactions. One of them is that the medium itself is excited by amounts which are substantial on the scale of the incident energy. In an electromagnetic cascade we mostly use the nucleus simply as a way to balance energy and momentum. The medium itself does not participate in the fundamental cascade processes such as

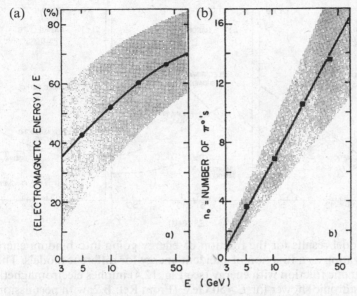

Fig. 12.4. (a) Model results for the growth of 'f_o' as a function of energy. (b) Total number of π° in the cascade as a function of incident energy. (From Ref. B.2, with permission.)

Bremsstrahlung and pair production and the nucleus is inert. The scale of energy transfer in a collision is $\sim m_e$ which is small on the scale of the nuclear binding energy (~ 8 MeV/nucleon, see Fig. 12.17). However, other processes do exist so that the medium does become activated when exposed to a photon shower. For example, photodisintegration, such as $\gamma + Fe^{56} \to Mn^{54} + p + n$.

For the case of hadrons, the interactions themselves are with the nucleons in the nucleus and the energy transfer that we quoted in Eq. 12.3 is the characteristic energy transfer. Hence, we see that the nucleus is disrupted. Therefore, substantial amounts of energy will go into 'binding energy losses' where the nucleus is excited. Detailed Monte Carlo results for three different models are shown in Fig. 12.5. The fraction of energy going into electromagnetic showers rises with energy, and the fraction going into binding energy effects and nuclear excitation or $\pi^\pm$ ionization ('charged particles' in Fig. 12.5) falls. The electromagnetic piece dominates at high energy, $E \gtrsim 50$ GeV. The next most important effect is excitation of the medium, followed by ionization losses. Although the results depend in detail on the model, the non-electromagnetic parts of the cascade are of order $^1/_2$ of the total at $E \sim 50$ GeV. The fact that the model results differ is an indication that the hadronic cascade is not very well modeled. A new generation of computer simulations (see Appendix K) has recently dramatically improved the situation.

This binding energy ultimately appears in the calorimeter when the nucleus

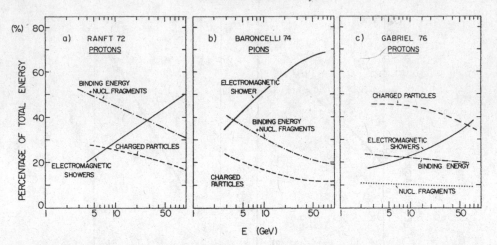

Fig. 12.5.   Model results for the fraction of energy going into binding energy, neutrals, and ionization as a function of incident energy for different models. The rise of the electromagnetic fraction with energy (see Fig. 12.4) implies electromagnetic dominance of the hadronic shower for $E > 50$ GeV. (From Ref. B.2, with permission.)

de-excites emitting a slow neutron, photon, or other fragments. It may well be, however, that the calorimeter is not sensitive to the deposit or that the deposit comes late enough that it is not detected because we are trying to do rapid pulse shaping in a high rate environment. We expect that a smaller fraction of the energy of an incident particle is detected in a hadronic calorimeter than in an electromagnetic calorimeter where, as we said, the medium itself does not participate and the time for energy deposit is consequently rapid. We also expect that there are stochastic fluctuations in the amount of energy given to the medium and that those fluctuations will limit the attainable energy resolution in a hadronic calorimeter.

12.4 Energy resolution

In analogy to the discussion in Chapter 11 on electromagnetic calorimetry, we expect that the stochastic contribution to the resolution for the hadronic case is going to be partially defined by the total number of particles in the cascade which in turn is controlled by the energy cutoff in the multiplication process. The electromagnetic shower is terminated at E_c while the hadronic cascade is cut off at the threshold energy for π production, E_{TH}.

$$dE/E = a/\sqrt{E} \oplus b \equiv \sqrt{a^2/E + b^2}$$

$$a_h/a_e \sim \sqrt{E_{TH}/E_c} \quad \text{(Chapter 11)} \tag{12.8}$$

$$\sim 6$$

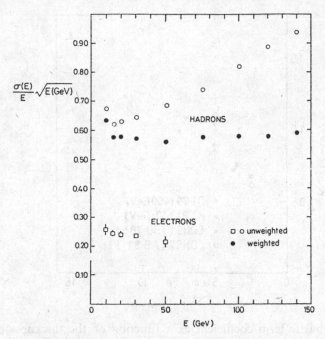

Fig. 12.6. Stochastic term coefficients for incident electrons and hadrons as a function of energy for iron sampling calorimeters. (From Ref. 12.9, with permission.)

We expect the ratio of hadronic to electromagnetic stochastic coefficients to be of order 6. Data is plotted for a specific calorimeter in Fig. 12.6 showing stochastic term coefficients for both electromagnetic and hadronic incident particles as a function of energy. We see that this particular device is fairly coarse sampling, with ∼20% stochastic coefficient, a, for electrons and ∼60% for hadrons. There are also other physical processes going on such as binding energy and nuclear excitation in the medium, which have their own inherent fluctuations. Thus, hadronic calorimetry will never be as precise as electromagnetic calorimetry because the underlying physical processes have larger intrinsic fluctuations.

On the subject of the stochastic coefficient, we argued that it should depend on the square root of the sampling thickness, and showed some data for electromagnetic calorimetry which confirmed that idea. Exactly the same arguments carry through for hadronic calorimetry. In Fig. 12.7 we show the stochastic coefficient for hadronic calorimetry using steel sampling plates as a function of the thickness of the plates. Note that the expectation that the coefficient increases as $\sqrt{\Delta v}$ is roughly obeyed. However, extrapolation to $\Delta v = 0$ does not yield an infinitely good resolution because there are non-sampling fluctuations. Compare the electromagnetic case, Fig. 11.5, to the hadronic case. We can never obtain the very good resolution that can be obtained in a fully

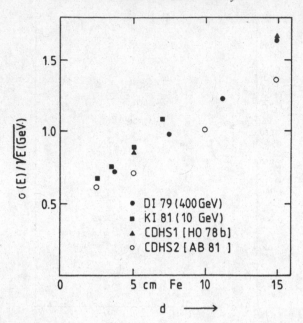

Fig. 12.7. Stochastic term coefficient as a function of the thickness of the steel plates. A functional dependence $dE/E \sim \sqrt{E_{TH} \Delta v/E}$ is expected. (From Ref. B.1, with permission.)

active electromagnetic calorimeter such as $PbWO_4$. One of the mechanisms which cause the worsened resolution is called 'non-compensation' which will be discussed later in this chapter.

12.5 Profiles and single cascades

We can, in analogy to the electromagnetic profile, Chapter 11, write down an energy profile for hadronic interactions.

$$\left(\frac{dE}{E}\right) = \left[\frac{u^a e^{-u}}{\Gamma(a+1)} f_o du + \frac{\omega^c e^{-\omega}}{\Gamma(c+1)}(1-f_o)d\omega\right]$$

(12.9)

$$\omega = dv, \quad d \sim 1 \quad \text{(Eq. 11.7)}$$

There are two pieces, as we might expect, in the energy deposition. There is a piece which is basically the electromagnetic fraction, f_o, developing from the first interaction as an electromagnetic shower. The second piece refers to the energy transport due to the charged particles with a characteristic length scale which is roughly equal to the nuclear absorption length. Although the actual energy deposition is mostly due to photons, energy transport by charged pions leaves $v \sim 1$ as the relevant scale.

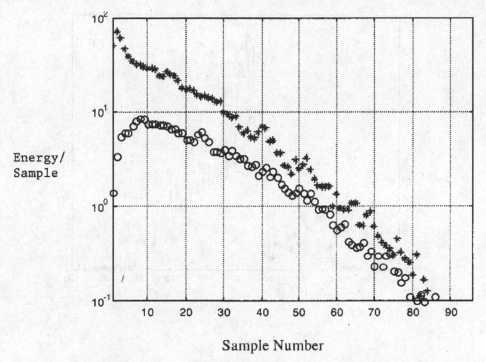

Energy/
Sample

Sample Number

Fig. 12.8. 'Profile' for 270 GeV pions incident on a stack of 90 Pb plates each of $^3/_4''$ thickness. ○ – Summed over many events. ∗ – Longitudinal profile with variable inter-action point subtracted.

Data for a 270 GeV pion beam incident on a stack of $^3/_4''$ lead plates sampled by scintillation counters is shown in Fig. 12.8. One plot is simply the samples of single events summed over many hadrons. We observe a rather smooth featureless curve. It is important to be careful and not jump to conclusions but to look at the profile with the first interaction point location subtracted. Remember that there is a fluctuation in the location of the first interaction point which occurs on the average at a depth of $1\,\lambda_I$ but with a fluctuation equal to the mean or $\pm\lambda_I$. The profile with the first interaction point subtracted is also shown. Now we see the two characteristic length scales. The initial electro-magnetic part is evidently due to the π^0s produced in the first collision which carry off about $^1/_3$ of the energy. The component with a characteristic length of order the nuclear interaction length in lead has an energy area roughly $^2/_3$ of the total.

Therefore the shape given in Eq. 12.9 appears to be a plausible representa-tion of the behavior of the energy deposition summed over many interactions, i.e. the energy profile. In the electromagnetic case, we showed individual events, Fig. 11.3, and argued that the main fluctuation was the fluctuation in the inter-action point. In Fig. 12.9 are shown the energy depositions of six single events

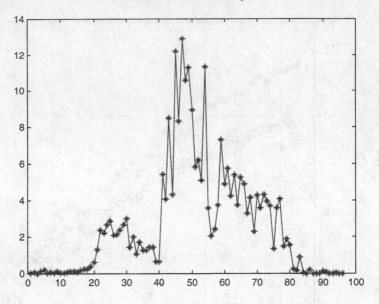

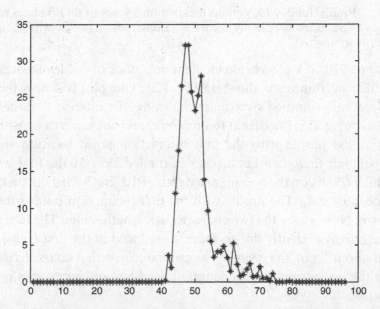

Fig. 12.9.   Longitudinal energy profiles of six individual 270 GeV pions incident on a stack of 40 lead plates of $^1/_8''$ followed by 55 iron plates of $1''$. The fluctuations inherent in hadron cascades are very evident. Vertical scales show energy/sample, horizontal scales show sample numbers.

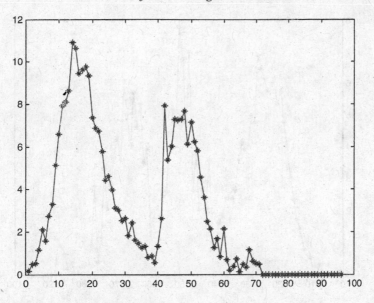

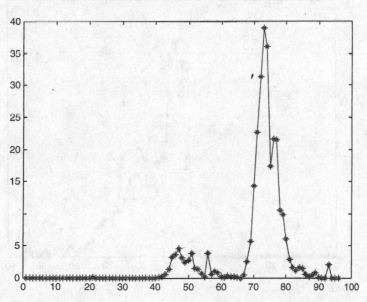

Fig. 12.9. (*cont.*)

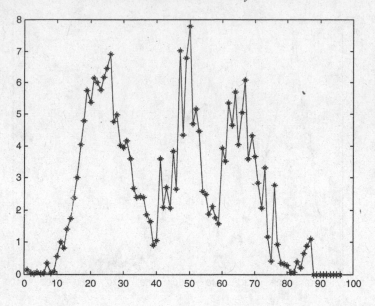

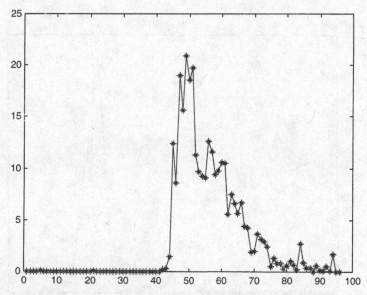

Fig. 12.9.   (cont.)

in a calorimeter for 270 GeV incident pions. Notice that the fluctuations inherent in the cascades are very large. Thus, the concept of a well reproduced shower shape developing in the same way that electromagnetic cascades develop is simply not valid. The profile given in Eq. 12.9 is sufficient to describe the average behavior of a hadronic shower but extremely misleading if thought to apply on an event by event basis.

A better model of the event by event behavior of a hadronic cascade can be derived using some data on individual pions incident on a stack of $^3/_4''$ lead plates. Six events of the data set are shown in Fig. 12.10. The data can be understood to consist of a sum of 'electromagnetic clusters' of fitted energy characterized by a fixed shape extending in depth by $\sim 20\ X_o$ ($\sim$six samples). The electromagnetic cascades develop and die out in five or six plates, see Fig. 12.8. The recurrence of such electromagnetic clusters at different depths indicates the transport of energy by charged particles with a length scale equal to λ_I ($\sim$nine samples). Data with a lead calorimeter was chosen because lead has a large ratio of $\lambda_I/X_o \sim 30$ which serves to separate the electromagnetic clusters in depth.

12.6 *e/h* and the 'constant term'

Let us consider now the situation for a calorimeter where the response to electrons and the response to hadrons are different. We might expect this to naturally be the case because of the large amount of energy lost in hadronic interactions to the binding energy. The electron response is defined to be *e*, while the response to hadrons is *h*. Since the π^0 decays rapidly into photon pairs and since photon showers are similar to electron showers (see Chapter 10), the π^0 response is about the same as the electron response.

Consider the case of an incident hadron with effective neutral fraction 'f_o'. Due to statistical fluctuations, 'df_o' (see Appendix J), in the neutral fraction there will be a limit to the achievable resolution of the hadronic measurements. If the *e/h* response of the calorimeter is not equal to 1, the neutral fluctuations cause an irreducible fractional energy resolution.

$$E \sim [e'f_o' + h(1 - 'f_o')]\ E_{IN}$$

$$\left(\frac{dE}{E}\right)_{df_o} \sim |e - h|'df_o' \sim |e/h - 1|'df_o' \tag{12.10}$$

As an example, for a 200 GeV hadron with mean total pion multiplicity of 9, the neutral multiplicity would be 3. Therefore, the fluctuation of the neutral

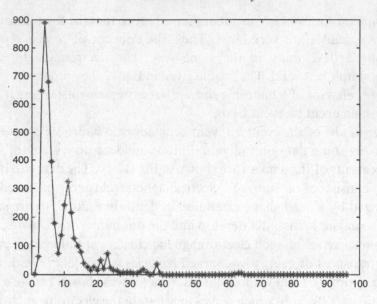

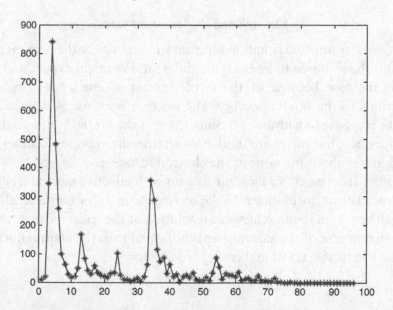

Fig. 12.10.   Longitudinal energy profiles of six individual 270 GeV pions incident on a stack of 90 plates of ³/₄″ thickness. The electromagnetic clusters with energy transport by charged particles are a prominent feature of the plots. Vertical scales show energy/sample, horizontal scales show sample number.

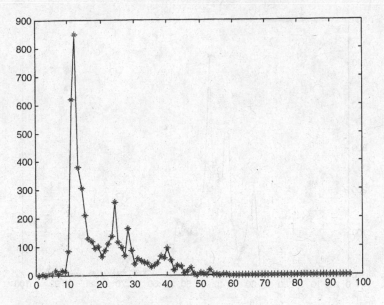

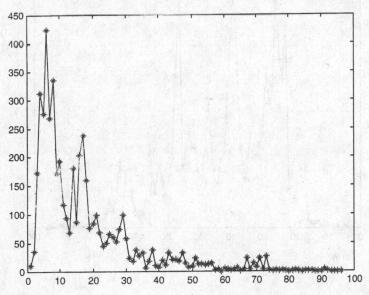

Fig. 12.10. *(cont.)*

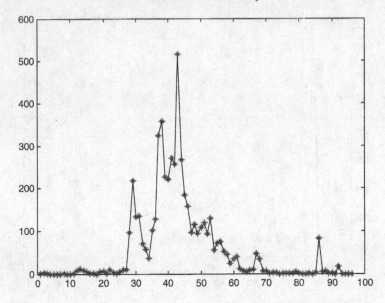

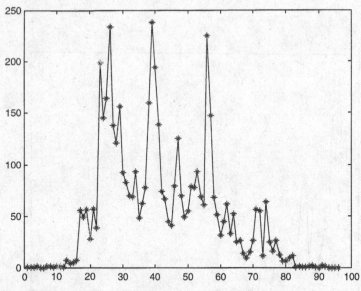

Fig. 12.10.   (*cont.*)

particle fraction, 'df_o', if taken to be equal to the fluctuation in the first inter-action, df_o, is $\sim 17\%$. The result is in good agreement with the results of a complex Monte Carlo program (Ref. 12.3).

$$df_o' \sim df_o \sim \sqrt{\langle N^o \rangle}/\langle N \rangle \sim 0.17$$

$$\left(\frac{dE}{E}\right)_{df_o} \sim |e/h - 1| \left[\sqrt{\langle N^o \rangle}/\langle N \rangle \right] \tag{12.11}$$

Hence, a calorimeter with a 20% different response to electrons and hadrons, $e/h = 1.2$, which is of the same size as the binding energy losses, would have a fractional energy error or 'constant term' of 3.5%. This means that the calorimeter could not measure energy to better than 3.5% even though the sto-chastic term at very high energies $\rightarrow 0$.

It is conventional to write the resolution of a calorimeter as in Eq. 12.8. The stochastic coefficient, a, is due to sampling fluctuations. The 'constant term', b, is due to intrinsic defects in the calorimetry either in the physics, $e/h \neq 1$, or in the non-uniformity of the manufacture of the medium. The fine sampling value of $\sim 50\%$ for $dE/\sqrt{E}$, Fig. 12.7, can be reduced to $\sim 30\%$ in hadron calorimeters designed to have $e/h = 1$.

We have made the approximation that the fluctuations in the effective neutral fraction, 'df_o', are dominated by the fluctuations in the first generation, df_o, which is plausible because that is where most of the energy is deposited (see Table 12.1). We further assumed that the fluctuations in the neutral particle fraction, df_o, are Gaussian with the number of produced neutral particles which we estimate using the data of Fig. 12.2. Note that $(dE/E)_{df_o}$ is not rigor-ously a constant error since as the energy goes to infinity 'f_o' $\rightarrow 1$. If the neutral fraction goes to 1, then the error will go to 0. (Eq. 12.10.)

$$\left(\frac{dE}{E}\right)_{df_o} \sim 1/\sqrt{\ln E} \rightarrow 0 \text{ as } E \rightarrow \infty \tag{12.12}$$

At the low end of the energy scale we expect that the different thresholds cutting off the multiplication processes for electromagnetic and hadronic inter-actions will cause the intrinsic e/h ratio to change. Since the binding energy losses reduce the hadron response, and since they in turn are reduced at low energies, we expect the hadronic response to increase, thus reducing e/h as the energy decreases.

Data on the e/h ratio as a function of incident energy are shown Fig. 12.11. The low energy e/h ratio is about 0.6 rising to a ratio near 1. The transition region occurs over a range of incident energy from 0.5 GeV to 3 GeV, with the low energy scale set by the pion production threshold energy, E_{TH}, of 0.28 GeV.

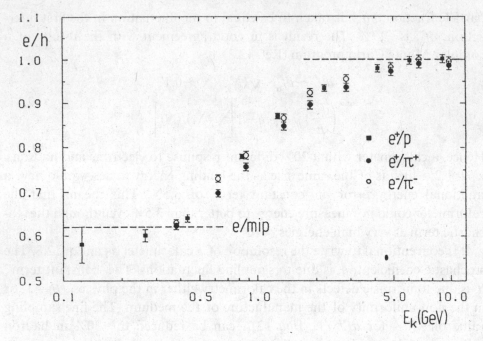

Fig. 12.11.   *e/h* ratio as a function of incident kinetic energy where $e/h \to 1$ as $E \to \infty$. (From Ref. 12.3, with permission.)

The data show that there is an energy dependent hadron energy nonlinearity at low energies even when the *e/h* ratio is approximately 1 at energies of 5 GeV and above. Clearly, the low energy hadron response of a specific calorimeter needs to be well measured because it is difficult to predict. We note that this effect is due to intrinsic physics and cannot be evaded.

The fact that $e/h \neq 1$ also makes the calorimeter a device which responds non-linearly to the energy deposit of a pion. From Eq. 12.10, the π/e response ratio is $\pi/e = ['f_o' + h/e\,(1 - 'f_o')]$. This is only 1, or linear, when $'f_o' \to 1$ at $E \to \infty$ or if $h/e = 1$. For h/e 0.7 we have $\pi/e = 0.8$ or a 20% differential nonlinearity between π and e.

12.7 Transverse energy flow

In the chapter on electromagnetic calorimetry, Chapter 11, we discussed the characteristic transverse size for an electromagnetic shower and defined the Moliere radius. For a hadronic shower we proceed by analogy. We know that the energy per particle at the end of the shower is roughly the threshold energy for pion production, E_{TH}. We also know that the mean transverse momentum of the produced secondary particles $\langle p_T \rangle_h$, is roughly, 400 MeV, Eq. 12.3. The angle for produced secondary particles at hadronic shower maximum would then be the mean transverse hadronic energy over the threshold energy.

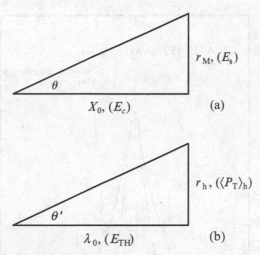

Fig. 12.12. Definitions of characteristic transverse sizes of cascades: (a) electromagnetic cascade, (b) hadronic cascade.

$$\langle \theta \rangle \sim \langle p_T \rangle_h / \varepsilon(v)$$

$$\langle \theta \rangle_{SM} \sim \langle p_T \rangle_h / E_{TH} \qquad (12.13)$$

We define the transverse distance for a hadronic shower particle, r_h, as the distance which is traveled going the last interaction length, λ_I. Since $\langle \theta \rangle_{SM}$ is ~ 1, r_h is the same size as the interaction length itself which is, for example, 16.7 cm in iron (Table 1.2). The definitions for the characteristic transverse sizes of both electromagnetic and hadronic cascades are shown in Fig. 12.12. Note that r_M is driven by multiple scattering while r_h is driven by the transverse momentum of the secondary particles. The reason for the difference is that $E_s > m_e$ but $\langle p_T \rangle_h \sim E_{TH} \gg E_s$

$$r_h \sim \lambda_I \langle p_T \rangle_h / E_{TH}$$

$$\sim \lambda_I \left[\frac{\langle p_T \rangle_h}{2 m_\pi} \right] \sim \lambda_I \qquad (12.14)$$

Data are shown for a one dimensional projection of the transverse position distribution in a uranium sampling calorimeter in Fig. 12.13. Indeed, the distribution of transverse coordinates is on the scale of λ_I. In fact, as we might expect from the previous discussion, things are slightly more complicated. There is a 'core' of electromagnetic energy at small transverse distances due to π^0s depositing a fraction $\sim f_o$ of the energy. In addition there is a distribution with a larger transverse size due to the energy transport of subsequent generations by the charged pions. The transverse shape of a hadronic shower has 'two components'. Data integrated over all depths for 150 GeV pions in a lead sampling

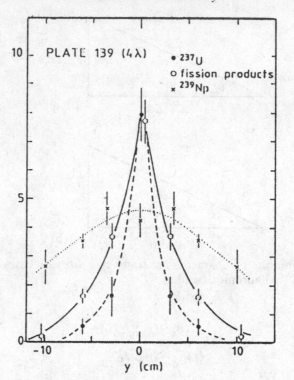

Fig. 12.13.   Data on the one dimensional projected (y) distribution (in cm) of transverse energy in a uranium sampling calorimeter at a fixed depth $4\lambda_I$. (From Ref. 12.3, with permission.)

calorimeter show that one component has a characteristic size of 3.7 cm while the second component has a size of 14.3 cm.

The transverse containment of the electromagnetic and hadronic showers allows calorimeters to be used as position measuring devices as well as energy measuring devices. The shower development is approximately one dimensional. We expect the transverse position resolution to depend on the fluctuation in the number of shower particles if, for example, the energy weighted center of gravity method is used as an estimator of the incident particle transverse position. Therefore, we expect the error in the position measurement to go as $1/\sqrt{E}$. Indeed, for typical electromagnetic calorimetry the error in the transverse position is 5 cm/$\sqrt{E}$ if E is expressed in GeV. For pions the corresponding coefficient is 31 cm reflecting the larger hadron transverse shower.

12.8 Radiation damage

As the new accelerators and space missions come on line, we get to situations where there are large radiation doses. We need, then, to consider radiation damage. Radiation damage will usually reduce the response of the calorimeter. More importantly, since the radiation damage is not uniform over the calorimeter active area, there is an 'induced' non-uniformity of response in the calorimeter. Because there are fluctuations in both electromagnetic and hadronic showers, errors are induced in the energy measurements. Individual showers develop differently in depth and sample different parts of the calorimeter, which have different radiation damage. We can try to evade that non-uniformity by a variety of techniques. This is an interesting topic all by itself and is a crucial aspect of detector design. We refer the interested reader to the references, e.g. Ref. 12.6.

12.9 Energy leakage

The fact that all calorimeters are of finite length, so that energy leaks out the back, must be faced. Fluctuations in shower development lead to fluctuations in the energy leakage and therefore measurement errors. Often 'inert' material is put in front of calorimeters, such as tracking detectors and the like (Chapter 13). For example, a solenoid magnet coil may be placed in front, causing the electromagnetic and hadronic showers to sometimes develop in this inert material. It is a very interesting topic to see how you can 'weight', or over sample, the energy deposit at the exit point of this inert material in order to regain some of the energy resolution. Similarly, with leakage, you can also 'weight' the energy exiting the calorimeter. A plot of the unweighted containment depth for hadrons of different energies is shown in Fig. 12.14. For an incident hadron of 1 TeV energy we need ~ 11 λ_l in order to contain 99% of the cascade energy on average. The required depth only increases as $\ln(E)$, so that calorimeters for high energy experiments need grow only modestly.

Plots of the energy responses of a 7 and 11 λ_l deep calorimeter are shown in Fig. 12.15. Clearly, a low energy 'leakage tail' of the thin calorimeter is seen in which significant energy is lost out the back. We can reduce the tail, Fig. 12.15c, by overweighting the samples at the rear of the calorimeter. By using this technique we can restore some of the resolution of a thin calorimeter which has financial implications. Note the induced 'high side tail' in Fig. 12.15 where we err since we overcorrect some showers, due to the fluctuations in shower development.

Typically, muon detectors are placed behind calorimeters (see Chapter 13) since low energy $(E < E_c^\mu)$ muons only ionize while all other particles (except neutrinos) interact more strongly. However, there is an intrinsic limit placed on the ability to filter out unwanted particles. Shown in Fig. 12.16 is the probability for a hadron cascade to 'punch through' and deposit energy behind a

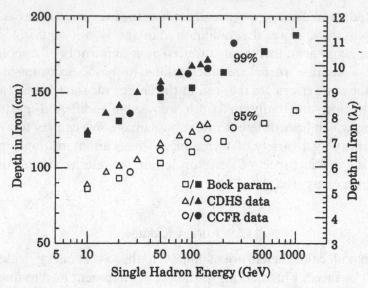

Fig. 12.14.   Required depth in iron for 95%, open, and 99%, closed, average energy containment of a hadronic shower as a function of energy. The slow logarithmic dependence is evident. (From Ref. 1.1, with permission.)

given depth of steel for different energies. After $1\,\lambda_I$ the probability begins to fall exponentially with a characteristic length $(\sim\lambda_I)$ which grows slowly (logarithmically) with energy. This behavior is expected given our previous discussion of the cascade mechanism, $\nu_{max} = \ln(E/E_{TH})/\ln(N)$ (Eq. 12.7).

However, at the few tenths percent level there is another component visible in Fig. 12.16. This component falls off slowly with depth and is of magnitude roughly proportional to E. It is due to the decay of pions to muons in the cascades. These decays occur before the pion is absorbed and they are intrinsic.

Note that on the earth's surface cosmic rays are mostly muons (see Chapter 6). The earth's atmosphere is a thick but diffuse 'calorimeter' of depth ~ 1000 g/cm^2 $\sim 11.4\lambda_I$ of N$_2$ gas (see Table 1.2) which 'leaks' muons since the diffuse 'absorber' allows decay, with a lifetime at rest of $(c\tau)_\pi = 7.8$ m, to compete with cascade absorption due to hadronic interactions. The deep component visible in Fig. 12.16 has a similar origin but its fraction is greatly reduced since in that case the calorimeter is a dense solid absorber. Still, at sufficient depth, the majority of particles remaining in a shower are muons both for ground level cosmic rays and deep in solid steel calorimeters. Fortunately for us, living as we do deep in our atmospheric 'calorimeter', muons only deposit ionization energy in us. It is this leakage which defines the irreducible annual dose of 0.2 rad due to cosmic rays which we are all subject to.

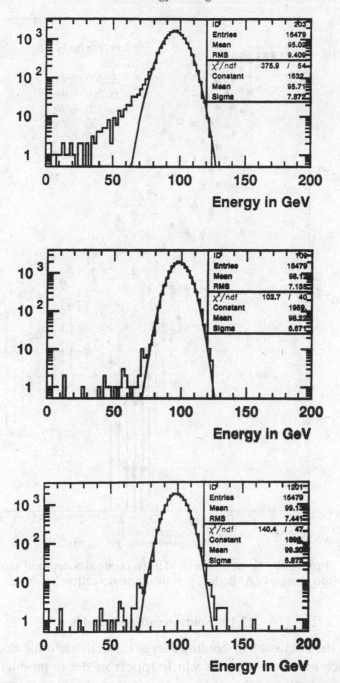

Fig. 12.15. Energy sampled in a calorimeter for 100 GeV incident pions. (a) 7 λ_o deep calorimeter, (b) 11 λ_o deep calorimeter and (c) 7 λ_I deep calorimeter with exit layer overweighting.

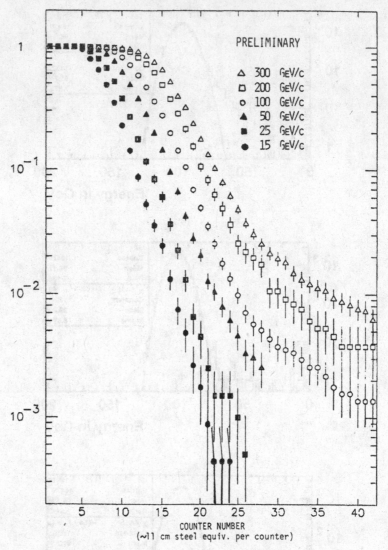

Fig. 12.16.   Probability for at least one MIP to occur at a depth of steel for different incident hadron energies. (A. Bodek, private communication.)

12.10 Neutron radiation fields

Implicit in the discussion of binding energy losses in hadronic calorimetry was the presence of many neutrons which appear as decay products during the nuclear de-excitation of the medium. A plot of the binding energy per nucleon as a function of A is shown in Fig. 12.17. The value is roughly constant with A for $A \gtrsim 15$.

$$B \sim 8 \text{ MeV/nucleon} \tag{12.15}$$

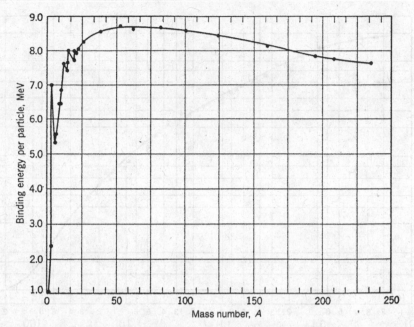

Fig. 12.17.   Binding energy per nucleon in MeV as a function of atomic weight A. (From Ref. 12.1, with permission.)

A transfer of energy appropriate to hadronic cascades, $\langle p_T \rangle_h$, implies that the absorptive medium is excited. These excited nuclei later decay, often with the emission of soft neutrons or photons with energies of order B.

As an aside, we mention that neutrons are unstable in vacuum, decaying into a proton plus an electron plus a massless anti-neutrino. The mass difference is positive, $m_n - m_p - m_e = 0.782$ MeV. The question naturally arises, why are stable atoms made of electrons and nuclei, where nuclei consists of protons and neutrons? The answer is that neutrons in nuclei are bound with negative energy, 8 MeV/nucleon, so that the bound mass difference is negative and the neutron decay reaction is not energetically possible. Thus, in nuclei neutrons are stable. The simplest example is the deuteron, a bound state of a proton + neutron, which requires 2.2 MeV to break apart.

A very crude rule of thumb is that 5 n/GeV are released in a shower and localized near hadronic shower maximum.

$$N_n \sim 5E \text{ (GeV)} \tag{12.16}$$

These neutrons will obviously have kinetic energies, T_n, near B. At these low energies they can only elastically scatter or participate in exothermic neutron capture reactions. For example, the mean free path for a 5 MeV neutron in steel is 9.4 cm.

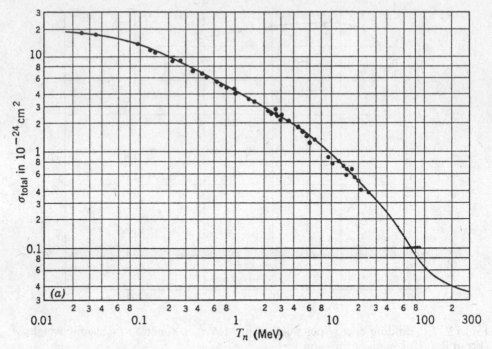

Fig. 12.18.   n–p elastic cross section as a function of n kinetic energy. (From Ref. 12.1, with permission.)

Scattering off a heavy nucleus transfers little energy (see Appendix A). Thus the n become 'transparent' to the medium and diffuse out of the calorimeter as a 'neutron gas'. Typical energies at the point of transparency are low.

$$\langle T_n \rangle \sim 1 \text{ MeV} \tag{12.17}$$

The elastic n–p scattering cross section as a function of T_n, from 10 keV to 300 MeV, is shown in Fig. 12.18. At 1 MeV the cross section is 4 b, and a rough $\sigma \sim 1/\sqrt{T_n} \sim 1/v_n$ behavior is observed, where v_n is basically the incoming flux (see Chapter 1).

12.11 Neutron detection

The kinematics for n + A elastic scattering imply a recoil neutron kinetic energy, T, which has maximum and minimum limits. Algebraic details are given in Appendix A.

$$\left(\frac{A-1}{A+1}\right)^2 \leq \frac{T}{T_n} \leq 1 \tag{12.18}$$

Clearly for large A, $T \sim T_n$ and little energy is transferred to the medium. Basically, the neutrons just bounce their way out of a heavy material with a constant energy, suffering no energy loss in collisions.

It is also clear that for $A = 1$, or n–p elastic scattering, the neutron can lose a substantial fraction of its energy. This loss mechanism can be exploited by detecting the recoil proton from the initially almost free proton in the hydrocarbons of a plastic scintillator, for example. This mechanism works well for $T_n \sim 1$ MeV neutron detection. The large cross section, evident in Fig. 12.18, means that the scintillator need not be very thick. A 1 cm, x, thick counter of scintillator has a fractional mean free path of $x/\langle L \rangle = N_o \rho \sigma / A \sim 15\%$ scattering probability for a 1 MeV neutron. The average energy transfer of 0.5 MeV can be detected efficiently (Chapter 2). It is also clear that scintillator based sampling calorimetry will efficiently absorb the 1 MeV neutrons leaking out of a high Z absorber. Kinematics also explains why hydrogenous, low Z, material is used to shield against slow neutrons.

The 'neutron gas' will be absorbed in the light sampling layers of a calorimeter if they contain hydrogenous material. The gas also diffuses while the neutrons slow down by elastic collisions. This gas may be treated using the diffusion equation, which we already have used for a molecular gas (Chapter 8).

$$n\mathbf{v} = \mathbf{j} = -D(\boldsymbol{\nabla} n) \tag{12.19}$$

The neutron number density is n and the velocity of the neutron 'fluid' is $\mathbf{v}$. The methods used in Chapter 8 may be adopted in the present case, where D is the neutron diffusion coefficient.

The slowing down process ends when the neutrons are in thermal equilibrium with the medium, $T_n \sim kT$. How can we detect neutrons with such a small energy? A plot of the cross section of n on various nuclei down to thermal ($kT \sim 1/40$ eV) energies is shown in Fig. 12.19. In this figure the unitary limit $4\pi/k^2$ $4\pi a_N^2$ is indicated (see Chapter 1, S wave unitarity), as is the geometric cross section. These two values for σ bracket the experimental values on the low and high ends respectively, which however span six orders of magnitude. Carbon has an approximately constant cross section, ~ 6 b, while boron has a $1/v_n$ behavior.

The thermal neutrons can be detected by capture reactions. Typical exothermic reactions are n+p$\rightarrow$D+γ (2.2 MeV) and n+^{6}Li$\rightarrow$^{3}H+α (4.76 MeV). The emitted photon can be detected via Compton recoil, while the α must be directly stopped in a lithium doped solid state detector due to its short range. These and other capture reactions are the basis for dosimetry in the case of slow (thermal) neutrons.

The design of detectors in some environments must take into account the

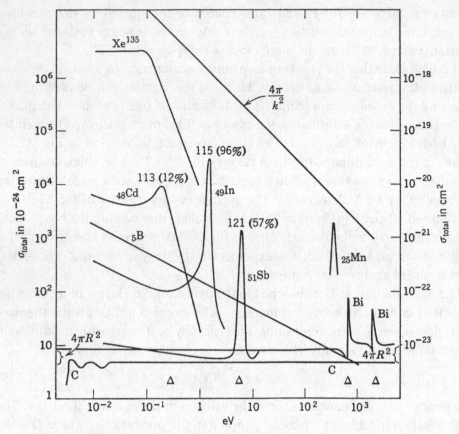

Fig. 12.19.   n–A reaction cross sections for capture of thermal neutrons. (From Ref. 12.1, with permission.)

existence of a large flux of neutrons. Detectors which will survive in the next generation of hadron colliders must live in a sea of neutrons with energies from ~1 MeV down to thermal scales and must operate in a fashion insensitive to this neutron 'bath'. That task is a major challenge for the new generation of detector designers.

Exercises

1. Consider a 2 TeV (2000 GeV) pion. Assume $\langle N \rangle = 9$. Work out the analogue to Table 12.1 for the four generations. What is 'f_o'?
2. For Exercise 1 what is the total number of charged particles in the cascade? What is the fractional fluctuation on that number?
3. For Exercise 1 estimate the number of MIPs contributed by the neutral energy by using the methods given in Chapter 11.

4. Use Eq. 12.6 to find the neutral energy deposited by a pion with 2000 GeV energy and four generations. Check your result against the explicit number worked out in Exercise 1.

5. Use Eq. 12.7 to find 'f_o' for a 2000 GeV pion. Check your result against the explicit number worked out in Exercise 1.

6. Consider a 25 GeV pion. Assume $\langle N \rangle = 9$. Work out the analogue of Table 12.1 for the two generations. What is 'f_o'? Compare to Fig. 12.4. Assume the charged particles ionize the deposit $(dE/dx)[dx = \lambda] = 0.2$ GeV per particle. Estimate the ionization energy in the cascade and compare to the neutral energy. Compare to the leftmost graph shown in Fig. 12.5.

7. What is the expected stochastic coefficient for $^1/_2$ absorption length sampling?

8. Show that unequal electromagnetic and hadronic response leads, with fluctuations in the neutral fraction, to a fractional energy error as given in Eq. 12.10. Remember that the hadronic energy is $E = hE_{IN}$. For $d'f_o' = 0.17$ and $e/h = 1.3$ show that the fractional energy error is 5.1%.

9. Use Fig. 12.2 to estimate $\langle N \rangle$ for a 1000 GeV and a 10000 GeV pion, using the approximate expression $s = 2m_p E$ for pion–proton collisions at high energies. Use Eq. 12.11 to estimate $d'f_o'$ for the two energies.

10. Find the maximum number of generations for a 1 TeV and a 10 TeV pion (1 TeV = 1000 GeV) assuming $\langle N \rangle = 9$ in both cases.

11. What is the ratio of the decay length to the interaction length for a 25 GeV pion? Recall that the lifetime of 7.8 m is stretched by a time dilation factor of γ.

12. Use Eq. 12.16 for a 25 GeV pion. Assume each neutron takes $B = 8$ MeV. What is the estimate for 'binding energy' in this case? Compare to the rightmost graph displayed in Fig. 12.5.

13. What is the mean free path of a thermal neutron in carbon?

References

[1] *Nuclear Interactions*, S. DeBenedetti, John Wiley & Sons, Inc. (1964).
[2] R. Bock, *et al., Nucl. Instrum. Methods* **134** 27 (1976).
[3] R. Wigmans, 'Advances in hadron calorimetry', *Annu. Rev. Nucl. Part. Sci.* **41** 133 (1991).
[4] T.A. Gabriel, *et al., Nucl. Instrum. Methods* **134** 271 (1976).
[5] C. Fabjan and R. Wigmans, *Rep. Prog. Phys.* **52** 1519 (1989).
[6] *Proceedings of the Workshop on Calorimetry for the Supercollider*, R. Donaldson and M.G.D. Gilchriese, World Scientific (1990).

Part IV

The complete set of measurements

We have now examined most of the techniques available to designers of particle detectors. Many detection systems are built using a single technique, for example a medical scanner using only NaI crystals and PMT transducers. However, in high energy physics, nuclear physics, and space physics the recent stress is on general purpose detectors using many detection technologies in concert. It is these composite devices which we examine in this last summary chapter.

13

Summary

The ideal of beauty is simplicity and tranquility.

Goethe

We have attempted to provide a sketch of the physics needed to understand the main types of particle detectors in use today. Our goal was to motivate, not rigorously derive, the operation of each type of detector from first principles in a single self-contained volume. The level of electromagnetism, quantum mechanics, and mechanics which was used should make this treatment accessible to advanced undergraduates and graduates. Chapter 1 introduced many of the concepts and the numerical constants. Velocity measurements via time of flight (TOF), Cerenkov, or transition radiation were the topics for Chapters 2, 3, and 4. Elastic scattering and ionization were introduced in Chapters 5 and 6. Non-destructive tracking and 'vertexing' with momentum and position determination was the subject of Chapters 7, 8 and 9. Radiative processes were examined in Chapter 10, leading to an exposition of destructive energy determination for electrons/photons and hadrons as described respectively in Chapters 11 and 12. Algebraic details were largely relegated to the Appendices.

The 'nuts and bolts' of actual devices have been stinted. That lack is easily remedied by exploring the set of references provided at the end of each chapter. The general physics discussions are illustrated with many numerical examples, which should establish the order of magnitude of the important parameters of a class of devices and thus motivate some of the approximations which are made. The choice of examples reflects the author's biases and specific experimental experiences.

13.1 Fundamental particles

As a final topic, we sketch a general purpose detector used in studying high energy collisions. In the world of present day research, the devices described in this note are often combined together into a powerful composite detector. As in an orchestra, the result is often greater than the mere sum of the parts. We

Table 13.1.   *The basic constituents of the 'Standard Model'*

Matter (spin $^1/_2$)[b]	Generations			Charge Q^a	
	$\begin{pmatrix} e \\ \nu_e \end{pmatrix}$	$\begin{pmatrix} \mu \\ \nu_\mu \end{pmatrix}$	$\begin{pmatrix} \tau \\ \nu_\tau \end{pmatrix}$	$\begin{pmatrix} 1 \\ 0 \end{pmatrix}$	Leptons
	$\begin{pmatrix} u \\ d \end{pmatrix}$	$\begin{pmatrix} c \\ s \end{pmatrix}$	$\begin{pmatrix} t \\ b \end{pmatrix}$	$\begin{pmatrix} ^2/_3 \\ -^1/_3 \end{pmatrix}$	Quarks

	Quanta	Force	Coupling	Quanta	Symbol
Interactions	Gluons	Strong	$\alpha_s = g_s^2$	8	g
(spin 1)	Photons	EM	$\alpha = e^2$	1	γ
	Weak bosons	Weak	$\alpha_w = g_w^2$	3	$W^- Z^0 W^+$

Notes:
[a] Units are electron charge $|e|$.
[b] Units are $\hbar$.

can draw an analogy between each chapter and the delights of chamber music while the full detector reflects the power and majesty of a symphonic performance. Specifically, the ability to make multiple redundant measurements makes for a powerful and versatile detector capable of studying extremely rare processes.

Consider our present knowledge of the elementary particles. These are listed in Table 13.1. The fermions (spin $^1/_2$) are either quarks, possessing strong interactions, or leptons, which only have electromagnetic and weak interactions. The leptons, in turn, are either charged or neutral. The neutral leptons, neutrinos, can only interact weakly.

In the list of spin 1 force carriers, the photon is familiar, while the gluons are the force carriers of the strong interactions. The electroweak gauge bosons, W and Z, unify electromagnetism with weak interactions and can be identified through their leptonic decays, as we will see. They are heavy, with masses ~ 90 GeV, while the photon and gluon are massless.

13.2 Detection of fundamental particles

How can these elementary particles be observed? The basic methods are listed in Table 13.2. All strongly interacting quarks and gluons decay 'virtually' and very rapidly into hadrons ('fragmentation') and appear as collimated 'jets' of hadrons (π, K, p, etc.) which can be observed in a calorimeter (Chapters 5, 11, 12, 13). For jets, the physics sets a transverse scale for the hadronic fragments of the quark jets, $\langle p_T \rangle \sim 0.4$ GeV. The longitudinal scale is set by the sharing of

Table 13.2. *Detection identification methods*

Signature	Detector	Particle
Jet of hadrons, λ_I	Calorimeter	u, c, t$\rightarrow$Wb
		d, s, b
		g
'Missing' energy	Calorimeter	ν_e, ν_μ, ν_τ
Electromagnetic shower, X_o	Calorimeter	e, γ, W$\rightarrow$ev
Only ionization interactions, dE/dx	Muon absorber	μ, $\tau\rightarrow\mu\nu\bar\nu$
		Z$\rightarrow\mu\mu$
Decay with $c\tau \geq 100$ μm	Si tracking	c, b, τ

the quark momentum among N fragments where $\langle N \rangle$ vs. $(p_T)_{jet}$ is similar to Fig. 12.2 for $\langle N \rangle$ vs. $\sqrt{s}$. For a $p_T = 100$ GeV jet, we find that $\langle N \rangle$ is ~ 14. Assuming the fragments share momentum equally, $p_{||}$ is ~ 7 GeV and the fragment makes an angle $\theta \sim p_T/p_{||} = 0.028$ or $3.4°$ with respect to the quark direction. Thus, high p_T 'jets' are collimated and easily recognized and measured. An example of a detected jet is shown in Fig. 13.1 both as a 'lego plot', where the detection sphere (θ, ϕ) is unfolded and as an azimuthal projection (r, ϕ). The collimated jets are quite apparent.

The neutrinos carry off energy without interacting, and therefore their existence and energy can be inferred by measuring the total final state energy in comparison to that of the well prepared initial state. The 'missing' energy is approximately equal to the neutrino energy. The neutrino cross section, for 10 GeV neutrinos off protons, is about $\sigma_{\nu N}$ (10 GeV) $\sim 9 \times 10^{-39}$ cm^2. Thus, for $L = 2$ m of iron the probability of interaction is $\sim 8 \times 10^{-10}$, indicating that neutrinos are unlikely to interact even in a thick detector. For example, a large flux of MeV neutrinos from nuclear reactions in the sun impinges on our bodies continuously but without causing noticeable effects.

A 'lego plot' of both electromagnetic and hadronic energy transverse to the incident beams in a scattering event is shown in Fig. 13.2. Note that the positron and the four jets do not balance E_T as the two jets did in Fig. 13.1. The presence of an unseen neutrino is indicated in Fig. 13.2 as having the θ, ϕ and E_T needed to make the total E_T of the final state system be zero.

The calorimetry itself is commonly segmented longitudinally beginning with electromagnetic detectors (Chapter 11) backed up with hadronic detectors (Chapter 12). The electrons and photons give large energy in the electromagnetic segment, and little energy in the following hadronic segment. As mentioned before, hadrons put little energy into the electromagnetic compartment since $\lambda_I \gg X_o$. Shading in Fig. 13.1 and Fig. 13.2 indicates this segmentation.

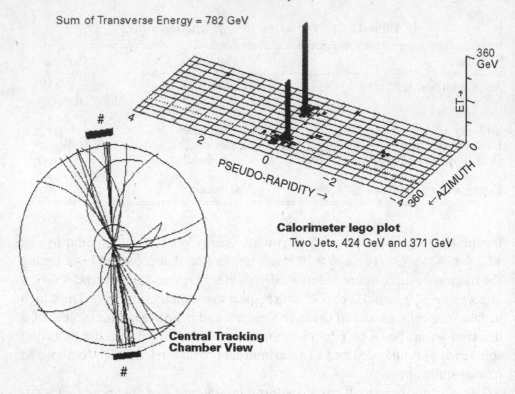

Sum of Transverse Energy = 782 GeV

Calorimeter lego plot
Two Jets, 424 GeV and 371 GeV

Central Tracking Chamber View

Fig. 13.1. The scattering of the quarks inside the proton leads to a 'jet' of particles traveling in the direction of, and taking the momentum of, the parent quark. Since there is no p_T in the initial state, by assumption the two quarks in the final state are 'back to back' in azimuth. (CDF event, with permission.)

A distinctive W→eν signature is shown in Fig. 13.3. The electron is identified as a track in the ionization tracking (shown in Fig. 13.1) with **p** matching the energy deposit in the calorimetry (redundant measurements). The calorimeter deposit in the direction of the electron track is approximately all in the electromagnetic compartment. The visible transverse energy is ~40 GeV, indicating a comparable missing energy. The W mass is inferred to be ~2 E_T (see Appendix A) or ~80 GeV.

All charged tracks first have their momentum measured by non-destructive measurements of their trajectory (Chapter 8) in a magnetic field (Chapter 7). The longer lived heavy quarks and leptons (c, b, τ) first have their secondary decay vertices measured and decays reconstructed using silicon detectors (Chapter 9). An expanded tracking view of the trajectories in the event shown in Fig. 13.2 is displayed in Fig. 13.4. Clearly jets 1 and 4 contain secondary ver-

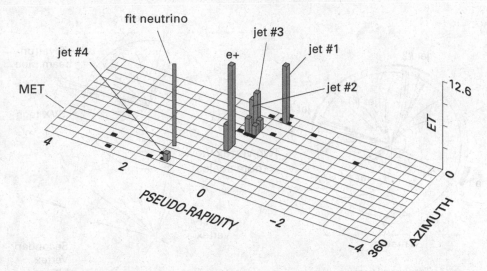

Fig. 13.2. 'Lego plot' for an event containing a positron and four jets. The existence of a neutrino is inferred from the large missing E_T in the event. (CDF event, with permission.)

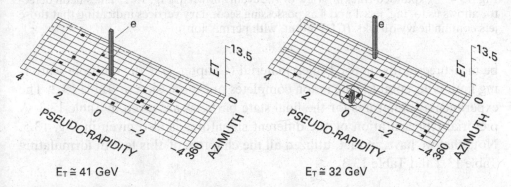

$E_T \cong 41$ GeV $E_T \cong 32$ GeV

Fig. 13.3. 'Lego plot' for W $\rightarrow e\nu$ events. The electromagnetic calorimetry shows a clean electron signal (associated track with the same momentum and angles in the ionization tracking). The missing energies are $\sim M_W/2$ indicating the two-body nature of the W decay. (CDF event, with permission.)

tices in the silicon detectors indicating that they contain heavy quarks. In addition, in many cases, particle identification done on the final state hadrons (e.g. K) can be used to infer the initial produced quark (e.g. s) by the use of time of flight (Chapter 2), Cerenkov detectors (Chapter 3) or transition radiation detectors (Chapter 4).

Finally, by forcing all other particles to cascade in the calorimetry, we can (see Chapter 12) identify muons. The muons, if below critical energy (Chapter 10), simply ionize the detection medium (Chapter 6) and their trajectories can

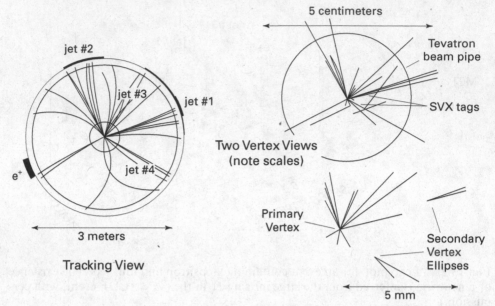

Fig. 13.4.   Expanded tracking view of the event shown in Fig. 13.2. The silicon detector allows us to 'tag' jets 1 and 4 as possessing secondary vertices indicating that those jets contain heavy quarks. (CDF event, with permission.)

be measured in magnetized thick material (Chapter 7) if the multiple scattering limit is acceptable. The muon completes our discussion of Table 13.2. The experimental signatures for the final state particles are given in Table 13.3. A pictorial representation of the different signatures is also given in Fig. 13.5. Note that we have, in fact, utilized all the chapters of this text in formulating Table 13.2 and Table 13.3.

13.3 General purpose detectors

An event of the type $W \rightarrow e\nu$ detected in a large general purpose collider detector is shown in Fig. 13.6. Note the ionization tracks, the calorimeter segmented azimuthally as 'towers' pointing to the interaction point, and the longitudinal segmentation into EM and hadronic 'compartments'. There is, unshown, muon tracking surrounding the thick iron toroidal magnets which exists outside the calorimetry.

A three dimensional plot of a dijet event where quarks in the initial protons scatter and appear as two jets in the final state, appears in Fig. 13.7. The debris of the initial protons appears in the forward direction where the shattered proton 'decays' into the calorimetry.

Lastly, in Fig. 13.8 the azimuthal display of the data available in detecting

Table 13.3. *Methods of identifying basic constituents*

Constituent	Si vertex	Track	TOF Č, TRD dE/dx	Cal EM	Cal Had	M U
e	Primary	✓	✓	✓	—	—
γ	Primary	—	—	✓	—	—
u, d, g	Primary	✓	—	✓	✓	—
ν	—	—	—	—	—	—
s	Primary	✓	✓	✓	✓	—
c, b, τ	Secondary	✓	✓	✓	✓	—
μ	Primary	✓	—	MIP	MIP	✓

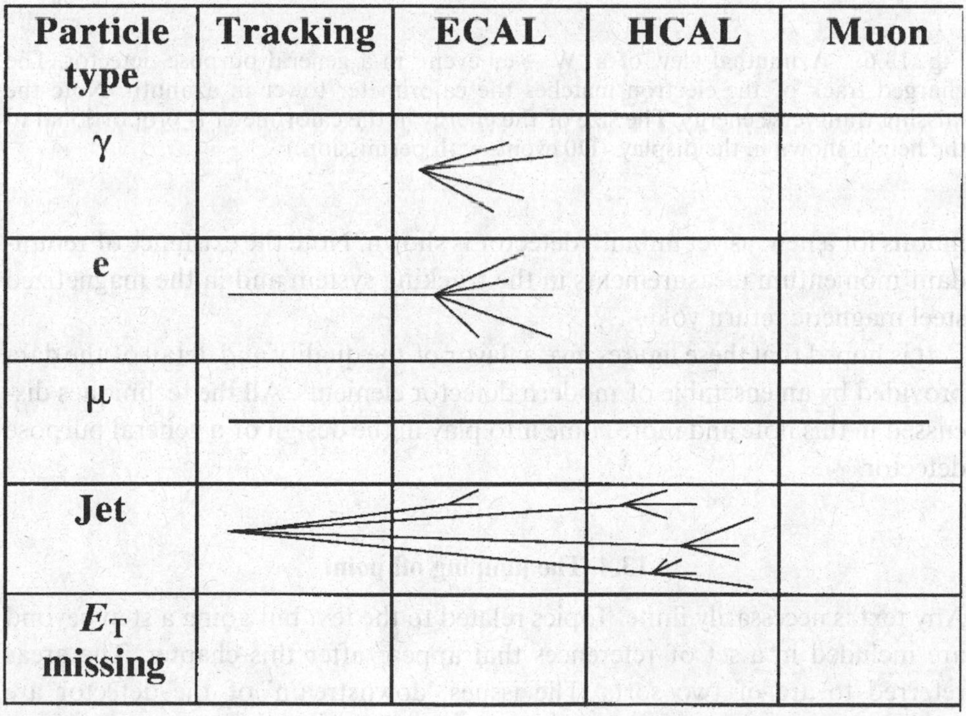

Signature (detector system)

Particle type	Tracking	ECAL	HCAL	Muon
γ				
e				
μ				
Jet				
E_T missing				

Fig. 13.5. Pictorial identification of particle topologies in the subsystems of a general purpose detector.

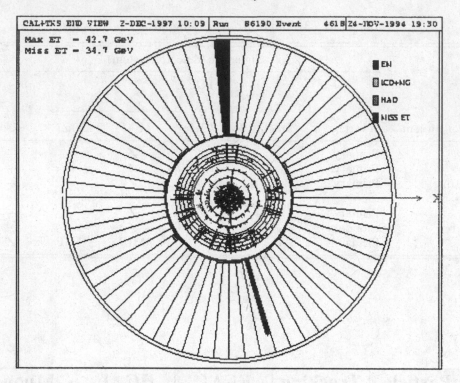

Fig. 13.6.   Azimuthal view of a 'W → eν event' in a general purpose detector. The charged track of the electron matches the calorimeter tower in azimuth. Note the missing transverse energy. The size of the energy in the calorimeter is proportional to the height shown in the display. (D0 event, with permission.)

muons for a new, as yet unbuilt, detector is shown. Note the existence of redundant momentum measurements in the tracking system and in the magnetized steel magnetic return yoke.

It is hoped that these figures give a flavor of the quality and detail of the data provided by an ensemble of modern detector elements. All the techniques discussed in this note and more come into play in the design of a general purpose detector.

13.4 The jumping off point

Any text is necessarily finite. Topics related to the text but going a step beyond are included in a set of references that appear after this chapter. The areas referred to are of two sorts. The issues 'downstream' of the detector are touched upon; triggering, data acquisition, data analysis and detector modeling. In addition, statistics is briefly introduced in Appendix J, and Monte Carlo

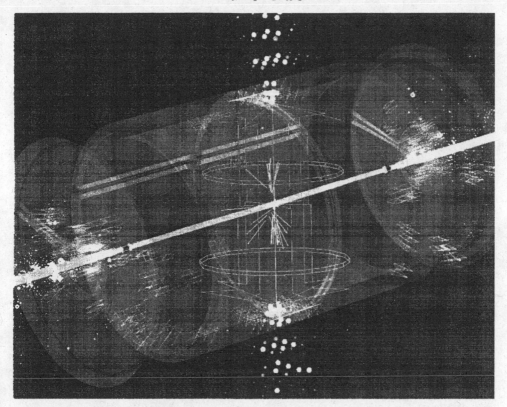

Fig. 13.7. Three dimensional view of the tracks and energy deposit in a dijet event. The initial state quarks are scattered by large angles and emerge at right angles to the beams. The initial protons are shattered by the interaction and 'decay' into angles near the direction of the initial colliding proton and antiproton. (D0 event, with permission.)

model creation in Appendix K. The second area deals with the specifics of hardware. The purpose of this set of references is to entice the student to go beyond the confines of this text and jump into a great sea of new information.

> If the doors of perception were cleansed, everything would appear
> to man as it is, infinite.
>
> *William Blake*

> Earth's crammed with Heaven, and every common bush afire
> with God.
>
> *Elizabeth Barrett Browning*

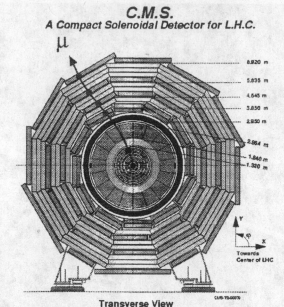

C.M.S.
A Compact Solenoidal Detector for L.H.C.

Fig. 13.8. Conceptual plot for muon identification and measurement. The muon track is measured in a solenoidal field, gives only MIP energy deposit in an electromagnetic and hadronic calorimeter, and is then measured again in the solid iron magnetized return yoke of the solenoid, in four measuring stations. (CMS figure, with permission.)

References

Topics outside the scope of this note (An invitation to go beyond)
[1] F. James, 'Monte Carlo Theory and Practice', *Rep. Prog. Phys.* **43** 1145 (1980).
[2] *Simulation and the Monte Carlo Method*, R.Y. Rubinstein, John Wiley & Sons (1981).
[3] N. Ellis and T.S. Virdee, 'Experimental Challenges in High-Luminosity Collider Physics', *Annu. Rev. Nucl. Part Sci.* **44** 609 (1944).
[4] H.H. Williams,'Design Principles of Detectors at Colliding Beams'. *Annu. Rev. Nucl. Part. Sci.* **36** 361 (1986).
[5] *Advances in Experimental Methods for Colliding Beam Physics*, W. Kirk, North-Holland Publishing Company (1988).
[6] *Physics of the Superconducting Supercollider – Snowmass 1986*, R. Donaldson and J. Marx, World Scientific Publishing Company (1986).
[7] *Future Directions in Detector R&D in High Energy Physics in the 1990s – Snowmass 1988*, S. Jensen, World Scientific Publishing Company (1988).
[8] *Statistical Methods in Experimental Physics*, W.T. Gadie, D. Oryard, F.E. James, M. Roos, B. Sadoulet, North-Holland Publishing Company (1971).
[9] *Data Reduction and Error Analysis for the Physical Sciences*, P.R. Bevington, McGraw-Hill Book Company (1969).
[10] J. Orear, *Notes on Statistics for Physicists*, revised Cornell University, UCRL-8417 (1982).
[11] *Mathematical Methods of Statistics*, H. Cramer, Princeton University Press (1958).
[12] Advanced Technology and Particle Physics, Como, Italy, 1990.
  (a) *Nucl. Phys. B* (Proc. Suppl.) **23A** (1991).
  (b) *Nucl. Phys. B* (Proc. Suppl.) **32** (1993).

Appendices

Men who wish to know about the world must
learn about it in its particular details.

Heraclitus

Appendix A

Kinematics

Kinematics refers to the constraints implied by energy and momentum conservation. Dynamics refers to the appropriate forces that define the detailed particle motion consistent with those constraints. We confine ourselves here either to the kinematics of two body decays of a single particle or to two body reactions.

Consider the two body reaction $a+b \rightarrow c+d$. The decay of a 'particle' of mass $\sqrt{s}$ into $c+d$ is a special case. Let us take the 'laboratory' frame with b at rest, and make the simplifying assumption that $m_c = m_d = m$. The center of momentum, CM, motion and the mass of the initial two body system, $\sqrt{s}$, follow from the expressions of special relativity, taking $a+b$ as the initial system; $M^2 = \varepsilon^2 - p^2 = p_\mu \cdot p^\mu$, $\gamma = \varepsilon/M$, $\beta = p/\varepsilon$. The particle energy is ε and momentum is $\mathbf{p}$, so that four dimensional momentum is $p_\mu = (\mathbf{p}, \varepsilon)$. The CM mass squared, γ and β factors are

$$s = (p_a + p_b)^\mu \cdot (p_a + p_b)_\mu \equiv M_{ab}^2$$

$$= m_a^2 + m_b^2 + 2p_a^\mu \cdot p_{b\mu}$$

$$= m_a^2 + m_b^2 + 2\varepsilon_a m_b \qquad (A.1)$$

$$\gamma \equiv \gamma_{ab} = \Sigma \varepsilon / \sqrt{s} = (\varepsilon_a + m_b)/\sqrt{s}$$

$$\beta \equiv \beta_{ab} = \Sigma \mathbf{p}/\Sigma \varepsilon = |\mathbf{p}_a|/(\varepsilon_a + m_b)$$

In the CM system, we can view the situation as either $2 \rightarrow 2$ scattering or as the two body 'decay' of a system of mass $\sqrt{s}$. The scattering angle in the CM frame is θ^* as illustrated in Fig. A.1. Energy and momentum balance requires (if $m_c = m_d = m$)

$$\varepsilon_c^* = \varepsilon_d^* = \sqrt{s}/2$$

$$\mathbf{p}_c^* = \mathbf{p}_d^* = p^* \qquad (A.2)$$

The Lorentz transformation back to the laboratory frame yields a momentum for the final state particles which depends on the scattering (decay) angle θ^*, $\gamma = 1/\sqrt{1 - \beta^2}$, $c\beta = v$.

$$p_T = p^* \sin\theta^*$$

$$p_\parallel = \gamma(p^* \cos\theta^* + \beta\varepsilon^*) \qquad (A.3)$$

$$\varepsilon = \gamma(\varepsilon^* + \beta p^* \cos\theta^*)$$

305

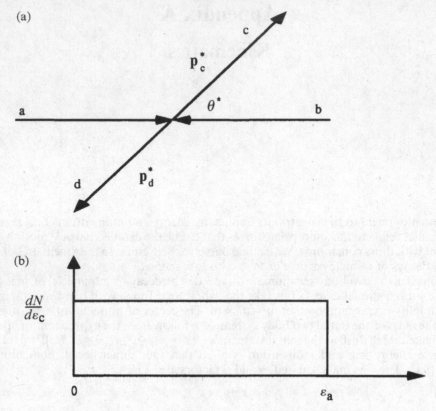

Fig. A.1.   (a) The CM scattering angle θ^*. (b) The distribution of the energy of particle c for an isotropic angular distribution in the case where c is light ($\sqrt{s} \gg m_c$).

The angular distribution, if isotropic (S wave scattering), is $d\Omega^* = d(\cos\theta^*)d\phi^* = 1/4\pi$, which implies an energy distribution in the laboratory which is uniform. In the limit of large $\sqrt{s}$ or light secondary particles ($\beta^* \sim 1$) ε_c ranges over the interval $(0, \varepsilon_a)$.

$$d\varepsilon = \gamma\beta p^* d(\cos\theta^*) = \gamma\beta p^*/2$$

$$\langle\varepsilon\rangle = \gamma\varepsilon^* \to \varepsilon_a/2 \tag{A.4}$$

$$\varepsilon_c \to (p_{\parallel}/2)[1 + \cos\theta^*]$$

The minimum secondary particle energy is ~ 0, in the limit that m is small with respect to $\sqrt{s}$. The maximum is given by the incident particle energy.

Consider now the solid angle element in the lab and CM (indicated by a * superscript) frames. The Lorentz transformation, $p_T = p_T^*$, insures that we only need to consider θ, since $\phi = \phi^*$. We also assume azimuthal isotropy.

$$d\Omega = d\cos\theta d\phi = d(p_{\parallel}/p)d[\tan^{-1}(p_x/p_y)] \tag{A.5}$$

$$= d(p_{\parallel}/p)d\phi^*$$

Using Eq. A.3 we differentiate to find the relationship between $d\Omega$ and $d\Omega^*$. We give the result only for the case where $\beta^* \to 1$ and $\beta \to 1$.

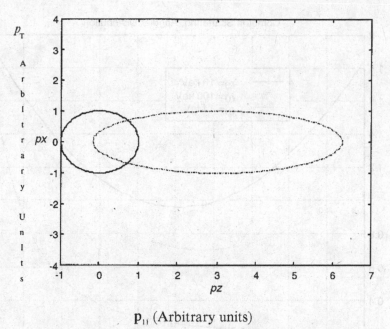

$\mathbf{p}_{||}$ (Arbitrary units)

Fig. A.2. Contours of momentum of photons in an isotropic decay. The CM contour (——) is spherical, while the lab contour (–·–·–) is highly peaked in the forward direction.

$$\cos\theta = p_{||}/p \sim p/\varepsilon \sim \frac{(\cos\theta^* + \beta)}{(1 + \beta\cos\theta^*)}$$

$$d\cos\theta = d\cos\theta^*[1/\gamma^2(1 + \beta\cos\theta^*)^2] \qquad (A.6)$$

$$\frac{d\Omega}{d\Omega^*} \cong (\varepsilon^*/\varepsilon)^2$$

In general, relativity implies a 'searchlight' effect, throwing the angular distribution forward by factors of γ. We show the contours of photons in the laboratory and CM frames in Fig. A.2. The isotropic CM contour is stretched along the direction of motion.

The resulting forward/backward asymmetry for relativistic motion is seen in Compton scattering at higher energies as shown in Fig. A.3. Part of this effect is purely kinematic, Eq. A.6.

An example of kinematics is Compton scattering. The basic equations are (see Chapter 10):

$$\omega_o + m = \omega + \varepsilon \qquad (A.7)$$

$$\mathbf{k}_o = \mathbf{k} + \mathbf{p}$$

We wish to find the outgoing photon energy, ω, in terms of the photon scattering angle, $\mathbf{k}_o \cdot \mathbf{k} = k_o k\cos\theta$. Clearly, we try to 'remove' the recoil particle. This is accomplished by placing ε and $\mathbf{p}$ on the right hand side of the equations,

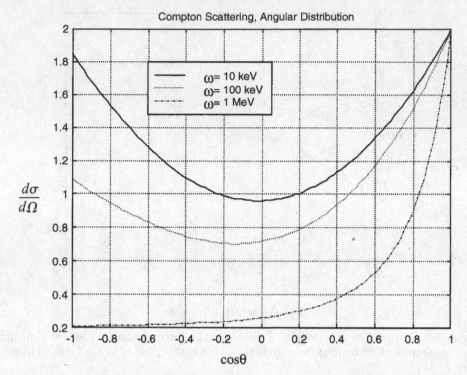

Fig. A.3. $d\sigma/d\Omega$ for Compton scattering, $\gamma + e \rightarrow \gamma + e$, at higher photon energies. The distortions causing a forward-backward asymmetry, which are partially due to Eq. A.6, are evident.

$$(\omega_o + m - \omega) = \varepsilon$$

$$(\mathbf{k}_o - \mathbf{k}) = \mathbf{p}$$

squaring and subtracting   (A.8)

$$\omega_o^2 + m^2 + \omega^2 + 2\omega_o m - 2m\omega - 2\omega_o\omega = \varepsilon^2$$

$$\frac{k_o^2 + k^2 - 2k_o k \cos\theta = p^2}{2m(\omega_o - \omega) + 2\omega_o\omega(\cos\theta - 1) = 0}$$

$$\left(\frac{1}{\omega} - \frac{1}{\omega_o}\right) = \frac{1}{m}(1 - \cos\theta)$$

The Compton angular distribution in the laboratory frame is given without proof to be $d\sigma/d\Omega = (\alpha\lambda)^2/2[\omega/\omega_o]^2[\omega/\omega_o + \omega_o/\omega - \sin^2\theta]$. We recognize the $(\omega/\omega_o)^2$ factor as being purely kinematic. The relation of ω/ω_o to $\cos\theta$ follows from Eq. A.8.

$$\omega/\omega_o = 1/[1 + y(1 - \cos\theta)], \quad y = \omega_o/m \tag{A.9}$$

This distribution is shown in Fig. A.3 for $\omega_o = 10$ keV, 100 keV and 1000 keV. The small y limit is the non-relativistic case, $\omega \rightarrow \omega_o$, $d\sigma/d\Omega \rightarrow (\alpha\lambda)^2/2[1 + \cos^2\theta]$ with integral, $\sigma = (8\pi/3)(\alpha\lambda)^2$ which is the Thomson result given in Chapter 10.

The most tedious calculation is this volume is perhaps that for the recoil angle as a function of recoil energy. The kinematics was shown in Chapter 5.

$$\mathbf{p}_o = \mathbf{p} + \mathbf{k}$$

$$\varepsilon_o + m = \varepsilon + e \tag{A.10}$$

We again 'remove' the scattered projectile variables by squaring and subtracting as in the case of Compton scattering.

$$p_o^2 + k^2 - 2p_o k \cos\Phi = p^2$$

$$\varepsilon_o^2 + m^2 + e^2 + 2\varepsilon_o m - 2\varepsilon_o e - 2em = \varepsilon^2 \tag{A.11}$$

$$2[m^2 + p_o k \cos\Phi + \varepsilon_o(m - e) - me] = 0$$

Define $e = T + m$ where T is the recoil kinetic energy. The energy and momentum constraints are then, $[p_o k \cos\Phi - T(\varepsilon_o + m)] = 0$. We now need to find the recoil momentum k in the terms of the recoil energy.

$$k = \sqrt{e^2 - m^2}$$

$$= \sqrt{(T + m)^2 - m^2} \tag{A.12}$$

$$= \sqrt{T^2 + 2mT}$$

This expression is used in Eq. A.11 and its square to solve for $T \equiv mQ$ as a function of the masses, the incident particle momentum, and the recoil angle Φ.

$$[(T^2 + 2mT)p_o^2 \cos^2\Phi] = T^2(\varepsilon_o + m)^2$$

$$T^2[p_o^2 \cos^2\Phi - (\varepsilon_o + m)^2] + T[2mp_o^2 \cos^2\Phi] = 0 \tag{A.13}$$

$$Q = \frac{2p_o^2 \cos^2\Phi}{[(\varepsilon_o + m)^2 - p_o^2 \cos^2\Phi]}$$

Using the relations $\varepsilon_o = \gamma M$ and $p_o = \gamma\beta M$ we can find Q as a function of recoil angle Φ, the incident particle mass M and velocity β, and the target mass m.

$$Q = \frac{2(\beta\gamma M \cos\Phi)^2}{[(\gamma M + m)^2 - (\beta\gamma M \cos\Phi)^2]} \tag{A.14}$$

This expression is quoted in Chapter 5. Clearly $Q_{max} \sim Q$ (cos $\Phi = 1$) or $Q_{max} = 2p_o^2/(M^2 + m^2 + 2\varepsilon_o m)$. For high energy incident particles, $\varepsilon_o \to p_o \to \infty$ and $Q_{max} \to p_o/m$ or $T_{max} \to p_o$ (remember, $c = 1$ is used so that $T_{max} \to p_o c$.).

Consider the special case of the scattering of slow neutrons by nuclei, $Q_{max} = 2p_o^2/[(\varepsilon_o + m)^2 - p_o^2]$ (Eq. A.13) where $m = AM$. In that case, $T_o + M = \varepsilon_o$ and $T_o = p_o^2/2M$, where M is the neutron mass.

$$Q_{max} = 4MT_o/[T_o^2 + (A + 1)^2 M^2 + 2MAT_o] \tag{A.15}$$

$$\to \frac{4MT_o}{(A + 1)^2 M^2}, \quad M \gg T_o$$

The scattered neutron has a minimum energy T_{min} when the recoil nucleus has a maximum kinetic energy.

$$T_{min} = T_o - Q_{max}(AM)$$

$$= T_o - \frac{4 M^2 A T_o}{(A+1)^2 M^2} \tag{A.16}$$

$$\frac{T_{min}}{T_o} = \left(\frac{A-1}{A+1}\right)^2$$

Thus, as intuition suggests, if $A \to \infty$, $T_{min} \to T_o$ and no energy is lost, while if $A \to 1$ (n + p elastic scattering) the neutron can give all its energy to the recoil proton, $T_{min} \to 0$. These results are quoted without proof in Chapter 12. Recall that in shooting pool, banking off a cushion ($A \to \infty$) means no loss of energy, while a head on collision of the cue ball and another ball ($A = 1$) leads to a complete loss of the kinetic energy of the cue ball.

The CM energy squared, s, is a relativistic invariant, being the square 'length' of the total four dimensional momentum of a system. Consider the reaction $\pi p \to (n\pi) p$ for multiple pion production. The initial state (p at rest) has $s \equiv (p_\pi + p_p)_\mu \cdot (p_\pi + p_p)^\mu \sim 2m_p E_\pi + m_p^2$, ignoring the mass of the light pion. The 'threshold' energy, $E_\pi = E_{TH}$, occurs at the lab energy where n pions can just be produced. The minimum s configuration is to have all the final state particles at rest in the CM, or $s = (nm_\pi + m_p)^2 \sim m_p^2 + 2nm_\pi m_p$. Therefore, the threshold lab pion energy for n pion production is

$$E_{TH}(n\pi) \sim nm_\pi \tag{A.17}$$

Appendix B

Quantum bound states and scattering cross section

The Schroedinger equation in quantum mechanics results from assigning differential operators to momentum and energy. These operators act on a wave function Ψ where $|\Psi|^2$ describes the probability density of finding a particle. The kinetic energy T plus the potential energy U is the total energy E. Note that this expression is non-relativistic, so that E really represents the energy $\varepsilon - m$.

$$(T + U)\Psi = E\Psi$$

$$\left(\frac{p^2}{2m} + U\right)\Psi = E\Psi \tag{B.1}$$

The differential operators are $\mathbf{p} = i\hbar\nabla$ and $E = i\hbar\partial/\partial t$ which have a compact relativistic representation, $p_\mu \to i\hbar\partial_\mu$

$$\left(\frac{-\hbar^2\nabla^2}{2m} + U\right)\Psi = i\hbar\frac{\partial\Psi}{\partial t} \tag{B.2}$$

The form of the solution for constant energy in three dimensions is the product of an angular function (spherical harmonics $= Y_\ell^m$) and a radial function, R, if the problem is 'central force', where U is a function only of r.

$$\Psi = RY_\ell^m$$

$$u \equiv rR \tag{B.3}$$

The radial wave function for u satisfies a differential equation with terms that can be identified as radial kinetic energy, total energy, central potential energy and centrifugal potential respectively.

$$-\frac{\hbar^2}{2m}\frac{d^2u}{dr^2} - \left(E - U - \frac{\hbar^2\ell(\ell + 1)}{2mr^2}\right)u = 0 \tag{B.4}$$

The solution to this equation is simple in the limits $r \to 0$ and $r \to \infty$. In the case $r \to 0$, the most singular part of the equation is the centrifugal potential (assuming U is well behaved).

$$r \to 0, \quad \frac{d^2u}{dr^2} - \frac{\ell(\ell + 1)}{r^2}u = 0$$

$$u = r^{\ell+1}$$

(B.5)

$$|\Psi|^2 \sim r^{2\ell}$$

The centrifugal barrier forces the wave function away from the origin, so that only S wave ($\ell = 0$) solutions have $\Psi(0) \neq 0$.

The solution for $r \to \infty$ is dominated by the energy term, since the potential U is assumed to fall as r increases. We have an oscillatory equation, whose behavior depends on the sign of E.

$$r \to \infty, \quad \frac{d^2u}{dr^2} + \left(\frac{2mE}{\hbar^2}\right)u = 0$$

(B.6)

$$k^2 \equiv (2mE/\hbar^2)$$

For $E < 0$ we have bound states, $|\Psi|^2$ localized, while for $E > 0$ we have scattering states, $|\Psi|^2 \sim$ constant. Note that both the small r and large r behavior of Ψ is 'generic', i.e. independent of the particular central force problem specified by $U(r)$.

$$E < 0, \quad u \sim e^{-r/a}$$

$$a = \hbar/\sqrt{2m|E|}$$

$$E > 0, \quad u \sim e^{\pm ikr}$$

(B.7)

$$\hbar k = \sqrt{2mE} = p$$

Consider the hydrogen atom with a radial quantum number n in addition to ℓ and m. The wave function is assumed to be $\Psi_n \sim r^\ell e^{-r/a_n}$ times some polynomial which interpolates between the small and large r behavior. As seen in Fig. B.1, the condition for a standing wave, representing a stable state, requires quantization of the de Broglie wavelength, which implies quantized momentum, and hence energy.

$$\lambda = 2\pi \, a/n = h/p \text{ (de Broglie wavelength)} = \lambda_{DB}$$

$$p = (\hbar/a)n$$

(B.8)

The Schroedinger equation for S waves, $\ell = 0$, assuming that the energy can be minimized with respect to radius in order to find the 'ground state', then becomes ($n = 1$)

$$(p^2/2m)\Psi + (U - E)\Psi = 0$$

$$\left(\frac{\hbar^2}{2ma^2} - e^2/a\right) = E$$

(B.9)

$$\partial E/\partial a = 0$$

$$a_0 = \frac{1}{\alpha m} = \lambda/\alpha \; (\lambda = \text{Compton wavelength} = \hbar/mc)$$

The 'virial theorem' relating T and U holds.

$$U = -e^2/a_0 = -\alpha^2 m$$

$$T = \alpha^2 m/2$$

(B.10)

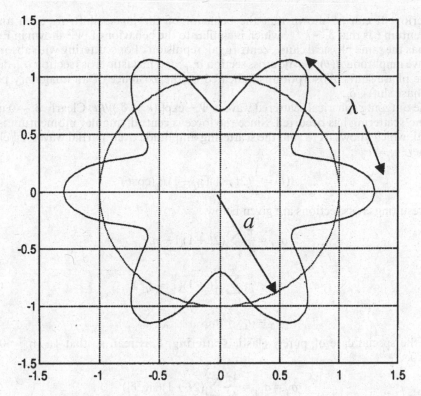

Fig. B.1.   Relationship of λ to a for the case of a standing wave solution for Ψ. The wavelength λ is the de Broglie wavelength $= h/p$.

Consequently $E_o = -T = U/2$. For excited states with $n > 1$, we use this result and Eq. B.8 to establish that $a_n = a_o n^2$ and then, with $E_n = U/2 = \alpha/a_n \sim 1/n^2$. The radius increases with n (not as deeply bound), and the $E_n \sim 1/n^2$ behavior indicates looser binding as n increases. The ionization continuum occurs at $n \to \infty$ or $E \to 0$.

$$E_o = T + U = -\alpha^2 m/2 < 0$$

$$a_n = a_o n^2, \quad n = 1, 2, 3 \tag{B.11}$$

$$E_n = E_o/n^2$$

In the case of a general power law potential, $U = g^2/r^d$, the radius analogous to Eq. B.9 can be shown by the same technique to be

$$a_n = (g^2 dm/\hbar^2 n^2)^{1/d-2} \tag{B.12}$$

Inserting back into the expression for energy, we find the dependence on m and n.

$$E_n \sim [n^2]^{(1+2/d-2)}[m]^{-(1+2/d-2)} \tag{B.13}$$

The electromagnetic results for $d = 1$, $a \sim n^2/g^2$ m $\sim \lambda n^2/\alpha$ and $E_n \sim m/n^2$, are recovered.

The scattering solutions, Eq. B.7, are oscillatory as , $r \to \infty$, $u \sim e^{\pm ikr}$, $\Psi \sim (e^{\pm ikr})/r$. The effect of scattering is to induce a change of phase, or 'phase shift', in the case of elastic scattering. It can be shown that the behavior of the phase shift, δ_ℓ, with angular momentum ℓ is that $\delta_\ell \sim k^{2\ell+1}$ which is similar to the behavior of $|\Psi|^2$ shown in Eq. B.5 and has the same physical cause, centrifugal 'repulsion'. For scattering with absorption we have amplitude $A(\theta)$, elastic cross section σ_{EL}, and inelastic cross section σ_{IN} defined by the phase shifts. Absorption is indicated by the existence of an imaginary part of the phase shifts, δ_ℓ.

The outgoing spherical scattered wave is $\Psi \sim \exp[i(kr+\delta_\ell)]/kr$. Clearly $\delta_\ell \to 0$ means that no scattering has occurred. Since the force is central, angular momentum is conserved, which allows us to sum the scattering amplitude over 'partial waves' labeled by ℓ, where $\eta_\ell \equiv e^{i\delta\ell}$:

$$A(\theta) = \frac{1}{2i}(2\ell+1)(\eta_\ell - 1)P_\ell(\cos\theta) \tag{B.14}$$

The resulting cross sections are given below.

$$\sigma_{EL} = \frac{\pi}{k^2}\sum_{\ell=0}^{\infty}(2\ell+1)|1-\eta_\ell|^2$$

$$\sigma_{IN} = \frac{\pi}{k^2}\sum_{\ell=0}^{\infty}(2\ell+1)(1-|\eta_\ell|^2) \tag{B.15}$$

$$\sigma_T = \sigma_{EL} + \sigma_{IN}$$

In the special case of purely elastic scattering, δ_ℓ is real, so that $1-|\eta_\ell|^2=0$, and $\sigma_{IN}=0$.

$$\sigma_T = \sigma_{EL} = \frac{4\pi}{k^2}\sum_\ell (2\ell+1)(\sin^2\delta_\ell) \tag{B.16}$$

Large phase shifts in a partial wave imply a large scattering cross section.

In general we find the phase shifts by matching the incoming and outgoing solutions. A simple example is a potential well with depth U_o and range a. The solutions for $r<a$ and $r>a$ can be seen from Eq. B.4. We consider only S waves ($\ell=0$). Clearly, since $\delta \sim k^{2\ell+1}$, if $ka \sim |\vec{L}|/\hbar \ll 1$, then only low angular momenta, $|L|$, are excited, and S waves will dominate. This situation typically obtains when the incoming particle wavelength, λ, is much larger than the size of the scattering system, $ka \sim a/\lambda \ll 1$.

$$\frac{d^2u}{dr^2} + k^2u = 0, \quad k^2 = 2mE/\hbar^2, \quad r>a$$

$$\frac{d^2u}{dr^2} + (K)^2u = 0, \quad (K)^2 = 2m(E+U_o)/\hbar^2, \quad r<a \tag{B.17}$$

The oscillatory scattering states are taken to be $\cos(kr)$ and $\sin(kr)$. We expand in that complete set to find the solutions u for $r<a$ and $r>a$.

$$u(r>a) = \sin(kr+\delta_o)$$

$$u(r<a) = A\sin(Kr) \tag{B.18}$$

The exterior wave is a mixture of sin and cos characterized by a phase shift δ_o. The interior wave must vanish at the origin (Eq. B.5) and has scattering amplitude A. The solution is obtained by matching u and du/dr at the $r=a$ boundary.

$$u : \sin(ka + \delta_o) = A\sin(Ka)$$

$$\frac{du}{dr} : k\cos(ka + \delta_o) = AK'\cos(Ka)$$

$$\frac{K}{k}\tan(ka + \delta_o) = \tan(Ka) \tag{B.19}$$

or

$$\delta_o = -ka + \tan^{-1}\left[\frac{k}{K}\tan Ka\right]$$

Consider the S wave elastic cross section, $\sigma_o = (4\pi/k^2)\sin^2\delta_o$. If the well is 'shallow' with respect to the energy of the incident wave, $U_o/E \ll 1$, then $K' \sim k$ and

$$\delta_o \sim -ka[1 - \tan(ka/ka)] \tag{B.20}$$

There are two limits depending on the size of ka. If $ka \gg 1$ the geometric cross section is obtained with $\delta_o \sim ka$. For $ka \ll 1$ the cross section is reduced with respect to the geometric cross section by a factor $\sim (ka)^4$.

$$ka \gg 1, \quad \sigma_o \to 4\pi a^2$$

$$ka \ll 1, \quad \sigma_o \to 4\pi a^2\left(\frac{(ka)^2}{3}\right)^2 \tag{B.21}$$

Let us now connect the wave equation for Ψ to a fictitious 'index of refraction'. Consider Eq. B.2 for energy eigenstates, or states of constant E, in one dimension, $d^2\Psi/dx^2 + [2m(E - U(x))/\hbar^2]\Psi = 0$. The oscillatory solutions go formally as $\exp(\pm ikx)$ with $T = p^2/2m = (E - U(x))$.

$$\hbar k = p = \sqrt{2m[E - U(x)]} \tag{B.22}$$

Thus the wave function of a free particle, $U(x) = 0$, is a plane wave. An interacting particle moves in a medium with an index of refraction

$$n = \sqrt{[E - U(x)]/E} \tag{B.23}$$

The energy E is a constant of the motion, by assumption.

In this fashion we can freely adopt the classical results of electromagnetic reflection to quantum mechanics. For example, consider photoelectric emission in metals. Inside the metal the electrons are bound by a negative potential energy, $U(x) = -U_o$. The emitted electron wave inside the metal is reflected when it encounters the interface to vacuum, $U(x) = 0$ at $x = 0$. We could match boundary conditions and solve the quantum mechanical problem. The wave functions are

$$\Psi_{IN} = be^{ik'x} + ce^{-ik'x}, \quad x < 0$$

$$\Psi_{OUT} = ae^{ikx}, \quad x > 0 \tag{B.24}$$

The wave vectors in the two regions, $x < 0$ and $x > 0$ are

$$k' = \sqrt{\frac{2m(E + U_o)}{\hbar^2}}, \quad k = \sqrt{\frac{2mE}{\hbar^2}} \tag{B.25}$$

Matching Ψ and $\partial \Psi \, \partial x$ at $x = 0$, we obtain the conditions

$$b + c = a$$

$$(b - c)k' = ak \tag{B.26}$$

In Eq. B.27 R is the reflection coefficient, the ratio of the intensity, $|\Psi|^2$, of the reflected wave to the incident wave. Removing a from Eqs. B.26 we find $b = c(k' + k)/(k' - k)$.

$$R = \left(\frac{c}{b}\right)^2 = \left(\frac{\sqrt{E + U_o} - \sqrt{E}}{\sqrt{E + U_o} + \sqrt{E}}\right)^2 \tag{B.27}$$

Note that we could also get this result by appealing to the classical result for optical reflection at a medium–vacuum interface, $R = [(n - 1)/(n + 1)]^2$. Note that this result also applies to waves traveling down a coaxial cable of impedance Z_1 striking another cable of impedance Z_2 (see Appendix D). Clearly, the same wave equation describes many different physical applications.

Appendix C

The photoelectric effect

The photoelectric effect derivation is sketched out in Chapter 2. In the interest of completeness, calculational details are given in this appendix. The kinematic quantities are given in Chapter 2. The transition rate, Γ, is given by the Fermi 'golden rule',

$$\Gamma = \frac{2\pi}{\hbar} \, \rho(p) |\langle f|H_I|i\rangle|^2,$$ where $\rho(p)$ is the density of final momentum states and H_I is

the Hamiltonian of the perturbation.

The essential physics is thus contained in the matrix element. The photon is specified by the vector potential, $\mathbf{A} \sim A_0 \boldsymbol{\varepsilon}[\exp(i\mathbf{k \cdot r})]$, where $\boldsymbol{\varepsilon}$, is the incoming photon polarization vector. The matrix element, $H_I \sim (e\mathbf{A \cdot p})/m$ was motivated in Chapter 2 and is

$$\langle f|H_I|i\rangle \sim \frac{eA_0}{m} \int \Psi^*_f (\boldsymbol{\varepsilon \cdot p}) e^{i\mathbf{k \cdot r}} \Psi_i \, d\mathbf{r} \tag{C.1}$$

Let us evaluate the matrix element. For a hydrogen-like atom, the inner electron 'sees' a charge Ze so that it is pulled in close to the nucleus, $a = a_0/Z$. The ground state S wave has $\Psi_i \sim e^{-r/a}$ (see Appendix B). Normalizing to $|\Psi_i|^2$ to 1, we note that the initial

electron is localized to a volume $\sim \frac{4\pi}{3} a^3$. Therefore the probability density $|\Psi_i|^2$ goes as

$1/a^3$ since $|\Psi|^2 \, d\mathbf{r}$ is a differential probability in three dimensions. The correctly normalized wave function is

$$\Psi_i(r) = \frac{1}{\sqrt{\pi a^3}} \, e^{-r/a} \tag{C.2}$$

When the photon energy is large, the ejected electron is quite energetic and its Coulomb interaction with the remaining ion can be ignored. Hence Ψ^*_f can be approximated as an outgoing plane wave, $\Psi_f(r) \sim \exp(i\mathbf{p \cdot r}/\hbar)$. Since Ψ_f is a momentum eigenstate, the momentum operator has a fixed value and we can pull $\mathbf{p}$ outside the integral.

$$\langle f|H_I|i\rangle \sim \frac{eA_0}{m} \, (\boldsymbol{\varepsilon \cdot p}) \, \frac{1}{\sqrt{\pi a^3}} \int e^{i(\mathbf{k}-\mathbf{p}/\hbar)\mathbf{r}} e^{-r/a} d\mathbf{r} \tag{C.3}$$

The integral can be obtained in a closed form in terms of the momentum transfer to the atom, $\mathbf{q} \equiv (\hbar\mathbf{k} - \mathbf{p})/\hbar$. Note that if qa is large, the phase factor $\exp(i\mathbf{q \cdot r})$ in the matrix element varies rapidly and destructively over the range $r < a$ where the integrand, $e^{-r/a}$, is large.

The angular integral is easily done first.

$$\langle f|H_{\mathrm{I}}|i\rangle \sim \frac{eA_{\mathrm{o}}}{m}\,(\boldsymbol{\varepsilon}\cdot\mathbf{p})\,\sqrt{\frac{1}{\pi a^3}}\,\int e^{i\mathbf{q}\cdot\mathbf{r}}e^{-r/a}d\mathbf{r}$$

(C.4)

$$\sim \frac{eA_{\mathrm{o}}}{m}\,(\boldsymbol{\varepsilon}\cdot\mathbf{p})\,\sqrt{\frac{1}{\pi a^3}}\left[4\pi\int\limits_{0}^{\infty} r^2\sin(qr)e^{-r/a}dr\right]$$

The radial integral involves the form of the bound state radial wave function.

$$\langle f|H_{\mathrm{I}}|i\rangle \sim \frac{eA_{\mathrm{o}}}{m}(\boldsymbol{\varepsilon}\cdot\mathbf{p})\,\sqrt{\frac{1}{\pi a^3}}\left[\frac{8\pi a^3}{(1+(qa)^2)^2}\right]$$

(C.5)

Ignoring the energy of the recoil atom, energy conservation reads, $\hbar\omega - \varepsilon = p^2/2m$. Thus, if the photon energy is well above the scale of inner electron binding energies, or $\hbar\omega > \dfrac{mc^2}{2}(Z\alpha)^2 = -\varepsilon$, then $\hbar\omega \sim p^2/2m$.

The controlling factor in the matrix element is qa. Note that $p \sim \sqrt{\omega} \sim \sqrt{k}$ so that $q \sim p/\hbar$ as long as the ejected electrons are non-relativistic. The momentum transferred to the atom is the momentum of the ejected electron. In this level of approximation, we have a matrix element that has a 'form factor' which leads to a cross section that is not characteristic of cross sections involving point-like structureless systems.

$$|\langle f|H_{\mathrm{I}}|i\rangle|^2 \sim \frac{1}{(pa/\hbar)^8}$$

(C.6)

The angular distribution, in the approximation $q \sim p/\hbar$, is simply a dipole due to the $\boldsymbol{\varepsilon}\cdot\mathbf{p} = \varepsilon p \sin\theta$ factor, where θ is the angle between the incoming photon, and the outgoing electron, and a polarization average is assumed. Thus, the photoelectrons are preferentially ejected transverse to the photon direction, along the direction of the accelerating transverse electric field associated with the photon.

Integrating over the angular distribution, $\int d\Omega \sin^2\theta = 2\pi\int d(\cos\theta)(1-\cos^2\theta) = 8\pi/3$. The cross section is derived from Γ, $d\sigma/d\Omega \sim \Gamma/(\hbar\omega A_{\mathrm{o}}^2) \sim p|\langle f|H_{\mathrm{I}}|i\rangle|^2/\hbar\omega A_{\mathrm{o}}^2$.

$$\frac{d\sigma}{d\Omega} \sim 2\pi\,\frac{mc^2\,p}{(2\pi\hbar)^2}\,\frac{\alpha\lambdabar^2}{\hbar\omega}\left[\left(\frac{1}{\pi a^3}\right)(p^2\sin^2\theta)\frac{(8\pi a^3)^2}{(qa)^8}\right]$$

(C.7)

$$\sigma \sim [\alpha\lambdabar^2]\left[\frac{mc^2}{\hbar\omega}\right](1/qa)^5$$

As in our discussion of Thomson scattering (Chapter 10), the A_{o}^2 factor in the square of matrix element is cancelled in $d\sigma/d\Omega$ since the cross section is normalized to unit incident flux.

Since we approximate $\hbar\omega$ as $p^2/2m = (q\hbar)^2/2m$, the basic behavior of σ is to go as $\sim 1/q^7$ or $\sim 1/\omega^{7/2}$. Note that the matrix element form factor behavior, $1/(qa)^8$, is softened to $1/(qa)^5$ by two other factors which are the $(\boldsymbol{\varepsilon}\cdot\mathbf{p})^2$ dipole factor and the density of final states factor, $\rho(p) \sim p$. The photon flux normalization of the cross section contributes $1/\omega$ the factor. Clearly, the resultant $(1/qa)^5$ behavior implies, $a = a_0/Z$, that σ goes as Z^5, strongly favoring high Z atoms for large photoelectric cross sections.

The mismatch of the scale of length between the initial state (a) and the final state ($1/q$) means that the overlap integral in the matrix element is small. That mismatch is minimized for high Z atoms since $a \sim 1/Z$.

Appendix D

Connecting cables

A key element of a complete detector system of great practical importance is the cables which are used to connect the parts together electrically. High speed operation dictates that a simple single wire connecting point to point is totally inadequate. For that reason a 'transmission line' environment should be provided for the signals. That environment can be provided by patterns etched on printed circuit boards, by twin leads (e.g. TV antenna lead, with 75 Ω impedance) or by coaxial cable (e.g. RG58 'Ethernet' cable with 50 Ω impedance). Coaxial cable is the method of choice, because it provides the best shielding against outside interference and thus the best noise immunity.

The geometry of some transmission line types is shown in Fig. D.1. In the coaxial case the characteristic impedance, Z_o, of the line is

$$Z_o = \left(\frac{1}{2\pi}\right)\sqrt{\frac{\mu}{\varepsilon}}\ln(b/a), \text{ MKS}$$

$$= \frac{60\ \Omega}{\sqrt{\varepsilon/\varepsilon_o}}\ln(b/a) \tag{D.1}$$

The impedance of the parallel wire configuration, Fig. D1, is twice that of the coaxial configuration for the same a and b.

We can think of the cable as the continuum limit of a discrete series of capacitors and inductors with impedance $Z_C \sim 1/i\omega C$, $Z_L \sim i\omega L$ and with the series–parallel combination having $Z \sim \sqrt{L/C}$ independent of frequency. Note that the 'impedance of free space' (MKS) sets the scale for Z up to dielectric constants and geometric factors as given in Eq. D.2.

$$Z = \sqrt{L/C}$$

$$Z_o = \sqrt{\mu_o/\varepsilon_o} = 377\ \Omega \tag{D.2}$$

The velocity of electromagnetic wave propagation is derived in Chapter 3. Typically, in coaxial lines solid dielectrics are used which have or $\varepsilon/\varepsilon_o \sim 2.3$ or $v \sim 0.67c = (5.0$ ns/m$)^{-1}$. The dependence on the geometry is weak (logarithmic). Typical values for the impedance of coaxial cables are 50 Ω. Some examples are given in Table D.1.

$$v = c/\sqrt{\varepsilon/\varepsilon_o} = 1/\sqrt{L\ C} = c/\sqrt{\mu\varepsilon} \tag{D.3}$$

$$= c/n$$

Table D.1 *Some examples of coaxial cables*

Type	Z_0 (Ω)	a(cm)	τ (n) (10 m)
RG58-U	50	0.147	0.2
RG8-U	52	0.362	0.038

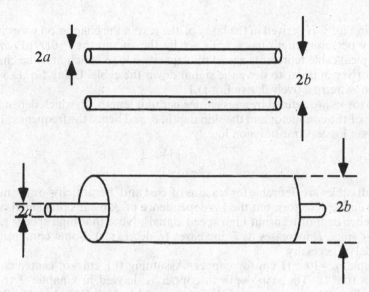

Fig. D.1.   Parallel wire and coaxial transmission lines.

A pulse propagating down a transmission line of impedance Z_1 which encounters a region of impedance Z_2 suffers a reflection with coefficient R for the power or $\sqrt{R}$ for the fields. The formalism is perhaps familiar from analogous situations in optics and quantum mechanics, with $n \sim \sqrt{\varepsilon}$ as quoted in Appendix B.

$$\sqrt{R} = \frac{(Z_2 - Z_1)}{(Z_2 + Z_1)} \tag{D.4}$$

If the impedance is matched (Fig. D.2), there is no reflection. If a short circuit, $Z_2 = 0$, appears, then $R = -1$ and the reflected pulse is inverted. If an open circuit appears, $Z_2 \to \infty$, $R = 1$, and the reflected pulse is non-inverted. Proper termination of a cable is shown in Fig. D.2.

For coaxial cable signal power is lost due to field penetration into the conductor by the skin depth effect caused by finite conductor resistance as discussed in Chapter 3. The capacity per unit length and inductance per unit length of the coaxial cable are $\underline{C}$ and $\underline{L}$.

$$\underline{C} = 2\pi\varepsilon/\ln(b/a), \text{ MKS}$$

$$\underline{L} = \left(\frac{\mu}{2\pi}\right)\ln(b/a) \tag{D.5}$$

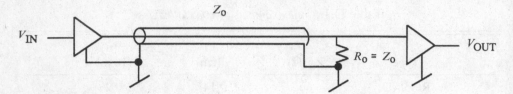

Fig. D.2. Layout for correct driving of a cable with characteristic impedance Z_o.

The expression for $\underline{C}$ is derived in the body of the text in the chapter on wire chambers. The capacity per unit length has a scale set by the vacuum, $\varepsilon_o = 0.09\,\mathrm{pF/cm}$. A 1 m length of typical cable represents a ~ 30 pF capacitive load which must be charged up by a buffer driver in order to drive the signal down the cable. Using Eq. D.5 and Eq. D.2 we can now retroactively derive Eq. D.1.

The cable losses are defined by a resistance per unit length, $\underline{R}$, which depends on the resistivity, ρ, of the conductor and the skin depth, δ, and hence the frequency. The cable is not a perfect lossless transmission line.

$$\underline{R} \cong \frac{\rho}{2\pi\delta}\left(\frac{1}{a}\right) \tag{D.6}$$

Clearly, small cables are desirable for reasons of cost and 'hermiticity' or the maximum coverage by active detectors, but the $1/a$ dependence of $\underline{R}$ means that small cables are lossy and hence cannot transmit high speed signals. Note that high speeds are preferentially lost since $\underline{R}$ increases as ω increases (δ decreases). Some compromises are usually made, of necessity.

For example, $\rho = 10^{-6}\,\Omega$ cm for copper. Assuming 0.1 cm $= a$, center conductor, $\rho/2\pi a = 1.6 \times 10^{-6}\,\Omega$. The skin depth in copper is derived in Chapter 3 to be 16.5 cm/$\sqrt{\omega}$ Hz. Thus, the resistance/unit length is roughly $10^{-7}\,\Omega/\mathrm{cm}\sqrt{\omega}$ Hz. The resistance of a 10 m cable to transmitting a frequency of 1 GHz is found to be $\sim 3\,\Omega$.

The measured dependence of cable loss as a function of frequency and length for several common coaxial cables is shown in Fig. D.3. The expected $1/\sqrt{\omega}$ behavior is clearly displayed.

The rise time of the cable to a step function, containing all frequencies, is defined to be τ, which is $\sim \underline{RC}$ times the square of the cable length (see Table D.1). For 10 m of cable, τ (for 1 GHz) is $\sim 3\,\Omega$ (300 pF) ~ 0.9 ns. Manufacturers quotes for rise times of RG58 ($a = 0.15$ cm) and RG8 (0.36 cm) are 0.2 ns and 0.038 ns for 10 m respectively confirming this rough estimate and showing the fall of τ with the conductor size, a. Note that the degradation of pulse shape fidelity goes as the square of the cable length. This argues to make the cable runs as short as possible. Oscilloscope traces of a pulse train at the driver and receiver ends (see Fig. D.2) of a 95-foot cable are shown in Fig. D.3b. Note the ~ 2 ns rise and fall times at the receiving end.

Note that we have simply ignored the 'quantized' nature of the waves propagating down the cable. The reason we can do this is that the minimum times to be considered are limited by the resistive cable losses to be $\tau \sim 1$ ns or $\omega_{max} \sim 1$ GHz. At that frequency the wavelength is still 30 cm which is much larger than the transverse size of the cables. We are not yet operating in the high frequency region where real r.f. techniques are required, nor will we explore r.f. 'plumbing'.

As an aside, if you look at the microwave oven in your home, there is a metal mesh screen embedded in the window. The metal will absorb electromagnetic waves with wavelengths much greater than the distance from metal to metal. The size of the grid is ~ 1 mm,

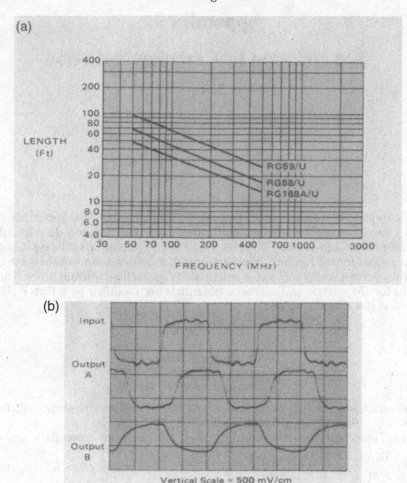

Fig. D.3. (a) Contours of constant loss for cables as a function of frequency and length. (b) Waveforms at the driver and receiver ends of 95 feet of RG58/U cable. (From Ref D.1 with permission.)

which is quite sufficient to block the wavelengths of the microwaves and thus protect the chefs from being cooked by their own ovens while allowing visible light to pass easily.

In the practical design of detectors, cables are actually quite crucial and the proper cable choice is of some importance. The cables must transmit the data signals well and with low noise. In a high radiation field, they must often also be small so as not to provide a leakage path to sensitive electronics components. A good working knowledge of the properties of cables is a useful tool in the kit of the detector designer.

Reference

[D.1] Motorola Semiconductor Products, Inc., *MECL System Design Handbook* 1971.

Appendix E

The emission of Cerenkov radiation

A truly complete derivation is well beyond the scope of this text. Nevertheless, it is useful to move beyond the heuristic motivations used in Chapter 3. In particular, we can relate radiation in general, Chapter 10, to the Cerenkov special case, Chapter 3. The expressions for the Lienard–Wiechert formulae relating a moving charge to the relativistically correct scalar and vector potentials are given here without proof. The static solutions for the electric and magnetic potentials are modified by a factor $1/1 - \hat{r} \cdot \boldsymbol{\beta})$ evaluated at the 'retarded time' t'.

$$V = [e/r(1 - \hat{r} \cdot \boldsymbol{\beta})]_{\text{RET}}$$

$$\mathbf{A} = [e\boldsymbol{\beta}/r(1 - \hat{r} \cdot \boldsymbol{\beta})]_{\text{RET}} = \boldsymbol{\beta} V \tag{E.1}$$

$$t' + r/c = t, \text{ retarded time } t'$$

The non-relativistic limit, $c \to \infty$, is that, $t \to t'$, $V \to e/r$ and $\mathbf{A} \to 0$ as expected. The definitions of r, t, β and t' are given in Fig. E.1.

We can then use the relationship, Table 6.1, between potentials and fields, $\mathbf{E} = \boldsymbol{\nabla} V - \dfrac{1}{c}\dfrac{\partial \mathbf{A}}{\partial t}$, to derive that the radiation fields, those which go as $E \sim 1/r$, are those given in Eqs. E.2 (See Ref. 3.1). Physically $\mathbf{r} - \boldsymbol{\beta} r$ is the 'virtual present position' as shown in Fig. E.1.

$$\mathbf{E} = \left(\frac{e}{c^2}\right)\left[\frac{1}{r(1 - \hat{r} \cdot \boldsymbol{\beta})^3}\right][\hat{r} \times (\hat{r} - \boldsymbol{\beta}) \times \mathbf{a}]_{\text{RET}}$$

$$\mathbf{B} = (\hat{r} \times \mathbf{E}) \tag{E.2}$$

At low velocity we find the familiar Larmor pattern, that of an electric dipole of moment ea/ω^2. In Chapter 10 we simply applied the dimensionless substitution (ar/c^2) to the static solutions in order to make the radiative solutions plausible. The non-relativistic, $\beta \to 0$, limits are

$$\mathbf{E} \to \left(\frac{e}{c^2}\right)\left[\frac{\hat{r} \times \hat{r} \times \mathbf{a}}{r}\right]$$

$$\mathbf{B} \to (\hat{r} \times \mathbf{E}) \tag{E.3}$$

$$P \to \frac{2}{3}(e^2 a^2/c^3)$$

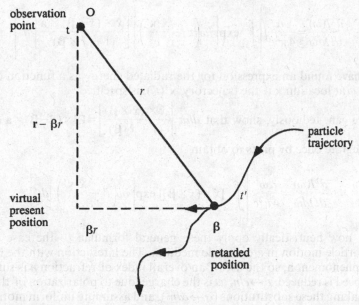

observation point O

t

r

r − βr

particle trajectory

virtual present position

βr

t'

β

retarded position

Fig. E.1. Definition of vectors at the retarded position, the observation point and the 'virtual present position'.

The general result for the power, $P = 2/3(e^2/c^3)\gamma^6[a^2 - (\boldsymbol{\beta} \times \mathbf{a})^2]$ was 'derived' in Chapter 10 using the correct relativistic expression for acceleration and writing P as a Lorentz invariant.

The expressions for the fields are used to find the energy radiated into solid angle element $d\Omega$ per unit frequency interval $d\omega$ where frequency is defined using the observer's clocks, t (see Fig. E.1). Assuming that there is only a finite duration of the acceleration, the radiated energy can be related to the Fourier transform of the potential, $\mathbf{A}(\omega)$. The non-relativistic limit is again familiar (see Chapter 10). Using Eq. E.2 for the electric fields,

$$\frac{d^2I(\omega)}{d\Omega\,d\omega} = 2|\mathbf{A}(\omega)|^2 \sim \frac{c}{2\pi}(rE)^2$$

$$\mathbf{A} = \sqrt{\frac{c}{4\pi}}[r\mathbf{E}]_{RET} \tag{E.4}$$

$$\frac{d^2I(\omega)}{d\Omega\,d\omega} \sim \left| \int_{-\infty}^{\infty} e^{i\omega t} \left[\frac{\hat{r} \times (\hat{r} - \boldsymbol{\beta}) \times \mathbf{a}}{r(1 - \hat{r}\cdot\boldsymbol{\beta})^3} \right]_{RET} dt \right|^2$$

If the observation point is very far away, then during the finite acceleration time, $\hat{r} \sim$ const. If the observation point is labeled by $\mathbf{x}$ from some arbitrary origin, as is the source point $\mathbf{x}'$, then during emission of the radiation, r can be approximated as $r = |\mathbf{x} - \mathbf{x}'| \sim x - \hat{x}\cdot\mathbf{x}'$. The expression for the radiated energy then simplifies considerably in terms of the time t' at emission, $t = t' + r/c \sim t' + x/c - \hat{x}\cdot\mathbf{x}'/c$.

$$\frac{d^2I(\omega)}{d\Omega d\omega} = \frac{\alpha}{4\pi^2 c^3} \left| \int_{-\infty}^{\infty} \exp\left[i\omega\left(t' - \frac{\hat{x}\cdot\mathbf{x}'}{c} \right) \right] \left[\frac{\hat{x} \times (\hat{x} - \boldsymbol{\beta}) \times \mathbf{a}}{(1 - \hat{x} \times \boldsymbol{\beta})^2} \right] dt' \right|^2 \tag{E.5}$$

Thus, we have found an expression for the radiated energy as a function of observed frequency ω at location $\mathbf{x}$ if the trajectory, $\mathbf{x}'(t')$, is specified.

Since we can, tediously, show that $d/dt'\left[\dfrac{\hat{x} \times (\hat{x} \times \boldsymbol{\beta})}{(1 - \hat{x}\cdot\boldsymbol{\beta})} \right] = [\hat{x} \times (\hat{x} - \boldsymbol{\beta}) \times \mathbf{a}/c](1 - \hat{x}\cdot\boldsymbol{\beta})^2$,

we can integrate once by parts to obtain

$$\frac{d^2I(\omega)}{d\Omega d\omega} = \frac{\alpha\omega^2}{4\pi^2 c} \left| \int_{-\infty}^{\infty} [\hat{x} \times (\hat{x} \times \boldsymbol{\beta})] \exp\left[i\omega\left(t' - \frac{\hat{x}\cdot\mathbf{x}'}{c} \right) \right] dt' \right|^2 \tag{E.6}$$

We can now heuristically apply these general formulae to the case of uniform charged particle motion in a dielectric medium. The interaction with the medium is a collective phenomenon, so the use of an overall index of refraction n is sufficient. The effective speed is reduced, $c \rightarrow c/n$, as is the charge (due to polarization of the medium), $e \rightarrow e/n$. Making these substitutions ($\alpha \rightarrow \alpha/n^2$), and assuming uniform motion, $\mathbf{x}' = \mathbf{v}t'$, we find

$$\frac{d^2I(\omega)}{d\Omega\, d\omega} = \frac{\alpha\omega^2}{4\pi^2 c^3} \left| \int_{-\infty}^{\infty} [\hat{x} \times (\hat{x} \times \mathbf{v})] \exp\left[i\omega\left(t' - \frac{\hat{x}\cdot\mathbf{x}'}{c} \right) \right] dt' \right|^2$$

$$\rightarrow \frac{\alpha\omega^2 n}{4\pi^2 c^3} \left| \int_{-\infty}^{\infty} [\hat{x} \times (\hat{x} \times \mathbf{v})] \exp\left[i\omega\left(t' - \frac{n\hat{x}\cdot\mathbf{v}t'}{c} \right) \right] dt' \right|^2 \tag{E.7}$$

$$\rightarrow \frac{\alpha n}{c^3} |\hat{x} \times \mathbf{v}|^2 \left| \frac{\omega}{2\pi} \int_{-\infty}^{\infty} \exp\left[i\omega t'\left(1 - n\frac{\hat{x}\cdot\mathbf{v}}{c} \right) \right] dt' \right|^2$$

Note that the expression which is squared is simply a delta function in the variable $1 - n\hat{x}\cdot\mathbf{v}/c$ or $1 - n\beta\cos\theta$. (Recall that the Fourier transform of a plane wave e^{ikx} (unique momentum) is $\delta(k)$.) Thus we have an infinite radiated power at an exact angle at this level of approximation.

$$\frac{d^2I(\omega)}{d\Omega\, d\omega} \rightarrow \frac{\alpha n\beta^2 \sin^2\theta}{c} |\delta(1 - n\beta\cos\theta)|^2 \tag{E.8}$$

A problem arises when the integral yields a delta function because we have assumed an infinitely long time for radiation. For a finite radiator length we would obtain a diffraction pattern peaked at $\cos\theta_c = 1/\beta n$, and the size of $I(\omega)$ would be proportional to the time interval over which radiation is emitted. Thus the energy radiated per unit path length, dx, is expected to be a constant. The number of emitted photons, N, per unit frequency is then very simply expressed.

$$\frac{d^2I(\omega)}{dx\, d\omega} = \frac{\alpha\omega}{c^2} \sin^2\theta_c$$

$$\frac{d^2 N(\omega)}{dx\, d\omega} = \frac{\alpha}{c^2} \sin^2 \theta_c \qquad\qquad (E.9)$$

Summarizing, the radiated power is related to the square of the Fourier transform of the fields. The $\exp(i\omega t)$ factor in the transform, given retardation, $t = t' + x/c - \hat{x}\cdot x'/c$, a trajectory of constant velocity, $x' = v t'$, and the heuristic $c \to c/n$ substitution leads to $\exp\left[i\omega t'\left(1 - \hat{x}\cdot v\, \dfrac{n}{c}\right)\right]$. The integral is a delta function in $1 - \beta n \cos \theta$ leading to the Cerenkov result, albeit in a formal fashion most properly relegated to this appendix. Further details are to be found in the references quoted at the end of Chapter 3.

Appendix F

Motion in a constant magnetic field

The Lorentz force equation (MKS) is $\mathbf{F} = d\mathbf{p}/dt = q(\mathbf{v} \times \mathbf{B})$. Since $\mathbf{F} \cdot \mathbf{v} = 0$, the magnetic force does no work on the particle, and hence $|\mathbf{p}| = p = \text{const}$, $\gamma = \text{const}$, $\nu = \text{const}$,. The arc length s can be used to parameterize the path instead of clock time t, and the general equation of motion is

$$ds = \nu dt$$

$$\mathbf{v} = \nu \hat{\alpha}, \quad \hat{\alpha} \equiv d\mathbf{x}/ds$$

$$\mathbf{p} = \gamma m \mathbf{v} = \gamma m \nu \hat{\alpha} \tag{F.1}$$

$$d\mathbf{p}/ds = q(\hat{p} \times \mathbf{B})$$

$$d\hat{\alpha}/ds = \left(\frac{\hat{\alpha} \times \hat{B}}{\rho}\right), \quad \rho = p/qB$$

Unit vectors are indicated by the ^ symbol. The instantaneous radius of curvature is indicated by the symbol ρ. This equation says that the momentum vector describes a circle of radius ρ about the direction of the magnetic field.

For the remainder of this appendix, we specialize to a uniform field, $\mathbf{B} = B_o \hat{k}$, along the z axis. Note that in the text x is usually considered to be along the direction of motion, e.g. dE_1/dx. However, to hew to the most common convention we adopt the z axis as the direction of the $\mathbf{B}$ field. We can always simply permute the axes if another Cartesian coordinate system is desired. Since there is no force along the z axis, we can integrate Eq. F.1 twice.

$$d^2 z/ds^2 = 0$$

$$dz/ds = \alpha_z = \alpha_{zo}$$

$$\Delta z = \alpha_{zo} \Delta s, \quad s = 0 \text{ at } Z_o \tag{F.2}$$

$$z - z_o = \alpha_{zo} s$$

Initial values are indicated by the o subscript. Note that initial means when clocks are started, at $t = s = 0$.

The forces act in the x and y directions. The equations of motion, from Eq. F.1 are

$$d\alpha_x/ds = \alpha_y/\rho$$

$$d\alpha_y/ds = \alpha_x/\rho \tag{F.3}$$

328

We then differentiate again in order to decouple the x and y equations of motion.

$$d^2\alpha_x/ds^2 = \frac{1}{\rho}\frac{d\alpha_y}{ds} = -\alpha_x/\rho^2$$

$$d^2\alpha_y/ds^2 = -\alpha_y/\rho^2 \tag{F.4}$$

The form of the equations indicates harmonic motion of α_x and α_y. The solutions are therefore sinusoidal. Thus, we assume an expansion in the complete set, $\sin\phi$ and $\cos\phi$, and impose the initial conditions, α_x at $s=0$ is α_{xo} and α_y at $s=0$ is α_{yo}.

$$\begin{pmatrix}\alpha_x\\\alpha_y\end{pmatrix} = \begin{pmatrix}\cos\phi & \sin\phi\\-\sin\phi & \cos\phi\end{pmatrix}\begin{pmatrix}\alpha_{xo}\\\alpha_{yo}\end{pmatrix} \tag{F.5}$$

$$\phi = s/\rho$$

Identifying $\phi_o = 0$ and $\phi = s/\rho$ we can easily show that Eq. F.5 satisfies Eq. F.4 by substitution. Physically ϕ is the angle through which the momentum vector is rotated perpendicular to **B** when the particle moves a total distance s in the field.

Having found the first integral, we integrate again to find the positions. The integral of Eq. F.5, $dx/ds = \alpha_{xo}\cos\phi + \alpha_{yo}\sin\phi$ with $\phi = s/\rho$ and initial conditions at $x = x_o$, $y = y_o$, $\phi = s = 0$ is;

$$\begin{pmatrix}\Delta x\\\Delta y\end{pmatrix}/\rho = \sin\phi\begin{pmatrix}\alpha_{xo}\\\alpha_{yo}\end{pmatrix} + (1-\cos\phi)\begin{pmatrix}-\alpha_{yo}\\\alpha_{xo}\end{pmatrix} \tag{F.6}$$

$$\Delta x = x - x_o, \quad \Delta y = y - y_o$$

Having found the trajectory in terms of the parameter of path length, we can also find alternative forms which eliminate explicit reference to the parameter s. Squaring Eq. F.6 we get the circular part of the trajectory perpendicular to the magnetic field direction, so that we can find x knowing y and the initial conditions. This formulation is useful for particles traversing dipole magnets to a fixed x or y boundary. Using Eqs. F.5 and F.6 we first find x in terms of α_y and y in terms of α_x

$$x - x_o = -\rho(\alpha_y - \alpha_{yo})$$

$$y - y_o = \rho(\alpha_x - \alpha_{xo}) \tag{F.7}$$

Squaring F.7 and adding we remove the parametric dependence on s (or ϕ) and find an expression for the trajectory.

$$[(x-x_o) - \rho\alpha_{yo}]^2 + [(y-y_o) + \rho\alpha_{xo}]^2 = \rho^2(\alpha_x^2 + \alpha_y^2) \equiv \rho_T^2 = \rho^2(\alpha_{xo}^2 + \alpha_{yo}^2) \tag{F.8}$$

$$\rho_T = p_T/qB_o, \quad p_T = p\sin\theta$$

Manipulating Eq. F.6 we can, alternatively, solve for the 'bend angle' ϕ as a function of points on the trajectory x and y.

$$\sin\phi = \frac{[(x-x_o)/\rho + \alpha_{yo}]\alpha_{xo} + [(y-y_o)/\rho - \alpha_{xo}]\alpha_{yo}}{(\alpha_{xo}^2 + \alpha_{yo}^2)}$$

$$\cos\phi = \frac{[(x-x_o)/\rho + \alpha_{yo}]\alpha_{yo} - [(y-y_o)/\rho - \alpha_{xo}]\alpha_{xo}}{(\alpha_{xo}^2 + \alpha_{yo}^2)} \tag{F.9}$$

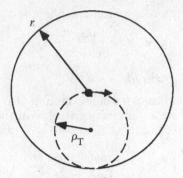

Fig. F.1.   The existence of a minimum transverse momentum required for particles to pass outside a radius r.

Clearly, the implication of Eq. F.2 and Eq. F.5 is that the momentum parallel to **B** is constant and the magnitude of the momentum perpendicular to **B**, p_T, is constant. If point (x,y) on the trajectory is known, then ϕ can be found. From $\phi = s/\rho$ and Eq. F.2 the point z can also be found.

It is often more convenient to use cylindrical coordinates because the basic physics is clearly cylindrically symmetric. For example, a solenoid has cylindrical symmetry and these coordinates are obviously the ones of choice. In the simplifying case $x_o = y_o = 0$, with ϕ' labeling the position at radius r, when $x = r\cos\phi'$, $y = r\sin\phi'$ is substituted into Eq. F.8 we find that

$$\sin(\phi' - \phi_o) = r/2\rho_T$$

$$\tan\phi_o = p_{yo}/p_{xo} \tag{F.10}$$

The angle ϕ_o is the initial direction of the momentum vector perpendicular to **B** as illustrated in Chapter 7. As shown in Fig. F.1, if $\rho_T < r/2$ then the track will curl up and never reach radius r.

Thus particles with transverse momentum $p_T < rqB_o/2$ do not reach a radius r. These low momentum tracks simply rotate in azimuth and drift down the solenoid in the $\hat{z}$ direction.

In order to find the exit vector we use Eq. F.7 to find ϕ'', the label for the exit angle. The geometry is given in the body of the text. (See Chapter 7.)

$$\alpha_y = \alpha_{yo} - x/\rho$$

$$\alpha_x = \alpha_{xo} + y/\rho$$

$$\tan\phi'' = \alpha_y/\alpha_x \tag{F.11}$$

$$= \frac{(\sin\phi_o - r/\rho_T\cos\phi')}{(\cos\phi_o + r/\rho_T\sin\phi')}$$

The solutions given in the appendix are exact in the case of a uniform field. Since any field can be locally taken to be uniform, a solution for an arbitrary field can be achieved numerically by pasting together a succession of these solutions. This technique has many applications in numerical recipes for solving much more complex problems.

Appendix G

Non-relativistic motion in combined constant **E** and **B** fields

Often we encounter situations where there are both electric and magnetic fields. We treat here the simplest case of motion in uniform fields. The electric field is taken to be along the z axis, while the magnetic field is assumed to be in the (y,z) plane with angle θ with respect to the electric field. The motion is assumed to be non-relativistic, but otherwise completely arbitrary. The Lorentz force equation, $\mathbf{F} = q(\mathbf{E} + \mathbf{v} \times \mathbf{B})$ (MKS), is, for $q = e$

$$\ddot{x} = \omega[\dot{y}\cos\theta - \dot{z}\sin\theta]$$

$$\ddot{y} = \omega[-\dot{x}\cos\theta]$$

$$\ddot{z} = \omega[\dot{x}\sin\theta] + a \tag{G.1}$$

$$\omega = eB/m$$

$$a = eE/m$$

$$\cdot = d/dt$$

For example, in a 40 kG field with a field of 3×10^4 V/cm we find that $\omega = 6.8 \times 10^{11}$ s^{-1} and $a = 5.9 \times 10^{19}$ cm/s^2. In order to solve the equations in a simple form the boundary conditions, $x(0) = y(0) = z(0) = 0$ and $\dot{x}(0) = \dot{y}(0) = \dot{z}(0) = 0$ are adopted, that is starting from the origin (choosable) at zero initial velocity. This situation would apply, for example, if the motion of photoelectrons emitted by a photocathode immersed in an arbitrary magnetic field were to be studied. The solutions are

$$x(t) = -(a/\omega^2)\sin\theta(\phi - \sin\phi)$$

$$y(t) = (a/\omega^2)\sin\theta\cos\theta[\phi^2/2 + (\cos\phi - 1)]$$

$$z(t) = (a/\omega^2)[\sin^2\theta(1 - \cos\phi) + \cos^2\theta(\phi^2/2)] \tag{G.2}$$

$$\phi = \omega t$$

as can be verified directly by differentiating twice.

A natural length scale appears which is (a/ω^2), while the natural rotational angle scale is ϕ. If there were no magnetic field the time t_o to fall to $z = d$ in a purely electric field and the final velocity, v_o, at $z = d$ would be

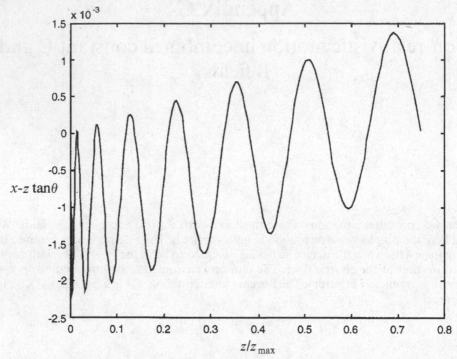

Fig. G.1.   The trajectory in uniform electric (50 kV/cm) and magnetic (40 kG) fields with a 30 degree angle between them. The particle starts at rest at $z = 0$. The trajectory is plotted with respect to motion along the magnetic field, $x = z \tan \theta$.

$$t_0^2 = (2d/a)$$

$$(v_o/c)^2 = (2eV_o)/m \qquad\qquad\qquad \text{(G.3)}$$

Where $E = V_o/d$ gives the potential V_o for a uniform electric field. Numerically, for $E = 3 \times 10^4$ V/cm we find $v_o/c = 0.2$ and, if $d = 3$ mm, $t_o = 0.1$ ns. These parameters are typical of the hybrid detectors mentioned in Chapter 9.

If there are no electric fields, then the radius of curvature about the axis of the **B** field at a velocity v_o is $a_o = mv_o/eB$. For $B = 40$ kG, $a_o = 0.084$ mm and $\phi = \omega t_o = 68 \cong 11/2\pi$ or 11 full electron rotations in the magnetic field.

For weak magnetic fields we expect mostly motion accelerated along the z axis, i.e. the electric field direction. However, for strong magnetic fields we expect that the electrons are captured along the magnetic field axis, with tight (i.e. small a_o) helical orbits about that axis. The general case, for example a PMT in an external magnetic field, can be studied using Eq. G.2. In Fig. G.1 is shown the trajectory in a strong magnetic field. The acceleration from $z = 0$ to $z = z_{max}$ is accompanied by rotation about the magnetic field direction. In this specific case, eight complete rotations are visible.

Appendix H
Signal generation in a silicon diode for point ionization

The equations of motion in a silicon diode are more complicated than those solved in the body of the text. We consider only point ionization as a simplification. Consider the creation of an electron–hole pair at x_o at time $t=0$, where $0<x_o<d$. The electric field in the p region between the diode electrodes is, just at depletion, $E=(E_o/d)(d-x)$ as shown in Fig. 9.4. If the diode is overdepleted, then the minimum field is not $E=0$ at $x=d$, but $E=\Delta E$ at $x=d$, or $E(x)=(E_o/d)(d-x)+\Delta E$. Thus, in this case there is a finite electric field in all locations in the p layer. Clearly this finite field will speed up the charge collection time which is why detectors are typically run overdepleted, if possible.

After creation, electrons are swept to $x=d$ by the electric field, while the holes are swept to $x=0$. Since $dx/dt=\mu E$, we find, solving the equation $dx=\mu(E_o/d)(d-x)dt$, that the electrons have a trajectory, just at depletion, $x_-(t)$.

$$x_-(t)=(x_o-d)e^{-t/\tau_-}+d$$

(H.1)

$$\tau_-=(d/\mu_-E_o)$$

Thus, $x_-(\infty)=d$ and $x_-(0)=x_o$, and it takes many time constants, τ_-, for the electrons to be swept to $x=d$ if the diode is operated just at depletion.

If, instead, the diode is run beyond depletion, the solution $x_-(t)$ is:

$$x_-(t)=(x_o+x_d-d)e^{-t/\tau_-}+(d-x_d)$$

(H.2)

Where $x_d \equiv d(\Delta E/E_o)$ defines a length incremental to the depletion depth which is in excess of the physical depth d. The solution, Eq. H.2, is derivable from Eq. H.1 via the substitution $d \to d-x_d$. As before, $x_-(0)=x_o$. However, the point $x=d$ is now reached in a finite time t_o.

$$\left(\frac{x_d}{x_d+x_o-d}\right)=e^{-t_o/\tau_-}$$

(H.3)

Clearly, if $x_d \to 0$, then $t_o \to \infty$ as we saw before in the just depleted case. In the limit where $x_d \gg d$, $t_o \to 0$ so that very fast response is obtainable if high fields can be sustained. In what follows we consider only the situation just at depletion.

The holes are swept in the opposite direction to $x=0$. The electric field increases during the motion, with the corresponding solution containing a positive exponential coefficient.

$$x_+(t)=(x_o-d)e^{t/\tau_+}+d$$

(H.4)

$$\tau_+=(d/\mu_+E_o)$$

333

The drift time for the holes, t_d, is defined by the $x=0$ condition, $d/(d-x_o)=e^{t_d/\tau_+}$. For $0<t<t_d$ the x position of the holes varies from x_o to zero. Note that there are two time constants, τ_- and τ_+, for the motion of the two charge carriers. The time for the holes to reach the electrode, t_d, is clearly $\sim\tau_+$. For $x_o=0$, $t_d=0$ since the holes are immediately sunk to the $x=0$ electrode. For $x_o=d$, $t_d\to\infty$ because the holes spend a lot of time in the zero field region.

The field in the p layer does work on the drifting charges. Energy conservation, as derived in the main text, implies that the work done on the charges is taken from the applied fields and this induces a change of charge, dQ, on the electrodes. Since the work done is $qEdx$, dQ is the same sign for both the electrons and the holes since q changes sign, but so does the sign of the drift motion dx. Thus the electron and hole currents reinforce one another.

For electrons $I_-(t)=dQ_-/dt$. We find that $dQ=qE\,dx/V$, $dx=\mu E\,dt$, or $dQ=q\mu E^2 dt/V$. Using the explicit formulae for $E(x)$ and $V(x)$ derived in Chapter 9;

$$I_-(t)=\left[1-\left(\frac{x_o}{d}\right)\right]^2\left(\frac{2e}{\tau_-}\right)e^{-2t/\tau_-} \tag{H.5}$$

The current is a maximum at $t=0$ when the fields are largest, $I_{max}=I_-(0)=\left(\frac{2e}{\tau_-}\right)\left[1-\left(\frac{x_o}{d}\right)\right]^2$. The current then decreases steadily as $t\to\infty$ when $I_-(\infty)\to0$. The current also depends on location in the gap. For $x_o=d$, $I_-=0$ since the electron moves no distance. For $x_o=0$ the maximum current is $I_-(0)|_{x_o=0}=(2e/\tau_-)$.

Integrating I to get the charge induced by the current of electrons, we obtain

$$Q_-(t)=e\left[1-\left(\frac{x_o}{d}\right)\right]^2[1-e^{-2t/\tau_-}] \tag{H.6}$$

The initial charge is defined to be $Q_-(0)=0$. The asymptotic collected charge is $Q_-(\infty)=e[1-(x_o/d)]^2$. The collected charge depends on where the point ionization occurs. For $x_o=d$, $Q_-(\infty)=0$ since the electron is immediately sunk to the electrode, while for $x=0$, $Q_-(\infty)=e$, and the entire electron charge is ultimately induced. Note that the basic $Q\sim x_o^2$ behavior occurs because $dQ\sim Edx$, $E\sim x$ and $dQ\sim x\,dx$, so that $Q\sim x^2$.

For the holes a similar situation obtains, but with $x_+(t)$ instead of $x_-(t)$ for the label of trajectory.

$$I_+(t)=\left[1-\left(\frac{x_o}{d}\right)\right]^2\left(\frac{2e}{\tau_+}\right)e^{+2t/\tau_+} \tag{H.7}$$

The minimum hole current occurs at $t=0$, $I_+(0)=[1-(x_o/d)]^2[2e/\tau_+]$. The largest current occurs when the holes ultimately sink on the electrode at $x=0$, $x_+(t_d)=0$, $I_+(x=0)=2e/\tau_+$.

The induced charge due to the hole motion is, $\int I_+(t)dt$.

$$Q_+(t)=e\left[1-\left(\frac{x_o}{d}\right)\right]^2[1-e^{2t/\tau_+}] \tag{H.8}$$

Hence, $Q_+(0)=0$ and $Q_+(t_d)=e(1-(x_o/d))^2(1-(d/(d-x_o))^2)=e[(d-x_o)^2-d^2]/d^2$. For $x_o=0$ the maximum hole induced charge is zero. For $x_o=d$ the maximum hole induced charge is e since the hole traverses the whole inter-electrode gap. This behavior is com-

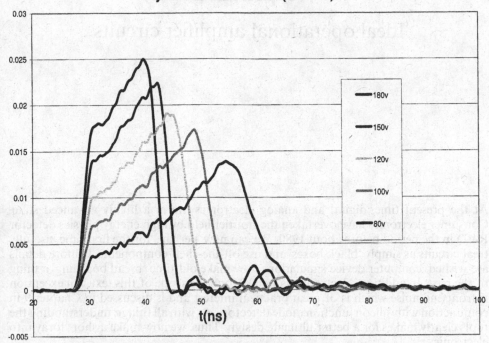

Fig. H.1. Hole signal data obtained from a silicon diode under point illumination at $x_o \sim d$. The diode is operated overdepleted so that the current begins at a finite value. The transit time of the $d \sim 300$ μm device is ~ 15 ns.

plementary to that displayed by the electron induced charge, as might be expected. The total current due to the joint motion of the holes and electrons is $I = I_+ + I_-$.

$$I(t) = \left[1 - \left(\frac{x_o}{d}\right)\right]^2 2e\left[\left(\frac{1}{\tau_-}\right)e^{-2t/\tau_-} + \left(\frac{1}{\tau_+}\right)e^{+2t/\tau_+}\right] \qquad \text{(H.9)}$$

The total charge induced by the joint hole and electron motions is $Q = Q_+ + Q_-$.

$$Q(t) = e\left[1 - \left(\frac{x_o}{d}\right)\right]^2 [(1 - e^{-2t/\tau_-}) + (1 - e^{2t/\tau_+})] \qquad \text{(H.10)}$$

The diode acts like a solid state ionization detector except that both the positive and negative charge carriers have a comparable mobility and contribute to induced currents and that the field is not uniform. This latter aspect leads to exponential time behavior rather than the linear time behavior for point ionization observed in an ionization chamber (see Chapter 8).

Data taken with a diode and point illumination with $x_o \sim d$ is shown in Fig. H.1. The electron current is ~ 0 since the electrons are immediately sunk at $x = d$. The holes move to $x = 0$. The diode is overdepleted, so that $I(0)$ is not zero. The exponential rise of the current, Eq. H.7, is clearly seen. The increased speed as the voltage is raised above depletion voltage is also very evident.

Appendix I

Ideal operational amplifier circuits

At the present time, digital and analog electronics are in a highly advanced state. Consumer electronics have overtaken the frontier held by high energy physics detector R&D in the period before about 1980. We can now, perhaps for the first time, usefully treat circuits as simply 'black boxes' and use off-the-shelf components. If more details are wished, computer device simulations now make old-time 'bread boarding' a thing of the past. In addition, electronics is well beyond the scope of this text. An exception is front end noise which is of great practical interest and is discussed in Chapter 9 in conjunction with silicon junction diode detectors. As with all things, understanding the tools clearly makes for a better ultimate design. Thus, we now make a short foray into electronics.

In order to get the flavor of possible circuit configurations, we adopt an ideal operational amplifier circuit model which is at the highest level of simplification. The 'op amp' is a high gain device with two inputs, one inverting, $(-)$, and one non-inverting, $(+)$. These inputs are assumed to have high input impedance. Together, the model requires that for stable operation $\Delta V = V_+ - V_- \sim 0$ and , see Fig. I.1, the amplifier output is not to be driven into a nonlinear operating point, because its gain (G) is taken to be very large, $G \to \infty$.

All the basic circuits shown below use negative feedback to reduce the gain G (assumed infinite) of the op amp and to stabilize operation. Feedback insures that the operation of the circuit depends only on the external passive components in the feedback loop. The input/output 'transfer' function then depends only on resistor values, for example.

The first such example is the non-inverting amplifier shown in Fig. I.2. Since $\Delta V \sim 0$ and $I_- = 0$ we have $IR_1 = V_I$, $IR_2 = V_o - V_I$, with a solution

$$V_o = V_I + IR_2$$

$$= V_I + (V_I/R_1)R_2 \tag{I.1}$$

$$V_o/V_I = (1 + R_2/R_1)$$

The device thus operates as an amplifier with a feedback defined effective gain, $G_{EFF} = (1 + R_2/R_1)$.

A specific example is a buffer-follower which might be used to drive/receive signals on cables without distortion. In this special case $R_1 \to \infty$, $R_2 \to 0$, and we have $V_o = V_I$. This 'unity gain' application is very often used to drive signals off circuit cards on cables over considerable distances (see Fig. D.2).

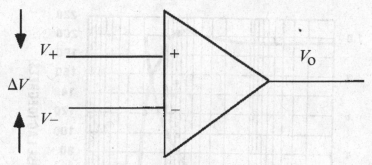

Fig. I.1. Ideal operational amplifier showing inverting and non-inverting inputs and output V_o.

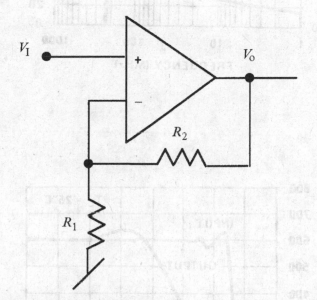

Fig. I.2. Basic non-inverting amplifier.

Any real amplifier has a finite gain and a finite bandwidth (finite noise). Shown in Fig. I.3 is the gain of a 'unity gain buffer' as a function of frequency. The response is quite flat out to ∼200 MHz. Also shown is the small signal response to a step function (rise time), which is about 2 ns. Such speeds are typical of current commercially available amplifiers. If our application is at frequencies less than 100 MHz, we can perhaps treat this device as an ideal buffer.

Turning to inverting amplifiers, the one shown in Fig. I.4 is easily analyzed using our basic assumptions about ideal op amps.

$$V_I + IR_1 = V_- \cong 0$$

$$V_o - IR_2 = V_- \cong 0$$ (I.2)

$$V_o/V_I = -R_2/R_1$$

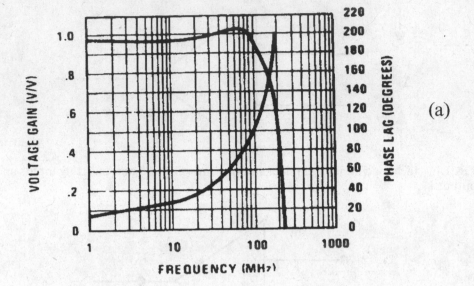

(a)

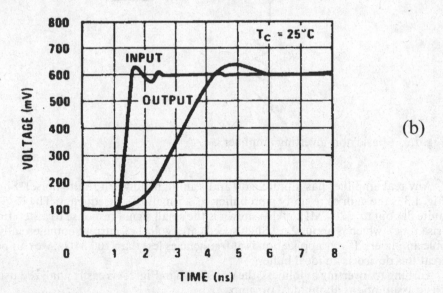

(b)

Fig. I.3.   (a) Frequency response of a commercial op amp operating as a unity gain
buffer driver. (b) Small signal time response to an input step function. (From Ref. I.1
with permission.)

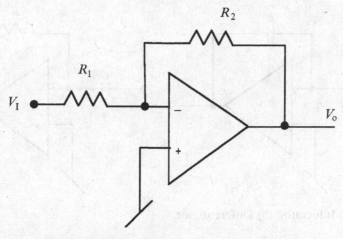

Fig. I.4. Basic inverting amplifier.

The effective gain is negative (i.e. inverting) and set by the two feedback resistors. If $R_1 = R_2$ the gain is -1.

There are many variations on the inverting amplifier which arise when driving it with current sources, using other circuit elements in place of R_1 and R_2, using both input legs, or using multiple inputs to a single leg (e.g. 'adders'). A few of the basic circuits will be indicated, but they are only a small representative sample.

First in Fig. I.5 we replace either R_1 or R_2 by a capacitor. That replacement allows us to either integrate or differentiate a voltage source ($V = Q/C = \int I dt/C$).

$$\left.\begin{array}{l} V_I + IR = 0 \\[4pt] V_o - V_c = 0 \\[4pt] V_I = -RC\, dV_o/dt \\[4pt] V_o = -(\int V_I dt)/RC \end{array}\right]$$

$$\text{integrator} \qquad\qquad (\text{I.3a})$$

$$\left.\begin{array}{l} V_I = (\int I dt)/C \\[4pt] V_o = IR \\[4pt] V_o = -\dfrac{1}{RC}\int V_I dt \\[4pt] V_o = -RC\, dV_I/dt \end{array}\right]$$

$$\text{differentiator} \qquad\qquad (\text{I.3b})$$

As noted in the text, most detector elements behave approximately as current sources with source currents i_s because they are typically taken off high impedance electrodes. High speed response or low noise performance is provided by the circuits shown in Fig. I.6. The 'transimpedance amplifier' simply converts the current to an output voltage, Eq. I.4a, while the 'charge sensitive preamp' integrates the current pulse to give an output voltage proportional to the source charge q_s, Eq. I.4b.

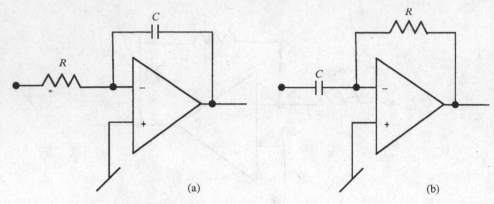

Fig. I.5.   (a) Integrator. (b) Differentiator.

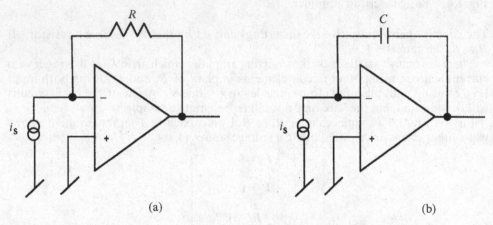

Fig. I.6.   (a) Transimpedance amplifier. (b) Charge sensitive preamplifier.

$$i_s R = V_o - V_- \sim V_o \qquad\qquad (I.4a)$$

$$q_s/C = V_o - V_- \sim V_o \qquad\qquad (I.4b)$$

Finally, it is often useful to exploit the fact that the 'op amp' is a true differential amplifier; $V_o = G(V_+ - V_-) = G\Delta V, G \to \infty$. Therefore, as shown in Fig. I.7a, when V_- exceeds $V_+ = V_T$, the output is rapidly driven to its maximum value yielding a sharp 'threshold' for the 'discriminator'. This behavior is assumed in describing a discrimination and coincidence circuit in Chapter 2. In this unstable mode of operation only two disconnected states are possible output = '0' or '1'.

$$V_o = \text{'1' if } V_- > V_T$$

$$V_o = \text{'0' if } V_- < V_T \qquad\qquad (I.5)$$

Noise is almost always an issue for detectors. If signals are to be driven off detectors on cables, see Appendix D, noise may be induced on the signal wire and on the 'grounded' outer coaxial layer or on both of the twin leads. One powerful technique

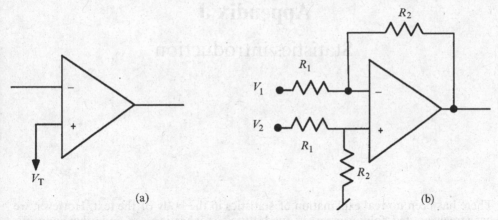

Fig. I.7. (a) A comparator or discriminator. (b) A fixed gain differential amplifier.

for reducing this 'pickup' noise is to use circuits which respond differentially, so that any common noise which is picked up by both inputs is rejected by the differential amplifier since it only, ideally, responds to the difference $V_+ - V_-$ ('common mode rejection'). A possible circuit is shown in Fig. I.7b.

$$V_2 - I(R_1 + R_2) = 0$$

$$V_+ = V_2 - IR_1 = V_2\left(\frac{R_2}{R_1 + R_2}\right) \cong V_- \tag{I.6a}$$

$$V_1 - I'(R_1 + R_2) = V_o$$

$$V_- = V_1 - I'R_1 \tag{I.6b}$$

$$V_o = (V_2 - V_1)(R_2/R_1)$$

This circuit can be thought of as a differential inverting amplifier. (See Eq. I.2 where with $V_2 \to 0$ in Eq. I.6b, the two expressions agree). Note that both inputs V_1 and V_2 'see' the same input impedance, $R_1 + R_2$. This fact allows us to make a good cable termination which enhances common mode rejection.

Reference

[I.1] *Linear Databook*, National Semiconductor Corporation (1982).

Appendix J

Statistics introduction

There has been no real explanation of statistics in the body of the text. However, we have freely quoted 'folding errors in quadrature', 'stochastic error' and other concepts. In order to give some minimal background, we expand a bit on the basic concepts of statistical error. References are provided at the end of the text (Chapter 13). The treatment here is completely without rigor.

Consider first a discrete series of N measurements of a quantity y, y_i, $i = 1, ..., N$. The mean and mean square deviation from the mean of those measurements are $\bar{y}$ and σ^2.

$$\bar{y} = \sum_i^N y_i / N$$

$$\sigma^2 = \sum_i (y_i - \bar{y})^2 / N \tag{J.1}$$

In the limit where $N \to \infty$ the distribution of results y approaches a continuous distribution function $\dfrac{dP}{dy}$ where the probability to observe y between y and $y + dy$ is defined to be $dP = \dfrac{dP}{dy} dy$. The expressions for $\bar{y}$ and σ^2 in the continuous case become

$$\bar{y} \to \int y(dP/dy)dy = \int y\, dP$$

$$\sigma^2 \to \int (y - \bar{y})^2 (dP/dy)dy = \int (y - \bar{y})^2 dP \tag{J.2}$$

$$\int dP = 1$$

The most commonly assumed distribution function is the Gaussian. It is often assumed to apply in experimental situations because, in the limit of large numbers of events, many distributions approach a Gaussian. The theoretical function is characterized by two parameters, a mean $<y>$ and a root mean square (RMS) deviation from the mean equal to Δy.

$$dP(y,\langle y\rangle,\Delta y) = \frac{1}{\sqrt{2\pi\Delta y^2}} \exp\left[-\frac{1}{2}\left(\frac{y - \langle y\rangle}{\Delta y}\right)^2 \right] dy \tag{J.3}$$

It is easy to show that the Gaussian given in Eq. J.3 is normalized to 1, and that the parameters $\langle y\rangle$ and Δy are $\bar{y}$ and σ respectively as defined in Eq. J.2. Note that $\bar{y}$ and σ refer to a large experimental data set, while $\langle y\rangle$ and Δy are parameters which define a theoretical distribution function, dP/dy.

Since it is used so extensively in the text, we quickly derive the Poisson distribution. Examples used in the body of the text include mean free path, decay lifetime, and phototube photoelectron statistics. Consider a case where the probability of interacting in traversing dx is $dx/\langle L \rangle$; the mean free path is $\langle L \rangle$. The probability of getting no events in traversing x is $\underline{P}(0,x)$.

$$d\underline{P}(0,x) = -\underline{P}(0,x)dx/\langle L \rangle$$

$$\underline{P}(0,x) = e^{-x/\langle L \rangle} \tag{J.4}$$

The exponential law for no interaction is already very familiar having been quoted in the discussion of cross sections. For getting N events in x, the appropriate probability is $\underline{P}(N,x)$, where we find none in dx and N in x (with the ordering of the events being assumed to be irrelevant). Assuming that joint probabilities multiply:

$$d\underline{P}(N,x) = \prod_i^N \frac{(dx_i/\langle L \rangle)}{N!} e^{-x/\langle L \rangle}$$

$$\underline{P}(N,x) = \frac{e^{-x/\langle L \rangle}}{N!} \left(\frac{x}{\langle L \rangle} \right)^N \tag{J.5}$$

$$= \frac{\langle N \rangle^N}{N!} e^{-\langle N \rangle}, \quad \langle N \rangle = x/\langle L \rangle$$

In terms of the mean number of events/traversal, $\langle N \rangle$, we recognize the Poisson photoelectron distribution, for example. The Poisson distribution approaches a Gaussian in the appropriate limit as N becomes large.

Suppose the N events yield measures y_i of some variable with measurement error σ_i and that we wish to make a hypothesis that they are described by a distribution function which predicts y as $\bar{y}$ when characterized by a single parameter α (for simplicity). Note that $\bar{y}$ is the theoretical prediction in what follows and not the sample mean as given in Eq. J.1. For example, we measure N decay times, t_i, and assume a distribution with a single lifetime, τ. The joint probability of the N independent measures defines a 'likelihood function' L. Recall that independent events have probabilities which multiply. For example, the chance to roll a five on a die is 1/6, while the change to roll a six and a five (in that order) is 1/36.

$$L(\alpha) = \prod_i^N d\underline{P}_i(y_i, \alpha) \tag{J.6}$$

The maximum of the likelihood would occur at a value of α, $\langle \alpha \rangle$ for which we have the largest joint probability. In the special case of a Gaussian function $d\underline{P}$, we minimize the χ^2 function i.e. the method of least squares. A maximum L, if $d\underline{P}$ is a Gaussian, means a minimum value of the argument of the exponential.

$$\chi^2 \sim \ln L$$

$$= \sum_{i=1}^N [y_i - \bar{y}(\alpha, x_i)]^2 / \sigma_i^2 \tag{J.7}$$

$$L(\alpha)|_{\max} \Rightarrow \chi^2(\alpha)|_{\min}$$

$$\partial \chi^2 / \partial \alpha|_{\langle \alpha \rangle} = 0$$

The minimum value of χ^2, in the special case $\bar{y} = \alpha$ occurs for

$$\frac{\partial \chi^2}{\partial \bar{y}} = 0$$

(J.8)

$$\dot{y} = \left(\sum_i y_i/\sigma_i^2\right) \Big/ \left(\sum_i \frac{1}{\sigma_i^2}\right)$$

In the special case that all errors are equal, $\sigma_i = \sigma$, $\bar{y} = \sum_i y_i/N$, and we recover Eq. J.1. The result Eq. J.8 is the best estimator of the mean of a set of measures assuming a Gaussian error distribution. We expect that it is related to the sample mean, as indeed we have shown.

The algebra for the error on the mean is straightforward. The value $\bar{y}$ occurs when χ^2 is minimized. The estimate of σ_y comes from looking at how fast we deviate from the minimum $(\partial^2\chi^2/\partial^2\bar{y})$, $\partial^2\chi^2/\partial_{\bar{y}}^2 = 1\sigma/_{\bar{y}}^2$.

$$\sigma_{\bar{y}}^2 = \frac{1}{\sum_i (1/\sigma_i^2)}$$

$$WT = \sum_i WT_i, \quad WT_i \equiv 1/\sigma_i^2$$

(J.9)

$$\bar{y} = \left(\sum_i y_i WT_i\right)/WT$$

If the measurements have different errors, σ_i, then they are 'weighted' as the inverse square of that error in finding the best estimation of the mean. This is called the weighted mean technique. If all individual measurement errors are the same, $\sigma_i = \sigma_o$, then the best estimate of the error to be attached to the ensemble of N measurements is, NWT_o or $\sigma = \sigma_o/\sqrt{N}$. We remark on this $1/\sqrt{N}$ behavior in the body of the text several times. We see that error estimators improve only slowly with the number of measures. It is also intuitively clear that poor measurements with large errors (small weights) should not strongly influence the estimate for $\bar{y}$ relative to good measurements with small errors (large weights).

The idea of estimating α is that for a large data set the likelihood clusters about its most probable value, $L(\langle\alpha\rangle) = L_{\max}$, has a width σ_α (see Eq. J.2).

$$\sigma_\alpha^2 = [\textstyle\int (\alpha - \langle\alpha\rangle)^2 L \, d\alpha]/[\int L \, d\alpha]$$

(J.10)

For the special case of Gaussian errors, and $L(\alpha) \sim \exp\left[-\frac{1}{2}\left(\frac{\alpha - \langle\alpha\rangle}{\sigma_\alpha}\right)^2\right]$ and $\chi^2 \sim \left(\frac{\alpha - \langle\alpha\rangle}{\sigma_\alpha}\right)^2$. Therefore, the best estimate of the error on α can be found using the expression for χ^2 and differentiating, $\partial\chi^2/\partial\alpha = (\alpha - \langle\alpha\rangle)/\sigma_\alpha^2$ which is minimized when $(\alpha - \langle\alpha\rangle) = 0$ by assumption. Differentiating again, $\partial\chi^2/\partial\alpha^2 = 1/\sigma_\alpha^2$, leading to an estimate for the error of α using the behavior of χ^2 as a function of α.

$$\sigma_\alpha^2 = \left(\frac{\partial^2 \chi^2}{\partial \alpha^2}\right)^{-1}_{\langle\alpha\rangle} \tag{J.11}$$

Clearly, if the hypothesis is good, then $\chi^2 \sim 1$ for a single variable, since in that case $\chi^2 \sim [(\alpha - \langle\alpha\rangle)/\sigma_\alpha]^2$. The test of how well a hypothesis fits the data depends on the size of χ^2. The precise idea of 'goodness of fit, and significance' can be explored in the references.

For multiple parameters, α_j, specifying the hypothesis, the error analysis becomes more difficult. Still, for N measurements at location x_j yielding values y_i of a variable y, the χ^2 is the same as in Eq. J.7, and we minimize χ^2 with respect to the set α_j, $j = 1$, M by simultaneously solving the set of M least square equations.

$$\partial\chi^2/\partial\alpha_j = 0 \tag{J.12}$$

An example might be chamber measurements sampling a trajectory in a magnetic field. The orbit, $\bar{y}$, is a function of the parameters $\mathbf{x}_o \hat{a}_o$, q, and p, as we have derived. A least squares fit to the helical path will yield the best fit value of the parameters defining the path of the track. In particular, we will determine the momentum of the track with some error. This outline of a procedure makes quite specific what we mean by using detectors to measure the momentum of a track.

With M parameters, the errors on the parameters are estimated from a straightforward generalization of Eq. J.11. However, in this case there is an 'error matrix' H^{-1}, of dimension $M \times M$, with off diagonal elements indicating a correlation between the parameters.

$$H_{ij} = \partial\chi^2/\partial\alpha_i\partial\alpha_j$$

$$\sigma_{ai}^2 = (H^{-1})_{ii} \tag{J.13}$$

$$\langle\alpha_i - \langle\alpha_i\rangle\rangle\langle\alpha_j - \langle\alpha_j\rangle\rangle = (H^{-1})_{ij}$$

The diagonal elements of the inverse H matrix provide an estimator of the errors on the M parameters α_j.

For example, assume N points at x_i measuring y_i with error σ_i. Assume no forces, so that the path $\bar{y}$ is a straight line. By hypothesis $M = 2$, $\bar{y} = ax + b$. The remaining degrees of freedom are $N - M$, so generalizing our previous discussion, we expect $\chi^2_{\min}/(N-M) \sim 1$ would indicate a reasonable fit to the hypothesis. The minimum value of χ^2 with respect to a and b occurs when

$$\chi^2 = \sum_i^N (y_i - ax_i - b)^2/\sigma_i^2$$

$$\frac{\partial\chi^2}{\partial a} = 2\sum_i^N (y_i - ax_i - b)(-x_i)/\sigma_i^2 = 0 \tag{J.14}$$

$$\frac{\partial\chi^2}{\partial b} = 2\sum_i^N (y_i - ax_i - b)(-1)/\sigma_i^2 = 0$$

This is two equations in two unknowns. The solution is left as an exercise for the reader.

$$b = \begin{vmatrix} \sum_i y_i WT_i & \sum_i x_i WT_i \\ \sum_i x_i y_i WT_i & \sum_i x_i^2 WT_i \end{vmatrix} / \Delta$$

$$a = \begin{vmatrix} \sum_i WT_i & \sum_i y_i WT_i \\ \sum_i x_i WT_i & \sum_i x_i y_i WT_i \end{vmatrix} / \Delta \qquad (J.15)$$

$$\Delta = \begin{vmatrix} \sum_i WT_i & \sum_i x_i WT_i \\ \sum_i x_i WT_i & \sum_i x_i^2 WT_i \end{vmatrix}$$

The error matrix on a and b follows from Eq. J.13. The algebraic details are again left to the reader.

$$\frac{\partial^2 \chi^2}{\partial a^2} = 2 \sum_i^N x_i^2 WT_i$$

$$\frac{\partial^2 \chi^2}{\partial a\, \partial b} = 2 \sum_i^N x_i WT_i$$

$$\frac{\partial^2 \chi^2}{\partial b^2} = 2 \sum_i^N WT_i \qquad (J.16)$$

$$\sigma_b^2 \sim \frac{1}{\Delta} \sum_i x_i^2 WT_i$$

$$\sigma_a^2 \sim \frac{1}{\Delta} \sum_i WT_i$$

These formulae serve us as a concrete example of the least squares method in a case of practical interest. They also illustrate the general method, Eq. J.12, Eq. J.13, in a particularly simple case. The reader is encouraged to fit some real data in order to gain experience.

Suppose we determine a variable Y which is a function of N variables y_i each with a different error σ_i. The maximum likelihood for $Y, \bar{Y}$ occurs, we assume, when the y_i attain their most likely values.

$$Y = Y(y_i)$$
$$\qquad (J.17)$$
$$\bar{Y} = Y(\bar{y}_i)$$

Using the definition, Eq. J.1, for mean square deviation, we can find the error on Y due to the errors on y_i. We assume that the fluctuations in the y_i are uncorrelated so that all cross terms, proportional to $y_i y_j$, $i \neq j$, average to zero. The series Taylor expansion in N dimensions is applied.

$$\sigma_Y^2 = (Y - \bar{Y})^2$$

$$\cong \left[\sum_i (y_i - \bar{y}_i) \frac{\partial Y}{\partial y_i} \right]^2$$

$$\cong \sum_i \left[(y_i - \bar{y}_i) \frac{\partial Y}{\partial y_i} \right]^2 \quad \text{uncorrelated} \qquad (J.18)$$

$$\cong \sum_i [\sigma_y \partial Y/\partial y_i]^2$$

This result allows us to 'propagate' errors. If we know the error in y, then we can find the error on any function of $y = Y$. For a single variable the error in Y becomes, $\sigma_Y = \sigma_y \, \partial Y/\partial y$. For a function of many variables, we can work out the result using Eq. J.18. For example, the product $Y = y_1 y_2 ... y_N$ has a fractional error which 'adds in quadrature'.

$$Y = y_1 y_2 ... y_N$$

$$(\sigma_Y/Y)^2 = \sum_i (\sigma_{y_i}/y_i)^2 \qquad (J.19)$$

Many volumes have been written on probability and statistics. The aim of this appendix has been only to introduce the concepts of a data set, the sample mean and standard deviation and its characterization by a distribution function $dP(y)$ defined by parameters α. The 'best' estimate of the mean is derived along with an estimate of the error on that best estimate assuming Gaussian errors and using the method of maximum likelihood/least sources. Propagation of errors is introduced as a final topic. Having provided this appendix, we have a self-contained explanation of the topics in statistics alluded to in the body of the text. More advanced references appear at the end of Chapter 13.

Appendix K

Monte Carlo models

Many of the problems encountered in making a realistic detailed model of a detector are so complex as to not be susceptible to analytic techniques. Nevertheless, we can break the problem of modeling a very complex system down into a series of choices for the relevant dynamical variables. The Monte Carlo method allows us to choose those variables and hence to construct such a model.

For example, suppose we want to model delta ray production in a medium by incident muons. The choices involved are first at what depth is the delta ray made. We choose x out of a distribution $e^{-(N_0\rho\sigma x/A)}$ where σ is the cross section for delta production. If there is indeed delta production within the active volume, the dynamics depends on a single variable, as has been mentioned previously. We can, for example, pick the recoil electron kinetic energy from $T=0$ to $T=T_{max}$ from a distribution $d\sigma/dT \sim 1/T^2$. Having the energy, the recoil angle follows, as derived in Appendix A, from the two body kinematics. We now have the position and momentum vectors of the electron at the point of production. We can find the final state muon as well if we are to trace it further through the system.

If it is desired that the delta ray be followed in its path, other choices must be made. Suppose there are no fields, for simplicity, so that the path is, at least locally, a straight line. We pick a distance which is short with respect to both the energy loss, dE_1/dx, and for which the multiple scattering angle is small. Over this distance the energy and angles of the electron can be considered to be constant. We extrapolate the electron path as a straight line. We pick a new energy T' by removing the ionization energy from the electron. We also pick new angles using a Gaussian distribution with $\mathrm{RMS}=\theta_0$ about the initial direction due to multiple scattering in the medium. Clearly, we can make a rather complicated model out of individual, fairly simple, choices. In fact, almost all the formulae appearing in the body of this text would, at some point, be used in a complete Monte Carlo simulation of the complex detectors used today. Therefore, we need to be conversant with all the dynamics discussed in this text if we want to be able to make a realistic model of a detector.

How do we actually go about choosing the individual dynamical quantities which define the evolution of the system? First, we need a 'random number generator' which produces a uniformly distributed number r in the interval $(0, 1)$. These generators are widely available on almost all computer platforms.

$$0 < r < 1, \text{ uniformly distributed} \tag{K.1}$$

Suppose we have a probability distribution $dP(x)/dx$ defined such that $dP(x)$ is the probability for x to occur between x and $x+dx$. If x is constrained to the interval

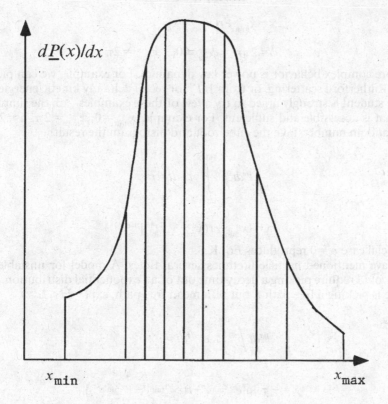

Fig. K.1. Visualization of intervals in x, Δx, over which the integral $\Delta \underline{P}$ is the same.

$(x_{\min}, x_{\max})$ then the integral probability is normalized to one, so that the cumulative probability from $x_{\min}$ to x should be uniformly distributed.

$$\int_{x_{\min}}^{x} d\underline{P}(t) \bigg/ \int_{x_{\min}}^{x_{\max}} d\underline{P}(t) = r = \int_{x_{\min}}^{x} d\underline{P} \qquad (K.2)$$

If $r = 0$ then $x = x_{\min}$, while if $r = 1$, $x = x_{\max}$.

A visual interpretation of $r = \int d\underline{P}$ is given in Fig. K.1. The discrete case is shown when we can perform the integral to find N intervals of x with equal probability, $\Delta \underline{P} = \int d\underline{P} \equiv g(x)$. If the inversion, $\Delta x = g^{-1}(\Delta \underline{P})$ is possible then x can be chosen by choosing index $1 < i < N$, with x between $x_{\min}$ and $x_{\max}$.

If this expression can be analytically inverted, we can 'pick x out of a distribution $d\underline{P}(x)/dx$'. Let us examine some possibilities. If $d\underline{P}(x)/dx$ is uniform, we can find a solution. An example arises in azimuthally uniform scattering, for example unpolarized Compton scattering. The azimuthal angle ϕ is chosen as,

$$\int_{x_{\min}}^{x} dt \bigg/ \int_{x_{\min}}^{x_{\max}} dt = r$$

$$x = x_{min} + r(x_{max} - x_{min}) \tag{K.3}$$

$$\phi = 2\pi r, \quad \phi_{min} = 0, \quad \phi_{max} = 2\pi$$

A more complex behavior is power law dynamics. For example, we can pick out of $1/\theta^4$ for Rutherford scattering, or from $1/T^2$ for recoil delta ray kinetic energies. By the way, the student is strongly urged to try a few of these examples – on the simplest computer that is accessible and sufficient. For example, $x_{min} = 0$, $x_{max} = 2\pi$, $\alpha = 2$, $x = r^{1/3}$, pick a random number, take the cube root and histogram the result.

$$\int_{x_{min}}^{x} t^\alpha dt \Big/ \int_{x_{min}}^{x_{max}} t^\alpha \, dt = r \tag{K.4}$$

$$x = [x_{min}^{\alpha+1} + r(x_{max}^{\alpha+1} - x_{min}^{\alpha+1})]^{1/(\alpha+1)}$$

The special case $\alpha = 0$ reproduces Eq. K.3.

We have mentioned particle lifetimes several times. A model for unstable particle decays would require picking a decay time out of an exponential distribution. Another example is picking a free path L out of a mean free path, $\exp(-L/<L>)$.

$$\int_{x_{min}}^{x} e^{-t/\tau} dt \Big/ \int_{x_{min}}^{x_{max}} e^{-t/\tau} dt = r \tag{K.5}$$

$$x = -\tau \{\ln[e^{-x_{min}/\tau} + r(e^{-x_{min}/\tau} - e^{-x_{min}/\tau})]\}$$

In the energy rather than the time domain any unstable particle is described by a Lorentzian energy spectrum. The Fourier transform of a decaying exponential with lifetime τ is a Lorentzian with energy full width $\Gamma = \hbar/\tau$. The uncertainty relation $\Delta E \Delta t = \hbar$ becomes $\Gamma\tau = \hbar$ in this special case. This type of behavior also occurs in any damped resonant system.

$$\frac{\int_{x_{min}}^{x} \left[\dfrac{1}{(t - x_o)^2 + (\Gamma/2)^2}\right] dt}{\int_{x_{min}}^{x_{max}} \left[\dfrac{1}{(t - x_o)^2 + (\Gamma/2)^2}\right] dt} \tag{K.6}$$

$$\phi_{min} = 2(x_{min} - x_o)/\Gamma$$

$$\phi_{min} = 2(x_{max} - x_o)/\Gamma$$

$$x = x_o + \Gamma/2\{\tan[\tan^{-1}\phi_{min} + r(\tan^{-1}\phi_{max} - \tan^{-1}\phi_{min})]\}$$

The most common distribution that is used is, perhaps, the Gaussian distribution. The integral leads to a non-analytic error function, so it seems that the technique fails. However, a trick avails: using the joint probability and displacing the mean to zero, $x = x' - \langle x' \rangle$, $y = y' - \langle y' \rangle$.

$$dP(x)dP(y) = e^{-x^2/2\sigma^2} e^{-y^2/2\sigma^2} \, dx \, dy$$

$$= e^{-r^2/2\sigma^2} r dr \, d\phi \qquad \text{(K.7)}$$

$$= e^{-u/2\sigma^2} \frac{du \, d\phi}{2}, \quad u = r^2$$

Therefore, we can pick ϕ from a uniform distribution as in Eq. K.3 and r^2 from an exponential distribution as in Eq. K.5. The result is, $x = r\cos \phi$, $y = r\sin \phi$, two Gaussianly distributed variables x and y with zero mean and rms $= \sigma$. The mean can be restored by addition, $x' = x + \langle x' \rangle$, $y' = y + \langle y' \rangle$.

Let us now consider angular distributions. In the case of isotropic dynamics, e.g. decay of a spinless particle, we can use the previous results for uniform distributions for ϕ and $\cos \theta$. The solid angle is then a constant.

$$d\sigma/d\Omega = \text{const} = 1/4\pi$$

$$d\Omega = d(\cos\theta)d\phi \qquad \text{(K.8)}$$

$$\phi = 2\pi r$$

$$\cos\theta = \cos\theta_{min} + r(\cos\theta_{max} - \cos\theta_{min})$$

Suppose we have a more complicated angular distribution. If, for example, $d\sigma/d\Omega = \cos^2 \theta$, then we can use the power law behavior, Eq. K.4.

$$d\sigma/d\Omega = \cos^2\theta \qquad \text{(K.9)}$$

$$\cos\theta = [\cos^3\theta_{min} + r(\cos^3\theta_{max} - \cos^3\theta_{min})]^{1/3}$$

What about the dipole behavior which is seen in non-relativistic radiation patterns? In that case the integral can be done, but it cannot be analytically inverted. In fact, for most distributions, an analytic inversion is not possible.

$$d\sigma/d\Omega = \sin^2\theta \qquad \text{(K.10)}$$

$$\int \sin^2 \theta \, d(\cos \theta) = \cos\theta - \frac{\cos^3\theta}{3}$$

How do we proceed? Clearly, by performing the integral given in Eq. K.2 numerically. Assume that x is contained in the interval (x_{min}, x_{max}). Assume also that within that interval there is a maximum value of $(dP(x)/dx)$ called $(dP(x)/dx)_{max}$. We use the 'rejection' method, see Fig. K.2.

$$\text{pick} r_1 : \text{distributing } x \text{ as } x = x_{min} + r_1(x_{max} - x_{min})$$

$$\text{pick} r_2 : \text{if } r_2 < (dP(x)/dP(x)dx)_{max} \text{ then accept } x \qquad \text{(K.11)}$$

$$\text{if } r_2 > (dP(x)/dP(x)dx)_{max} \text{ then reject } x$$

As can easily be seen in Fig. K.2, this procedure weights x by $(dP(x)/dx)$. Clearly, if $(dP(x)/dx) = 0$, then x is never accepted, while if $(dP(x)/dx) \sim (dP(x)/dx)_{max}$ then x is almost always accepted. For example, in the case $d\sigma/d\Omega = \sin^2 \theta = 1 - \cos^2 \theta$ we pick $\cos\theta$ uniformly between -1 and 1. If $1 - \cos^2 \theta < r_2$ then we accept that choice of $\cos\theta$ and continue. If not we repeat. The student should explicitly try this procedure and verify that it works.

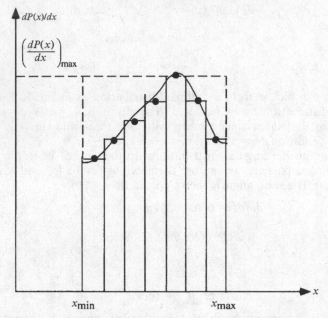

Fig. K.2.   Plot of distribution function $(dP(x)/dx)$ indicating how one approximates it by choosing uniform width strips in x and then accepting the strip x if $(dP(x)/dx)/(dP(x)/dx)_{max} <$ a random number.

We can also combine methods by using the fact that joint probabilities are multiplicative. Suppose we factor $(dP(x)/dx)$ into $g(x)$ and $h(x)$ where $0 < g < 1$ and $h > 0$, $\int h(x)dx = 1$.

$$dP(x)/dx = g(x)h(x) \qquad (K.12)$$

We might do this because $h(x)$ is analytically invertable, for example. We draw x from $h(x)$ and then accept x if $r < g(x)$. Clearly the probability to accept x is $g(x)$ by the rejection technique.

Several 'processes' with probability P_i can be used in a further generalization. For example, the total probability might be the total cross section while the individual probabilities would refer to the possible reactions making up the total reaction rate.

$$dP(x)/dx = \sum_i^N P_i g_i(x)h_i(x) \qquad (K.13)$$

First choose i by picking a process using a weight $P_i / \sum_i^N P_i$. Then draw x from $h_i(x)$ either directly or by rejection. Then accept x if $r < g_i(x)$. Clearly very complex distribution functions for x can be built up this way.

Glossary of symbols

a	Radius of a system or acceleration or magnetic radius of curvature or stochastic term in calorimeter resolution.
a_o	First Bohr radius $= 0.53$ Å.
a_T	Transverse radius of curvature in a magnetic field.
A	Atomic weight or area or electromagnetic vector potential, $\mathbf{A}$.
Å	Angstrom $= 10^{-10}$ m.
$A(\theta)$	Scattering amplitude or transition amplitude.
A^μ	Four dimensional acceleration.
α	Fine-structure constant or first Townsend coefficient or $\hat{a}$ direction cosines.
α	Helium nucleus.
α_s	Strong interaction coupling constant.
b	Impact parameter. Point of closest approach in a collision or constant term in calorimeter resolution.
b	barn $= 10^{-24}$ cm^2.
$\mathbf{B}$	Magnetic induction field, or binding energy per nucleon.
β	Velocity with respect to $c = v/c$.
β	Electron from nuclear beta decay.
$\boldsymbol{\beta}_d$	Drift velocity in $\mathbf{E}$ and $\mathbf{B}$ fields.
c	Speed of light.
C	Counting rate or capacitance.
C_s	Source capacity.
C'	Capacitance per unit length.
CB	Conduction band in a solid.
χ^2	The chi-squared function.
d	Distance between detector electrodes or distance between wires in a PWC.
$d\bar{V}^2$	Mean square noise voltage frequency distribution.
$d\Omega$	Element of solid angle $= d(\cos\theta)d\phi$.
D	Diffusion coefficient or displacement electric field.
δ	Phase shift or skin depth or difference of refractive index from one or Dirac delta function or gain/dynode in a PMT or optical phase difference.
Δt	Sample thickness of an EM calorimeter in X_o units.
Δy	RMS of a distribution function.

ΔE	Energy deposited in a calorimeter sample.
Δv	Sampling thickness of a hadronic calorimeter in λ_I units.
e	Electronic charge.
e/h	Ratio of the response of a calorimeter to EM and hadronic components of a cascade.
e_n	Voltage noise in volts/$\sqrt{Hz}$ for a transistor.
eV	Electron volt.
E	Electric field strength or total system energy.
E_g	Band gap in a solid between the top of the VB and the bottom of the CB.
E_n	Bohr energy in a quantum state n.
E_o	Rydberg ground state energy $= 13.6$ eV.
E_c	Critical energy where ionization and radiative losses are equal.
E_s	Characteristic multiple scattering energy, 21 MeV.
E_{TH}	Threshold for pion production in a hadronic interaction.
ENC	Equivalent Noise Charge or noise referred to the input compared to source charge.
ε	Particle energy or detection efficiency or electrical permitivity or polarization vector.
$\varepsilon(t), \varepsilon(v)$	Particle energy in a cascade as a function of depth t or v.
f	Frequency of a wave or focal length of a lens system.
f_o	Fraction of electromagnetic energy in a single hadronic collision, $\sim 1\sqrt{3}$.
'f_o'	Fraction of electromagnetic energy in a hadronic cascade.
$f(\omega)$	Transfer function or filter function in noise source studies.
fm	Fermi $= 10^{-13}$ cm.
F	Force on a particle.
g_m	Transconductance of a front end transistor.
GeV	10^9 eV.
γ	Ratio of particle energy/mass $= 1/\sqrt{1 - \beta^2}$.
γ	Photon
Γ	Transition rate or inverse lifetime $=$ decay width.
H	System Hamiltonian or magnetic field.
$\hbar$	Planck constant, $\hbar = h/2\pi$.
i_s	Source current liberated in a detector by particle passage.
I	Intensity $=$ energy crossing unit with frequency ω, $I(\omega)$ or ionization potential $\langle I \rangle$ or electric current.
$I(t)$	Capacitively induced current flow between detector electrodes.
$\mathbf{j}$	Current density or current/area.
k	Wave number $= 2\pi/\lambda$ or particle momentum in $\hbar$ units ($\mathbf{p} = \hbar\mathbf{k}$) or Boltzmann's constant or quadrupole magnet gradient constant.
K	Kaon or K meson.
L	Mean free path, $\langle L \rangle$, or length of travel or total path length in a cascade or angular momentum or likelihood function.
L_z	Angular momentum about the z axis.
$\underline{L}$	Inductance per unit length.
ℓ	Angular momentum quantum number or length.

λ	Wavelength or charge/length on an electrode or nuclear interaction length.
$\bar{\lambda}$	Compton wavelength of a particle $= \hbar/mc$.
λ_{DB}	deBroglie wavelength $= h/p$.
m	Mass of particle (light particles), or $L_z = m\hbar$ angular momentum projection quantum number.
M	Mass of a particle (used for ions) or system of particles or matrix for magnetic beam transport or matrix of multiple scattering error.
MeV	10^6 eV.
MIP	Minimum Ionizing Particle.
μ	Magnetic permeability or mobility of charge carriers in an electric field or electron magnetic moment.
n	Number density or principal Bohr quantum number or index of refraction or magnet current turns per unit length.
N	Number of occurrences or noise.
$<N>$	Mean number of particles produced in a hadronic collision.
N_o	Avogadro's number.
$N^o(\nu), N^\pm(\nu)$	Number of neutral and charged particles at depth ν in a hadronic cascade.
$N(t), N(\nu)$	Number of particles in a cascade as a function of depth t or ν.
ν	Path length in nuclear interaction length units.
ω	Circular frequency, $\omega = 2\pi f$, or photon energy, $\varepsilon = \hbar\omega$.
ω_p	Plasma frequency.
Ω	Solid angle.
p	Particle momentum or number density of positive charge carriers.
p^*	Momentum in the CM frame.
p^μ	Four dimensional momentum.
p_T	Transverse momentum.
P	Probability of occurrence or gas pressure or 'pitch' (electrode spacing) of a silicon detector.
P	Power of a system or probability distribution function.
PMT	Photomultiplier tube.
P_ℓ	Legendre polynomial.
ϕ	Azimuthal angle or bend angle in a magnetic field. (Dipole ϕ_B or quadrupole ϕ_Q.)
ϕ_L	Lorentz angle in combined electric and magnetic fields.
π	Pion, or π meson.
Ψ	Wave function, where $|\Psi|^2 d\mathbf{r}$ is the probability of finding a particle in a volume element $d\mathbf{r}$.
q	Particle charge or momentum transfer in a collision.
q_p	Charge induced on 'pads' of a wire chamber cathode.
q_s	Source charge liberated in a detector by particle passage.
$q(t)$	Charge in a detector appearing between the electrodes.
Q	Ratio of recoil kinetic energy to mass in a collision.
$Q(t)$	Charge induced on detector electrodes.
r	Radius or random number.
r_h	Characteristic transverse size of a hadronic shower.
r_M	Moliere radius. Characteristic transverse EM cascade size.

R	Radial wave function or electrical resistance or reflection coefficient.
RMS	Root mean square of a distribution.
$\underline{R}$	Range of a particle or resistance per unit length.
ρ	Mass density or charge density of a system or electrical resistivity $= 1/\sigma$ or density of quantum states or magnetic radius of curvature.

s	Four dimensional interval between two space–time events or path length for orbit in a magnetic field or spin quantum number.
s_z	Spin angular momentum component about the z axis.
$\sqrt{s}$	Energy in the Center of Momentum frame.
STP	Standard Temperature and Pressure.
σ	Cross section or electrical conductivity or charge per unit area (surface charge density) or root mean square deviation of a set of measurements.

t	Time label of events or path length in radiation length units.
T	Kinetic energy or wire tension or absolute temperature.
τ	Lifetime of process or mean time of occurrence or decay lifetime.
θ	Spherical angle. $\langle \theta^2 \rangle$ = mean square scattering angle.
θ_{MS}	Multiple scattering angle.

u	Energy density of a field or radial wave function (bound state).
U	Energy of an electrical system or total potential energy of a system.
U^μ	Four dimensional velocity

v	Particle velocity.
v_d	Drift velocity.
v_T	Thermal velocity.
V	Electric potential, voltage ($U = qV$).
$\underline{V}$	Volume of a system.
$\langle V^2 \rangle$	Mean square output noise voltage after frequency filtering.
V_B	Magnetostatic potential.
V_T	Threshold voltage.
$V(t)$	Voltage as a function of time induced on detector electrodes.
VB	Valence band in a solid

W	Ratio of sampled energy in a cascade to the total energy or sampling fraction.
WT	'Weight' of a measurement $= 1/\sigma^2$.

x	Distance traveled along the path of a particle, or **x** vector labeling particle position.
x'	Transverse 'velocity' dx/ds (dimensionless).
x^μ	Four dimensional position.
X_o	Radiation length.

y	Energy in units of the critical energy or frequency in units of $\gamma \omega_p$ or transverse multiple scattering displacement or ratio of time to drift time.
$\bar{y}$	Mean of a set of measurements of y.
$\langle y \rangle$	Mean of a distribution function.
Y_ℓ^m	Spherical harmonic, angular solution of the Schroedinger equation for central forces.

Z	Atomic number or electrical impedance of a circuit element.
ξ	Formation zone length in transition radiation.

Index

Page numbers in italic, e.g. *214*, signify references to figures. Page numbers in bold, e.g. **7**, denote entries in tables.

357

Printed in the United States
By Bookmasters